AF358718

Effect of Protein and Peptide Supplementation on Physical Performance and Health Status

Effect of Protein and Peptide Supplementation on Physical Performance and Health Status

Editors

Lei Zhao

Liang Zhao

Basel • Beijing • Wuhan • Barcelona • Belgrade • Novi Sad • Cluj • Manchester

Editors

Lei Zhao	Liang Zhao
Beijing Engineering and	Beijing Engineering and
Technology Research Center	Technology Research Center
of Food Additives	of Food Additives
Beijing Technology and	Beijing Technology and
Business University	Business University
Beijing	Beijing
China	China

Editorial Office
MDPI
St. Alban-Anlage 66
4052 Basel, Switzerland

This is a reprint of articles from the Special Issue published online in the open access journal *Nutrients* (ISSN 2072-6643) (available at: https://www.mdpi.com/journal/nutrients/special_issues/8XR32PU4QJ).

For citation purposes, cite each article independently as indicated on the article page online and as indicated below:

Lastname, A.A.; Lastname, B.B. Article Title. *Journal Name* **Year**, *Volume Number*, Page Range.

ISBN 978-3-7258-0983-7 (Hbk)
ISBN 978-3-7258-0984-4 (PDF)
doi.org/10.3390/books978-3-7258-0984-4

Contents

Review

Carnosine and Beta-Alanine Supplementation in Human Medicine: Narrative Review and Critical Assessment

Ondrej Cesak [1,2], Jitka Vostalova [3], Ales Vidlar [1,2], Petra Bastlova [4] and Vladimir Student, Jr. [1,2,*]

1 Department of Urology, University Hospital Olomouc, 775 20 Olomouc, Czech Republic; ondrej.cesak@fnol.cz (O.C.)
2 Faculty of Medicine and Dentistry, Palacky University, 775 15 Olomouc, Czech Republic
3 Department of Medical Chemistry and Biochemistry, Faculty of Medicine and Dentistry, Palacky University, 775 15 Olomouc, Czech Republic
4 Department of Rehabilitaion, University Hospital Olomouc, 775 20 Olomouc, Czech Republic
* Correspondence: vladastudent@gmail.com

Abstract: The dipeptide carnosine is a physiologically important molecule in the human body, commonly found in skeletal muscle and brain tissue. Beta-alanine is a limiting precursor of carnosine and is among the most used sports supplements for improving athletic performance. However, carnosine, its metabolite *N*-acetylcarnosine, and the synthetic derivative zinc-L-carnosine have recently been gaining popularity as supplements in human medicine. These molecules have a wide range of effects—principally with anti-inflammatory, antioxidant, antiglycation, anticarbonylation, calcium-regulatory, immunomodulatory and chelating properties. This review discusses results from recent studies focusing on the impact of this supplementation in several areas of human medicine. We queried PubMed, Web of Science, the National Library of Medicine and the Cochrane Library, employing a search strategy using database-specific keywords. Evidence showed that the supplementation had a beneficial impact in the prevention of sarcopenia, the preservation of cognitive abilities and the improvement of neurodegenerative disorders. Furthermore, the improvement of diabetes mellitus parameters and symptoms of oral mucositis was seen, as well as the regression of esophagitis and taste disorders after chemotherapy, the protection of the gastrointestinal mucosa and the support of *Helicobacter pylori* eradication treatment. However, in the areas of senile cataracts, cardiovascular disease, schizophrenia and autistic disorders, the results are inconclusive.

Keywords: carnosine; beta-alanine; zinc-carnosine; supplementation; human diseases; health benefits

Citation: Cesak, O.; Vostalova, J.; Vidlar, A.; Bastlova, P.; Student, V., Jr. Carnosine and Beta-Alanine Supplementation in Human Medicine: Narrative Review and Critical Assessment. *Nutrients* **2023**, *15*, 1770. https://doi.org/10.3390/nu15071770

Academic Editors: Lei Zhao and Liang Zhao

Received: 15 March 2023
Revised: 30 March 2023
Accepted: 31 March 2023
Published: 5 April 2023

1. Introduction

Carnosine is a dipeptide composed of the non-proteogenic amino acid beta-alanine and the essential amino acid L-histidine (Figure 1). Carnosine is found in the human body not only in relatively high millimolar concentrations in excitable tissues (skeletal muscle and brain [1]) but also in smaller amounts in other tissues (gastrointestinal tract, kidney, liver, adipose tissue and heart) [2]. The bioavailability of beta-alanine limits the synthesis and amount of carnosine [3].

Several biological effects have been described for carnosine and its precursor beta-alanine. At the cellular level, carnosine reacts with reactive oxygen species (ROS) and reactive nitrogen species (RNS), and oxidative damage products of biomolecules [4]. In this way, it protects other biomolecules from modification and damage [5]. Carnosine reduces not only the level of reactive carbonyls (RC) but also the formation of advanced glycation end products (AGEs) and advanced lipid peroxidation end products (ALE) [6,7]. Carnosine also has the ability to scavenge nitric oxide (NO) [8] and is involved in the chelation of transition metals [9]. Further, carnosine affects the regulation of calcium levels in muscle cells [10]. Carnosine can directly react with protons and, thus, participates in maintaining pH balance and reducing muscle fatigue caused by intense muscle activity [11]. Furthermore, carnosine is associated with the ability to modulate the endogenous antioxidant system by activating the signaling pathway controlled by the transcription factor Nrf2, which is involved in the removal and detoxification of oxidative modification products of biomolecules [11]. Carnosine also affects telomerase activity and slows down cell senescence [12], affects the resistance of proteins to heat or chemical stress [13] and has immunomodulatory effects [14]. It is also significantly involved in the regulation of intracellular metabolism, inhibiting glycolysis and increasing mitochondrial activity (the production of adenosine triphosphate, ATP) [15]. Finally, in vitro studies have revealed that carnosine has antineoplastic effects and affects the metabolism and senescence of tumor cells [15–22]. Figure 2 summarizes the published functions of carnosine.

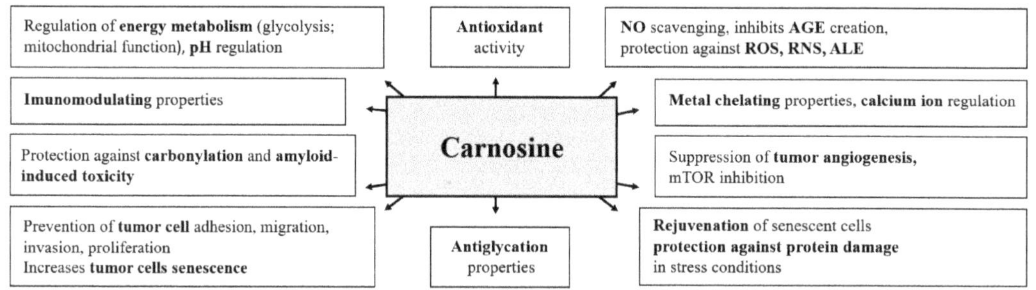

Figure 1. Chemical structure of carnosine, zinc-L-carnosine and beta-alanine.

Figure 2. Biological activities of carnosine, simplified. ALE: advanced lipoxidation end products; AGE: advanced glycation end products; mTOR: mammalian target of rapamycin; NO: nitric oxide; RNS: reactive nitrogen species; ROS: reactive oxygen species. Additional information on carnosine's molecular functions can be found in Supplementary Table S1.

Beta-alanine is one of the most used sports supplements worldwide [23], as it improves muscle performance in active athletes, specifically by increasing the concentration of carnosine in the muscles [24]. It is recommended for this purpose by the International Society of Sports Nutrition (ISSN) [25]. Besides carnosine and beta-alanine, *N*-acetylcarnosine (NAC) and the synthetically prepared derivative chelate compound zinc-L-carnosine (ZnC), also called Polaprezinc® (containing 23% zinc and 77% L-carnosine), have biologically significant effects on the human body [26]. For a long time, the supplementation of carnosine and beta-alanine has been associated only with active athletes, but is now gaining popularity among persons with chronic diseases.

Current publications present recent data from a critical perspective, specifically from clinical trials in human medicine. This narrative review aims to increase knowledge of the clinical benefits of the supplementation with a focus on the primary and secondary prevention of diabetes mellitus; the prevention of sarcopenia; the improvement of cognitive abilities and neurodegenerative disorders; the treatment of schizophrenia, autism spectrum disorders and senile cataracts; and the improvement of prognoses of cardiovascular diseases, oral mucositis, taste disorders and other gastrointestinal diseases.

2. Search Strategy and Methodology

A narrative review was conducted between December 2021 and February 2023. Peer-reviewed journal articles were located from inception up to January 2023. The search strategy was aimed at evaluating clinical studies on the role of carnosine and beta-alanine supplementation in human clinical studies. Scientific articles were searched for using PubMed, Web of Science, the National Library of Medicine and the Cochrane Library databases. A search strategy using database-specific keywords was employed. The search terms used included "carnosine", "beta-alanine", "zinc-carnosine", "polaprezinc", "supplementation" combined with "cataract" and/or "neurodegenerative" and/or "mucositis" and/or "diabetes mellitus" and/or "sarcopenia" and/or "gastroenterology" and/or "psychiatry" and/or "cardiovascular system" and/or "schizophrenia" and/or "ADHD". Only papers with English translations were reviewed. The search retrieved approximately 800 articles and 118 were used in the present review.

3. Synthesis and Degradation

Carnosine is synthesized in the human body from beta-alanine and histidine [27]. The synthesis is catalyzed by ATP-dependent carnosine synthetase and the rate of the intracellular synthesis is greatly limited by the level of beta-alanine [28]. Beta-alanine itself is synthesized in the liver (during the catabolism of polyamines, pyrimidines and coenzyme A; the bacterial catabolism of aspartate; and the transamination of malonic acid semi aldehyde [27]) and then transported to muscle and brain cells, where it is utilized for carnosine synthesis [2]. It can be estimated that a 60 kg woman and a 70 kg man synthesize 427 and 606 mg of carnosine, respectively, per day [29].

The intracellular concentration of carnosine depends on the hydrolytic activity of carnosinases (CN 1 or CN 2) and the synthetic activity of carnosine synthetase, and is greatly limited by the dietary uptake of beta-alanine and essential amino acid histidine [10].

Dietary carnosine and beta-alanine are absorbed in the human small intestine on the apical side by specific transporters: carnosine by a peptide transporter, beta-alanine by a specific transporter for beta-amino acids and taurine, and histidine by a sodium dependent neutral amino acid transporter [2]. In enterocytes, carnosine is hydrolyzed by CN 2 [2]. Carnosine and histidine cross the basolateral membrane of enterocytes to the blood by a proton coupled transporter, and beta-alanine by a specific transporter for beta-amino acids [2]. Nearly all ingested carnosine enters the portal circulation [30]. In the blood, carnosine is hydrolyzed by CN1 and the half-life of carnosine in the human serum is under 5 min [31]. Histidine and beta-alanine cross the membrane of target cells in the extraintestinal tissues, mainly, muscle, the liver and the brain, by specific transporters [2], where they are used for the synthesis of carnosine.

The carnosine molecule is enzymatically hydrolyzed in the human body by CN1 and CN2 [31], which act as the regulators of carnosine level [32].

In the central nervous system (CNS) [33], carnosine is hydrolyzed, and the resulting histidine is enzymatically converted by histidine decarboxylase in specific parts of the brain to the neurotransmitter histamine [34].

The metabolism of carnosine and beta-alanine, along with their synthesis and degradation cycle, is summarized in Figure 3.

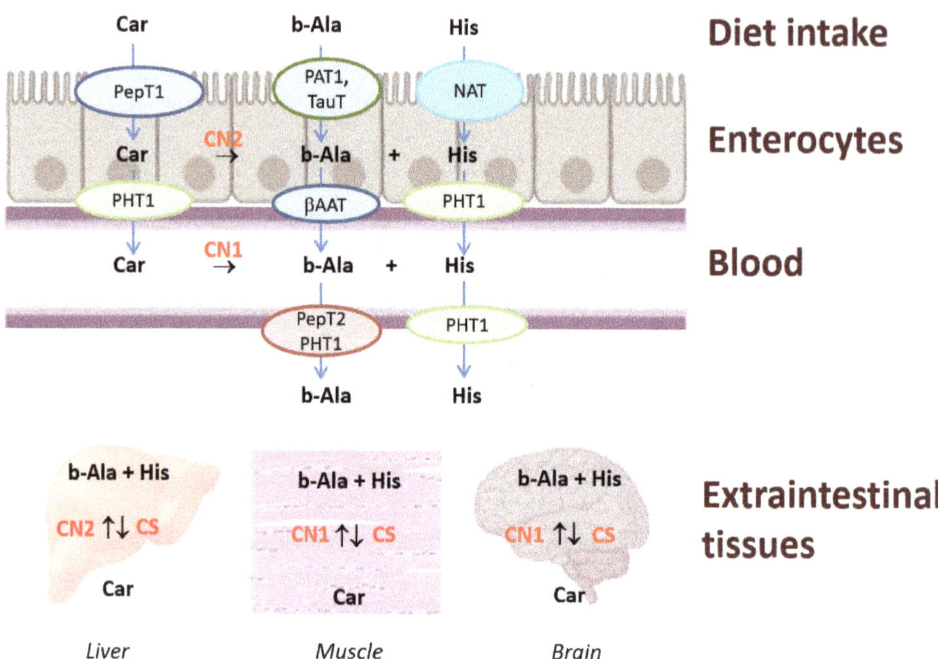

Figure 3. Metabolism of carnosine and beta-alanine. Carnosine (Car) enters the enterocyte by a peptide transporter (PepT1); beta-alanine (b-Ala) enters by a specific transporter for beta-amino acids, a proton-assisted amino acid transporter (PAT1) and a taurine transporter (TauT); and histidine (His) enters by a sodium-dependent neutral amino acid transporter (NAT). Proton-coupled peptide-histidine transporter 2 (PepT2) carnosine transporting is expressed in the membrane on majority extraintestinal tissue. b-Ala is transported from the circulation by beta-amino acid specific transporter (ßAAT) and His by PHT1. In the extraintestinal tissues, carnosine is cleaved by carnosinases CN1 or CN2 into b-Ala and His, respectively, and formed from these two amino acids through carnosine synthetase (CS).

4. Supplementation and Food Sources

Increasing the dietary intake of carnosine enhances its concentrations in, mostly, skeletal muscle, the brain and the heart [10]. An important source of carnosine is provided by foods such as chicken meat [35], fish and shrimp, as well as asparagus, green peas and white mushrooms [36]. Interestingly, it is suggested that some components in beef inhibit serum CN1 [10]. A daily dietary intake of 30 g of dried beef should be able to completely supply daily carnosine requirement of a 70 kg adult to ameliorate human nutrition and health [2]. Even though its limiting precursor beta-alanine is produced endogenously in the liver, its main source is from a person's diet. Humans acquire beta-alanine through the consumption of foods with large beta-alanine content, such as poultry, beef and fish [25]. The administration of beta-alanine seems to be more convenient and effective than carnosine, because the bioavailability of carnosine is reduced by CN activity [31]; however, beta-alanine is reused for carnosine synthesis by skeletal muscle, the heart, and the olfactory bulb of the brain.

In a meta-analysis of the adverse effects and risks of oral beta-alanine supplementation, no adverse effects on human health were observed (at doses from 4 to 6 g per day) [23]. The only adverse effect was paresthesia and a small increase in alanine aminotransferase activity was described (the increase was within reference ranges) [23]. There is no risk of overdose when carnosine is administered because carnosine is cleaved by CN1 into amino acids, which are then involved in metabolism [10].

5. Carnosine, Beta-Alanine and Diabetes Mellitus

The formation of AGEs and RC is one of the causes of diabetic complications. In patients with type 2 diabetes mellitus (T2DM), RC levels correlate with insulin resistance [37]. Carnosine can interact with these reactive molecules and, thus, prevent the development of adverse complications in patients with diabetes mellitus (DM) [38].

The potential of carnosine supplementation in the prevention of T2DM was described in a clinical trial by de Courten et al. [39]. The administration of carnosine (2 g/day for 12 weeks) led to a reduction in fasting insulin levels and a reduction in insulin resistance [39]. Furthermore, carnosine supplementation in a randomized placebo-controlled trial has shown a nephroprotective effect (the reduction in urinary transforming growth factor-beta) in 40 patients with diabetic nephropathy [40]. The effect of supplementation was also suggested by a different randomized trial, in which it reduced glycaemia in 82 patients with T2DM; however, this study combined carnosine supplementation and 2 other supplements, so this effect cannot be attributed to carnosine alone [41]. The benefits of beta-alanine administration (4 g/day in 3 doses, over 28 days) in combination with increased physical activity in patients with T2DM led to a reduction in glycaemia and a simultaneous improvement in physical capacity [42]. Another placebo-controlled randomized trial described the effect of carnosine supplementation (500 mg twice daily, over 12 weeks) on complications of type 1 diabetes (T1DM) in children and adolescents. The dose was well tolerated by the subjects, and a decrease in parameters capturing oxidative stress, an increase in antioxidant capacity and an improvement in glycated hemoglobin (HbA1c) and renal metabolism (alpha-1 microglobulin) were observed [43]. After carnosine supplementation (2×500 mg/day, over 12 weeks) in T2DM patients, fasting glycaemia, HbA1c, serum triglyceride (TAG) and tumor necrosis factor-alpha (TNF-α) levels were lowered [44].

A meta-analysis showed that supplementation with carnosine and beta-alanine led to reductions in fasting glycaemia, a decrease in HbA1c levels and insulin resistance [45]. Another meta-analysis including only randomized trials confirmed the effect of carnosine supplementation on DM patients, specifically on their HbA1c and fasting glucose levels [46]. This is supported by the results of yet another meta-analysis on 30 clinical trials of the effect of carnosine and related dipeptides in obese subjects, which showed reductions in fasting glycaemia and HbA1c, as well as reductions in obesity (waist circumference) [47]. Available trials on the effectiveness of supplementation are summarized in Table 1.

Table 1. Carnosine, beta-alanine and diabetes mellitus.

Author (Year)	Study Design	Intervention	Number of Patients	Effect
de Courten et al. (2016) [39]	double-blind RCT	Carnosine orally (2 g in 2 doses/day) or placebo; 12 weeks	30	Preserved insulin sensitivity and insulin secretion, normalized glucose intolerance and reduced 2-h insulin levels after o-GTT in a subgroup of individuals with impaired glucose tolerance
Siriwattanasit et al. (2021) [40]	RCT	Carnosine (2 g/day) or placebo; 12 weeks	40	Nephroprotective effect of oral supplementation to decrease urinary TGF-β.
Karkabounas et al. (2018) [41]	double-blind RCT	Alpha-lipoic acid (7 mg/kg bodyweight), carnosine (6 mg/kg bodyweight), thiamine (1 mg/kg bodyweight) or placebo; 8 weeks	82	Supplementation effectively reduced glucose concentration in patients with T2DM.
Nealon et al. (2016) [42]	double-blind RCT	Beta-alanine (4 g split into 3 doses/day) or placebo; 28 days	12	Beta-alanine supplementation can increase exercise capacity in individuals with T2DM.

Table 1. *Cont.*

Author (Year)	Study Design	Intervention	Number of Patients	Effect
Elbarbary et al. (2018) [43]	double-blind RCT	Patients with diabetic nephropathy received supplemented carnosine (1 g/day) or placebo; 12 weeks	90	Oral supplementation with L-carnosine for 12 weeks resulted in a significant improvement of oxidative stress, glycemic control and renal function.
Houjeghani et al. (2018) [44]	double-blind RCT	Patients with T2DM received carnosine (500 mg, 2×/day, capsules) or placebo	54	Carnosine supplementation lowered fasting glucose, serum levels of triglycerides, AGEs and tumor necrosis factor-α without changing sRAGE, IL-6 or IL-1β levels in T2DM patients.

AGEs: advanced glycation end products; IL-6: interleukin 6; IL-1β: interleukin 1-beta; o-GTT: oral glucose tolerance test; RCT: randomized controlled trial; sRAGE: soluble receptor for advanced glycation end products; TGF-β: transforming growth factor-β; T2DM: type II diabetes mellitus.

Findings from clinical trials show that supplementation with carnosine (ranging from 500 mg to 2 g per day) and beta-alanine (4 g per day) is important in preventing and slowing the progression of T2DM. At the same time, it reduces complications associated with obesity, T1DM and T2DM, including oxidative stress, and improves carbohydrate metabolism [45–47].

6. Carnosine, Beta-Alanine, and Neurological Diseases

In addition to muscle tissue, carnosine is also found in the CNS. Due to its properties, carnosine participates in the maintenance of homeostasis in the CNS, acting as a neuroprotective agent [48]. Through the mechanism of neuroprotection by reactions with ROS, RNS, AGEs and RC, it has the potential to promote cognitive maintenance and prevent the development of neurodegenerative diseases. Even the precursor of carnosine, beta-alanine, affects many neurocirculatory processes through competitive inhibition of taurine [49]. The activity of CN1 increases with age, leading to a lower accumulation of carnosine in muscle, the brain and other tissues [50]. This is also related to the clinical entity of dementia (specifically in the aging population), which is accompanied by increased production of ROS, RNS and AGEs and inflammation [51].

The increasing prevalence of neurological and neurodegenerative diseases is a social issue. These diseases are often associated with disability and their treatment is still inadequate. The potential of carnosine or beta-alanine supplementation in clinical trials was extensively described in the review by Schön et al., in which the authors describe promising results in the field of neurology, for diseases such as Alzheimer's disease (AD), Parkinson's disease (PD), and multiple sclerosis [52]. This section of the review presents the results of selected clinical trials focusing on the CNS, and available studies are summarized in Table 2.

Table 2. Carnosine, beta-alanine and neurological diseases.

Author (Year)	Study Design	Intervention	Number of Patients	Effect
Rokicki et al. (2015) [53]	double-blind RCT	Carnosine/anserine (1:3 ratio; 500 mg/day) or placebo; 3 months	31	Intervention group had better verbal episodic memory performance and decreased connectivity in the default mode network, the posterior cingulate cortex and the right frontal parietal network. A correlation between the extent of cognitive and neuroimaging changes was observed.

Table 2. *Cont.*

Author (Year)	Study Design	Intervention	Number of Patients	Effect
Szcześniak et al. (2014) [54]	RCT	Chicken meat extract containing anserine and carnosine (2:1 ratio; 1 g/day) or placebo; 13 weeks	51	Mean values of Short Test of Mental Status (STMS) scores increased in the intervention group (in the subscores of construction/copying, abstraction and recall), and promising effects on physical capacity.
Katakura et al. (2017) [55]	double-blind RCT	Anserine/carnosine (3:1 ratio; 1 g/day) or placebo; 3 months	60	Supplementation may preserve verbal episodic memory, probably owing to inflammatory chemokine CCL24 suppression in the blood.
Hisatsune et al. (2015) [56]	double-blind RCT	Anserine/carnosine (3:1 ratio; 1 g/day) or placebo; 3 months	39	MRI analysis showed a suppression in the age-related decline in brain blood flow in the posterior cingulate cortex area. Delayed recall verbal memory showed significant preservation in the intervention group.
Baraniuk et al. (2013) [57]	double-blind RCT	Carnosine (500, 1000 and 1500 mg/day, increasing at 4-week intervals) or placebo; 12 weeks	25	Supplementation may have beneficial cognitive effects. Fatigue, pain, hyperalgesia, activity and other outcomes were resistant to treatment.
Solis et al. (2015) [58]	double-blind RCT	Beta-alanine (6.4 g/day) supplementation or placebo; 28 days	19	Supplementation did not influence cognitive function before or after exercise in trained cyclists.
Masuoka et al. (2019) [59]	double-blind RCT	Anserine (750 mg/day) and carnosine (250 mg/day) or placebo; 12 weeks	54	Protective effects against cognitive decline in APOE4 (+) MCI subjects exist.
Boldyrev et al. (2008) [60]	two-arm, prospective	In addition to basic PD therapy, carnosine (1.5 g/day); 30 days	36	The combination of carnosine with basic therapy may be a useful way to increase efficiency of PD treatment.
Cornelli et al. (2010) [61]	two-arm, RCT	Carnosine (100 mg/day) with a mixture of antioxidants (beta-carotene, selenium, cysteine, ginko biloba and coenzyme Q10) and vitamins (B1, B2, B3, B6, B9, B12, C, D and E) on Alzheimer's disease (AD) patients treated with donepezil; 6 months	52	A reduction in oxidative stress parameters and an improvement in mini-mental state examination, version 2 (MMSE-2) scores were observed.

AD: Alzheimer's disease; MCI; Mild Cognitive Impairment; MMSE-2: mini-mental state examination, version 2; PD: Parkinson's disease; RCT: randomized controlled trial; STMS: Short Test of Mental Status.

The investigation of improving cognition and slowing the progression of neurodegenerative diseases was conducted in an interventional, double-blind, placebo-controlled trial by Rokicki et al. [53]. The results showed that supplementation with carnosine in combination with anserine (500 mg in a 1:3 ratio, for 3 months) had a positive cognitive effect in elderly people (40–78 years) [53]. Similarly, another double-blind trial in an elderly population (56 subjects aged 65+) combining carnosine with anserine (2.5 g of anserine and carnosine in a 2:1 ratio, for 13 weeks) revealed improvements in cognitive function and physical capacity [54]. Another study that investigated supplementation of carnosine with anserine (1 g of carnosine and anserine in a 3:1 ratio for 3 months) in elderly patients resulted in a significant improvement in verbal memory and correlated with the suppression of the inflammatory chemokine CCL24, which may be responsible for these cognitive symptoms [55]. However, none of the previous three trials could completely attribute the results to carnosine, given the ratio of active ingredients.

Hisatsune et al. observed a decrease in blood levels of pro-inflammatory cytokines and an increase in cerebral blood flow after carnosine supplementation [56]. The effect of carnosine was studied in a double-blind randomized trial on Gulf War veterans who suffered from complex cognitive and multisystemic symptoms (Gulf War Illness, GWI, or Chronic Multisymptom Illness). The etiology of GWI is thought to be associated with increased ROS production. Study participants were given increasing doses of carnosine (0.5–1 g daily at 4-week intervals for 12 weeks), which significantly improved their Wechsler Adult Intelligence Test scores [57]. Solis et al. did not confirm an effect on cognitive function after beta-alanine supplementation (6.4 g/day for 28 days), but this trial was limited by the small number of participants [58].

The beneficial properties of carnosine and its components (beta-alanine and histidine) have been described in the additional literature [53,56,57]. A meta-analysis of randomized trials confirmed the effects of carnosine supplementation on the Wechsler memory scale-revised logical memory delayed recall (WMS-LM2), specifically in doses of around 300 mg of carnosine per day [46].

Neurodegenerative Diseases

Globally, neurodegenerative diseases represent a significant burden on the healthcare system. Three of the major neurodegenerative diseases are AD, PD and amyotrophic lateral sclerosis (ALS). The prevalence and incidence of these diseases increase dramatically with age, and, therefore, the number of cases is expected to increase as life expectancy continues to rise [62].

Patients with probable AD (pAD) showed changes in plasma amino acids, including a reduction in carnosine levels, which may be related to reduced antioxidant capacity in AD patients [63].

After supplementation with anserine and carnosine (54 subjects; 750 and 250 mg/day for 12 weeks), no significant effect on cognitive function was observed in patients with a mild form of cognitive dysfunction; however, in a follow-up analysis, it was found that there was an improvement in cognitive function in participants who are Apo E4 carriers (Apo E4 functions as a risk factor for AD development). For these patients, supplementation had a protective function against cognitive deterioration in patients with mild cognitive impairment [59]. Only one randomized placebo-controlled trial investigated carnosine supplementation (1.5 g/day for 1 month) in 36 patients with PD, according to the basic treatment protocol. Carnosine led to improvements in neurological symptoms associated with improved antioxidant capacity. The administration of carnosine together with the basic therapy not only improved the clinical parameters but also improved the antioxidant status of the PD patients [60].

In another clinical trial, the effect of a mixture of antioxidants and vitamins, including carnosine (100 mg/day), on AD patients treated with donepezil was investigated. After a six-month treatment with this mixture, a reduction in oxidative stress parameters and an improvement in mini-mental state examination, version 2 (MMSE-2) scores were observed in the donepezil-treated group when compared to the placebo [61].

Slowing the progression and improving the symptoms of certain neurodegenerative diseases, specifically AD and PD, has been confirmed sporadically in clinical trials in specific patient groups [59–61,64]. A meta-analysis of randomized trials confirmed the positive effect of carnosine supplementation (in doses of up to 1.5 g per day) in the Alzheimer's Disease Assessment Scale and the Beck's Depression Inventory [46]. Unfortunately, the number of clinical trials is very small. Nevertheless, we suggest that preventive long-term supplementation with these substances in the aging population could slow down and delay the onset of degenerative changes in the CNS. High-quality clinical studies are needed to confirm this theory.

7. Carnosine, Beta-Alanine, and Psychiatric Diseases

The increasing prevalence of psychiatric illness is considered a major social problem. Recently, there has been an increase in the number of cases of psychiatric illnesses and associated rising demands on the healthcare system worldwide. For example, the negative symptoms of schizophrenia, with a prevalence of about 25%, are a challenging aspect of this disease, precisely because of the inconsistent efficacy of current antipsychotic medication [65]. Several mechanisms are thought to be responsible for the negative symptoms, such as, for example, the effect of the disease on glutaminergic synapses [66]. Carnosine and its effect on these synapses could lead to the clinical effect of reducing these symptoms of schizophrenia. The potential of supplementation in clinical trials has been extensively described in a review by Schön et al., in which the authors present promising results of carnosine or beta-alanine supplementation in clinical trials concerning psychiatric disorders, autism spectrum disorders and many others [52].

An 8-month randomized double-blind placebo-controlled trial of 60 patients with chronic schizophrenia treated with risperidone focused on studying the effects of carnosine (2 g/day). The administration of carnosine along with therapy resulted in a reduction in negative symptoms of schizophrenia without an increase in side effects [67]. This was confirmed in a randomized trial of 75 patients with chronic schizophrenia, in which carnosine supplementation (2 g/day for 3 months) improved executive functions [68].

In addition, the depletion of beta-alanine and histidine may not occur in active disease exclusively. Increase in the levels of these amino acids might show potential to monitor treatment effect [52]. In a study of patients with depressive disorders treated with selective serotonin reuptake inhibitors (SSRIs), plasma beta-alanine levels were decreased compared to healthy controls [69]. This has also been indicated in people suffering from depression. The use of the antidepressant quetiapine led to a decrease in carnosine levels. If the medication was not administered, there was a gradual increase in serum carnosine levels over the course of the depression (over 40 months). The authors assume that this is a defensive reaction of the organism to this condition [70].

Isolated published studies suggest a benefit of carnosine supplementation on schizophrenia symptomatology [67,68], but further high-quality studies are needed to confirm the benefit in practice. The compounds' ability to monitor treatment effect is only theoretically suggested, currently, without robust data to support it.

Autism Spectrum Disorders in Children

Autism spectrum disorders (ASD) are complex disorders accompanied by impairments in certain brain functions and represent a global healthcare issue. The etiology of ASD is multifactorial [71]. Studies show that imbalances in transit metals (increased levels of lead, cadmium, mercury and nickel and decreased levels of zinc) can disrupt mitochondrial function and antioxidant capacity; induce synapse dysfunction; impair myelination; and affect neurogenesis and neural differentiation, synapses, myelination and neuroinflammation development [72,73]. Carnosine could reduce oxidative stress in ASD sufferers due to its antioxidative properties and ability to chelate transit metals.

Most clinical trials evaluating levels of carnosine and similar peptides in neurological and neurodevelopmental disorders have been conducted in children with ASD [74]. Reduced levels of carnosine, including histidine and beta-alanine, have also been demonstrated in children and adolescents with ASD [75,76], suggesting the potential use of carnosine in the treatment of autistic patients.

Carnosine supplementation (800 mg/day for 8 weeks) resulted in improved communication skills and behavior in autistic children. The authors also reported improvements in receptive speech and social attention, a reduction in apraxia and an overall improvement in brain function [77]. Mehrazad-Saber et al. observed a reduction in the frequency of sleep disturbances in children with ASD with carnosine supplementation (500 mg daily for 2 months), but the effect on the course of the disease was not confirmed in the trial [78]. In a double-blind randomized trial of 70 children with ASD, carnosine supplementation (800 mg daily for 10 weeks) resulted in a reduction in hyperactivity and non-compliance [79], although these results were not confirmed in another randomized placebo-controlled trial on 63 participants [74]. The effectiveness and safety of carnosine has also been studied in children with ADHD (Attention-Deficit/Hyperactivity Disorder). In a randomized double-blind placebo-controlled trial, either carnosine or a placebo was administered in addition to methylphenidate (0.5–1.5 mg/kg/day). Treatment with carnosine (2 g/day for 8 weeks) and risperidone showed good tolerability and significant beneficial effects on negative symptoms in patients with stable disease [80].

Studies with carnosine have pointed to its potential use in improving sleep in children and adolescents with ASD; however, Esposito et al., in their review, stress the need for further research to confirm the efficacy of this molecule for the treatment of sleep disorders in patients with ASD [81]. Although it appears from the results of individual trials that the administration of carnosine, or its components beta-alanine and histidine, could be beneficial for children with ASD [77–80], carnosine was also included in a meta-analysis on the use of drugs and dietary supplements in ASD, and the available data were considered insufficient and, therefore, carnosine cannot be therapeutically recommended [82]. Based on the results of a clinical trials meta-analysis on the effect of carnosine supplementation (up to 2 g per day) on children with ASD (only 5 eligible studies with 215 participants were selected), it is not possible to recommend carnosine supplementation for children with ASD [74].

Further high-quality clinical trials with a larger number of patients will be needed to verify the effect of these agents on children's health and possibly recommend them as an add-on to conventional therapy. Available trials are summarized in Table 3.

Table 3. Carnosine, beta-alanine and psychiatric diseases and their monitoring treatment effects.

Author (Year)	Study Design	Intervention	Number of Patients	Effect
Ghajar et al. (2018) [80]	double-blind RCT	Carnosine (2 g/day in two doses) or placebo; 8 weeks	60	Administration of carnosine along with therapy resulted in a reduction in negative symptoms of schizophrenia without an increase in side effects.
Chengappa et al. (2012) [68]	double-blind RCT	Carnosine (2 g/day) or placebo; 3 months	75	Intervention group performed significantly faster on non-reversal condition trials of the set-shifting test. The strategic target detection test displayed improved strategic efficiency and fewer perseverative errors.
Chez et al. (2002) [77]	double-blind RCT	Carnosine supplementation (800 mg/day) or placebo; 8 weeks	31	Improved communication skills and behavior in children with ASD. The authors also reported improvements in receptive speech and social attention, a reduction in apraxia and an overall improvement in brain function.
Mehrazad-Saber et al. (2018) [78]	double-blind RCT	Carnosine (500 mg/day) or placebo; 2 months	43	Carnosine supplementation did not change anthropometric indices and showed no effect on autism severity, whereas it significantly reduced sleep duration, parasomnias and total sleep disorders scores.

Table 3. *Cont.*

Author (Year)	Study Design	Intervention	Number of Patients	Effect
Hajizadeh-Zaker et al. (2018) [79]	double-blind RCT	Carnosine (800 mg/day in 2 doses) or placebo in addition to risperidone; 10 weeks	70	Carnosine supplementation resulted in a reduction in hyperactivity and non-compliance in children with ASD.
Ghajar et al. (2018) [67]	double-blind RCT	Carnosine (2 g/day in two divided doses) or placebo; 8 weeks	63	Carnosine add-on therapy reduced the primary negative symptoms of patients with schizophrenia.
Ann Abraham et al. (2020) [83]	double-blind RCT	Carnosine (10–15 mg/kg in 2 divided doses/day) or placebo; 2 months	63	No statistically significant difference was observed for any of the outcome measures assessed.
Woo et al. (2015) [69]	two-arm prospective	SSRI; 6 weeks	68/22 (90 total)	A potential was shown to measure therapeutic response. Patients with MDD, after 6 weeks of SSRI treatment, had alterations of amino acids, including beta-alanine (and alanine, beta-aminoisobutyric acid, cystathionine, ethanolamine, glutamic acid, homocystine, methionine, O-phospho-L-serine and sarcosine).
Ali Sisto et al. (2023) [70]	two-arm, prospective	Antidepressant quetiapine; 40 weeks	99/253 (352 total)	The use of any antipsychotic medication was associated with lowered carnosine levels. Elevated serum levels of carnosine were also associated with a longer duration of the depressive episode.

ASD: autism spectrum disorders; MDD: major depressive disorder; RCT: randomized controlled trial; SSRI: selective serotonin reuptake inhibitors.

8. Carnosine, Beta-Alanine and Cataract

Cataracts are a global healthcare concern in both developed and developing countries. The main cause of cataracts is aging, but it is also related to other exogenous factors such as UV radiation and trauma, and endogenous factors such as DM or hereditary predisposition [84]. The potential of carnosine in slowing the process of degenerative changes in vision and in the treatment of senile cataracts itself is summarized in the article by Wang et al. [85]. Although cataract surgery is effective and safe, using an antioxidant in the form of topical carnosine eye drops can provide patients with another treatment option. Carnosine might have the potential, through its antioxidant and antiglycation properties [7], to delay visual impairment with aging, effectively preventing and treating senile cataracts [85]. The beneficial effect of NAC on senile cataracts was found in 49 elderly patients (mean age 65 years) with variously advanced cataracts in a randomized, placebo-controlled, double-blind trial with eye drops of 1% aqueous NAC solution (2 drops/2 times daily for 6 and 24 months) [86]. NAC's positive effect on eye health and its safety have also been described in other clinical trials from the same author [87,88].

Although the results of the abovementioned trials appear promising, an externally conducted analysis of randomized clinical trials (including only two randomized trials) concluded that there is no credible evidence at this time that NAC has preventive effects against, or that it slows the development of, cataracts. The primary reason for this decision was the lack of information on the methodology of the trials conducted [89]. Furthermore, well-designed clinical trials with a larger number of participants will be needed to confirm or refute the effect.

9. Carnosine, Beta-Alanine and Sarcopenia

Sarcopenia is a complex and multifactorial progressive muscle disorder that develops with age. In elderly patients, sarcopenia is associated with a greater likelihood of adverse events, including falls, fractures, physical disability and even mortality [90]. The deterioration of the human organism in old age may be associated with reduced tissue concentrations of carnosine, and, therefore, a lack of antioxidant capacities [91]. The activity of CN increases with age, leading to less accumulation of carnosine in muscle, the brain and other tissues [50]. It is believed that a diet rich in carnosine can slow the process of sarcopenia, aging and the development of age-related diseases [92,93]. Considering the confirmed functions of carnosine, increasing its levels in muscle could at least partially slow down the progression of sarcopenia.

In a double-blind, placebo-controlled trial of 18 subjects (60–80 years), supplementation with the carnosine precursor beta-alanine (1.6 g twice daily for 12 weeks) resulted in an increase in muscle carnosine content and improved physical performance [94]. In a double-blind placebo-controlled trial, when 2 doses of beta-alanine (800 mg and 1200 mg for 12 weeks) were administered, a significant increase in work capacity was seen compared to the placebo group [95]. Similar conclusions were reached in a trial that found a positive effect of beta-alanine (2.4 g/day) on exercise performance and subsequent muscle recovery in elderly subjects (60.5 ± 8.6 years) [96].

Supplementation with beta-alanine (in doses up to 3.2 g per day) could potentially represent one possible strategy to increase muscle activity in old age and slow down the sarcopenia process. This dietary intervention represents one option that could, therefore, lead to increased muscle performance and delay the problems associated with sarcopenia, thereby improving the quality of life of the elderly [94–96]. Available trials are summarized in Table 4.

Table 4. Carnosine, beta-alanine and sarcopenia.

Author (Year)	Study Design	Intervention	Number of Patients	Effect
del Favero et al. (2012) [94]	double-blind RCT	Beta-alanine (3.2 g divided into 4 doses/day) or placebo; 12 weeks	18	Supplementation is effective in increasing the muscle carnosine content in healthy elderly subjects, with improvement in exercise capacity.
McCormack et al. (2013) [95]	double-blind RCT, three-arm	(1) ONS (containing 8 oz; 230 kcal; 12 g protein, 31 g cholesterol, 6g fat); 12 weeks (2) ONS plus beta-alanine (800 mg, 2×/day); 12 weeks (3) ONS plus beta-alanine (1200 mg, 2×/day); 12 weeks	60	ONS fortified with beta-alanine may improve physical working capacity, muscle quality and function in older men and women.
Furst et al. (2018) [96]	double-blind RCT	Beta-alanine (2.4 g/day) or placebo; 28 days	12	Supplementation increased exercise capacity and eliminated endurance exercise-induced declines in executive function seen after recovery.

ONS: oral nutritional supplement; RCT: randomized controlled trial.

10. Carnosine, Beta-Alanine and Diseases of the Cardiovascular System

Cardiovascular disease (CVD) remains the leading cause of death globally. Endothelial dysfunction represents one of the earliest pathophysiological factors leading to the development of CVD [97]. Carnosine may also be useful in the treatment and prevention of CVD [97]. Both in vitro and experimental studies have documented the benefits of carnosine supplementation in reducing the risk of atherosclerosis [98] and have described anti-ischemic effects [99].

One of the few studies in human medicine to address carnosine's effect on CVD, a randomized placebo-controlled clinical trial, found that adding carnosine (500 mg/day for 6 months) to established treatments for chronic heart failure improved cardiopulmonary stress tests, 6-min walk test scores and quality of life [100]. In a randomized placebo-controlled trial of 50 patients after acute myocardial infarction and percutaneous coronary intervention, the anti-inflammatory effect of ZnC (75 mg twice daily for 9 months) was confirmed along with an improvement in cardiac function compared to the placebo group [101].

In some experimental studies, mostly performed in vitro, the added value of carnosine supplementation is suggested [98,99]. The results of very few human clinical trials are available [100,101] and the effect was not confirmed in the published meta-analysis [46]. High-quality clinical trials confirming these mechanisms are still lacking.

11. Zinc-Carnosine and Oral Mucositis, Loss of Taste and Gastrointestinal Tract

Zinc is a biogenic element that is essential for the proper function of a wide range of biomolecules. Zinc acts as a cofactor of a number of antioxidant enzymes and is essential for cell proliferation and cell repair [102]. ZnC has several biological effects, including maintenance, prevention and treatment of mucosal and epithelial tissue damage [26,103,104].

Loss of taste is a common side effect of chemoradiotherapy. Oral mucositis, oesophagitis and other gastrointestinal complications are common following radiotherapy or chemotherapy. Although these are not life-threatening conditions, they significantly reduce quality of life, and, therefore, it makes sense to investigate the possibilities of their prevention and treatment. Since Zn is involved in the healing processes of connective and epithelial tissue [105], the next part of the review is devoted to Zn.

Evidence supports the use of ZnC in the maintenance, prevention and treatment of mucosal and epithelial tissue damage, and in the treatment of oral mucositis and taste disorders in cancer patients undergoing radiotherapy and chemotherapy [104].

The efficacy of ZnC was confirmed in trials investigating taste receptors, in which ZnC supplementation (17 mg, 34 mg or 68 mg/day for 12 weeks) resulted in a faster return of taste [106]. Patients with taste dysfunction experienced faster symptom disappearance after ZnC supplementation (150 mg twice a day until symptom disappearance) [107]. A prospective trial on patients with radiation-induced oral mucositis confirmed the effect of an oral ZnC rinse, which led to a lower incidence of grade 3 oral mucositis, both by mucosal findings and subjective symptomatology [108]. Similar benefits were also described by Hayashi et al. [109,110]. In a trial of patients with hematological malignancies after chemoradiotherapy followed by hematopoietic stem cell transplantation, a benefit was also demonstrated. Patients rinsed their mouths with ZnC mouthwash containing sodium alginate (P-AG), which reduced the incidence of moderate to severe oral mucositis [109]. P-AG has also been shown to reduce the time to hospital discharge after radiotherapy in patients with head and neck cancer [111]. In a trial involving 10 patients, ZnC was found to reduce damage to the stomach and small intestine and contribute to gastrointestinal mucosal stabilization [112]. In one randomized study, ZnC oral rinse had a beneficial effect when it came to the use of analgesics [113]. Available trials on the effectiveness of ZnC supplementation are summarized in Table 5.

Table 5. The effects of carnosine on oral mucositis, taste disorders, gastritis and gastrointestinal dysfunctions.

Author (Year)	Study Design	Intervention	Number of Patients	Effect
Doi et al. (2015) [108]	two-arm, prospective	1 min ZnC mouth rinse (37.5 mg/10 mL, 4×/day)	32	Grade 3 mucositis was observed less frequently according to clinical findings and symptomatology. ZnC promoted recovery.
Watanabe et al. (2010) [113]	RCT	ZnC oral rinse	16/15 (31 total)	Use of analgesics was less frequent and the amount of food intake was significantly higher. Tumor response rate was not affected in patients receiving ZnC.
Hayashi et al. (2016) [110]	prospective, three-arm	ZnC lozenge (18.75 mg 4×/day), ZnC suspension (75 mg in 4 doses/day)	19/31/16 (66 total)	ZnC lozenge was highly effective for prevention of moderate to severe oral mucositis in patients receiving high-dose chemotherapy for HSCT. The efficacy of lozenge preparation was comparable suspension.
Hayashi et al. (2014) [109]	retrospective	ZnC (500 mg) in 20mL P-AG, mouth rinse	36	Reduced the incidence of moderate-to-severe oral mucositis, and pain was significantly relieved. Incidence of xerostomia and taste disturbance tended to be lowered, but not significantly.
Yanase et al. (2015) [114]	retrospective	60 mL P-AG and 150 mg ZnC (3×/day)	19/19 (38 total)	ZnC highly effective in suppressing the development of radiation esophagitis without affecting the tumor response rate.
Sakagami et al. (2009) [106]	double-blind RCT, multi-center	ZnC (17 mg, 34 mg or 68 mg/day; 12 weeks)	28/27/26/28 (109 total)	An amount of 68 mg of ZnC showed a significant improvement in gustatory sensitivity.
Fujii et al. (2018) [107]	retrospective	ZnC (150 mg; 2×/day), until symptom disappearance	80	The administration of 150 mg of ZnC to patients (with pancreatic cancer treated with fluoropyrimidines) with grade 2 dysgeusia significantly shortened its duration.
Mahmood et al. (2007) [112]	RCT	ZnC (37.5 mg; 2×/day) before and after 5 days of indomethacin treatment (50 mg; 3×/day)	10	ZnC, at concentrations likely to be found in the gut lumen, stabilized gut mucosa.
Tan et al. (2017) [115]	RCT, multi-center	ZnC (150 mg/day) combined with triple therapy; ZnC (300 mg/day) combined with triple therapy; triple therapy	113/108/111 (332 total)	Confirmed the effectiveness of the zinc compound in improving *HP* eradication rate.
Takaoka et al. (2010) [116]	two-arm, prospective	150 mg of ZnC orally	12/28 (40 total)	No significant correlation between improvement of VAS pain score and the zinc concentration in the serum after zinc supplementation.
Masayuki et al. (2002) [117]	two-arm, prospective	ZnC and 2% carmellose sodium	19	ZnC was found to have efficacy and safety as a preventive drug for radiation-induced stomatitis.
Suzuki et al. (2016) [111]	retrospective	P-AG oral rinse	104	P-AG was found to be effective in preventing severe oral mucositis and reducing the irradiation period and median time to discharge after completion of radiotherapy.
Baraniuk et al. (2013) [57]	RCT	Carnosine (500 mg, 1000 mg and 1500 mg increasing at 4-week intervals)	25	Decrease in diarrhea associated with irritable bowel syndrome.

Table 5. *Cont.*

Author (Year)	Study Design	Intervention	Number of Patients	Effect
Kitagawa et al. (2021) [118]	RCT, multi-center	ZnC lozenge (18.75 mg) 4×/day, until 35 days after transplantation	47/41 (88 total)	In patients receiving high-dose chemotherapy followed by hematopoietic stem cell transplantation, grade ≥2 oral mucositis was significantly reduced in the intervention group.
Jung et al. (2021) [119]	RCT	ZnC (150 mg/day), pantoprazole or rebamipide (300 mg/day), and pantoprazole	200	ZnC plus PPI treatment showed noninferiority to rebamipide, with PPI treatment of the ulcer healing rate at 4 weeks after endoscopic submucosal dissection.

HP: *Helicobacter pylori*; HSCT: hemopoetic stem cell transplantation; P-AG: polaprezinc suspension sodium alginate solution; PPI: proton pump inhibitor; RCT: randomized controlled trial; VAS: visual analogue scale; ZnC: zinc carnosine.

However, in another study, the effect of ZnC supplementation was not described [116]. ZnC supplementation reduced the incidence of esophagitis by more than two grades and delayed the onset of grade 2 esophagitis [114].

Supplementation of ZnC (4 × 18.75 mg/day) affected mucositis grade >2 in a trial involving 88 patients after high-dose chemotherapy before hematopoietic cell transplantation [118]. Supplementation also reduced diarrhea associated with irritable bowel syndrome [57], and reduced the incidence of radiation-induced stomatitis in a placebo-controlled trial [117].

ZnC as a supplement has been approved in the USA, Canada and Australia for use in the prevention and mitigation of post-radiation oral mucositis, loss of taste after chemotherapy and as a regulator of zinc release into tissue structures [120]. In Japan and South Korea, ZnC is prescribed for use in the treatment of surgical wounds after the removal of gastric ulcers and the eradication of *Helicobacter pylori* [115]. Its usability as an add-on therapy (omeprazole 20 mg, amoxicillin 1 g and clarithromycin 500 mg, each twice daily) for the eradication of *H. pylori* was confirmed in a randomized placebo-controlled prospective multicenter trial [115]. ZnC (150 mg/day) and proton pump inhibitor together showed non-inferiority to the standard treatment in the gastric ulcer healing rate [119]. ZnC is officially approved for this purpose in Japan [120]. A review by Hewlings et al. described the effect of ZnC on oral mucositis, taste disorders and the gastrointestinal system and presented more evidence in the area of promoting the mucosal integrity of the gastrointestinal tract [120].

ZnC has also been investigated by the European Food Safety Authority (EFSA), which mentioned limited bioavailability, since ZnC is insoluble in water at neutral pH, and also expressed reservations about the nature of the particles being absorbed [26]. ZnC is absorbed by the human body and effectively provides zinc to the tissues [26]. The molecule has also been reviewed for safety and use in humans by the US Food and Drug Administration (FDA) and was granted "new dietary ingredient" status in 2002 [121].

12. Conclusions

Data from clinical trials demonstrate the potential benefits of beta-alanine (4–6 g per day) and carnosine (up to 2 g per day) supplementation mainly in the prevention of sarcopenia, neuroprotection and cognitive preservation, neurodegenerative diseases and the improvement of diabetes mellitus parameters. The effectiveness has also been demonstrated for a synthetic derivative of zinc-L-carnosine (68–500 mg per day) to improve symptoms and clinical findings of oral mucositis, to treat and prevent taste disorders after chemotherapy, to regress symptoms of esophagitis, to protect the gastrointestinal mucosal layer and as an additive to the eradication treatment of *Helicobacter pylori*.

Conversely, in the areas of senile cataracts, cardiovascular disease, autistic disorders in children and schizophrenia in adults, the results to date are inconclusive and further clinical trials are needed to confirm a possible effect.

Supplementary Materials: The following supporting information can be downloaded at: https://www.mdpi.com/article/10.3390/nu15071770/s1, Table S1. Molecular functions of carnosine.

Author Contributions: O.C. contributed to the conception and design of the manuscript, to the acquisition, analysis, and interpretation of the data and drafted the written manuscript; V.S.J., contributed significantly to the design of the review and the interpretation of the data; A.V., J.V. and P.B. contributed equally to the data interpretation; All authors critically revised the manuscript, gave final approval, and agree to be fully accountable for ensuring the integrity and accuracy of the work. All authors have read and agreed to the published version of the manuscript.

Funding: The study was supported by the Internal Grant of Palacky University IGA_LF_2022_031. Supported by Ministry of Health, Czech Republic—conceptual development of research organization (FNOl, 00098892).

Institutional Review Board Statement: Not applicable.

Informed Consent Statement: Not applicable.

Data Availability Statement: Not applicable.

Acknowledgments: We thank Vilim Simanek, in memoriam, for his guidance on writing this paper.

Conflicts of Interest: The authors declare no conflict of interest.

References

1. Drozak, J.; Veiga-da-Cunha, M.; Vertommen, D.; Stroobant, V.; Van Schaftingen, E. Molecular Identification of Carnosine Synthase as ATP-grasp Domain-containing Protein 1 (ATPGD1). *J. Biol. Chem.* **2010**, *285*, 9346–9356. [CrossRef]
2. Wu, G. Important roles of dietary taurine, creatine, carnosine, anserine and 4-hydroxyproline in human nutrition and health. *Amino Acids* **2020**, *52*, 329–360. [CrossRef]
3. Xing, L.; Chee, M.E.; Zhang, H.; Zhang, W.; Mine, Y. Carnosine—A natural bioactive dipeptide: Bioaccessibility, bioavailability and health benefits. *J. Food. Bioact.* **2019**, *5*, 8–17. [CrossRef]
4. Haus, J.M.; Thyfault, J.P. Therapeutic potential of carbonyl-scavenging carnosine derivative in metabolic disorders. *J. Clin. Investig.* **2018**, *128*, 5198–5200. [CrossRef] [PubMed]
5. Hipkiss, A.R.; Chana, H. Carnosine Protects Proteins against Methylglyoxal-Mediated Modifications. *Biochem. Biophys. Res. Commun.* **1998**, *248*, 28–32. [CrossRef]
6. Dolan, E.; Saunders, B.; Dantas, W.S.; Murai, I.H.; Roschel, H.; Artioli, G.G.; Harris, R.; Bicudo, J.E.P.W.; Sale, C.; Gualano, B. A Comparative Study of Hummingbirds and Chickens Provides Mechanistic Insight on the Histidine Containing Dipeptide Role in Skeletal Muscle Metabolism. *Sci. Rep.* **2018**, *8*, 14788. [CrossRef] [PubMed]
7. Abdelkader, H.; Longman, M.; Alany, R.G.; Pierscionek, B. On the Anticataractogenic Effects of L-Carnosine: Is It Best Described as an Antioxidant, Metal-Chelating Agent or Glycation Inhibitor? *Oxid Med. Cell. Longev.* **2016**, *2016*, 3240261. [CrossRef]
8. Nicoletti, V.G.; Santoro, A.M.; Grasso, G.; Vagliasindi, L.I.; Giuffrida, M.L.; Cuppari, C.; Purrello, V.S.; Stella, A.M.G.; Rizzarelli, E. Carnosine interaction with nitric oxide and astroglial cell protection. *J. Neurosci. Res.* **2007**, *85*, 2239–2245. [CrossRef] [PubMed]
9. Corona, C.; Frazzini, V.; Silvestri, E.; Lattanzio, R.; La Sorda, R.; Piantelli, M.; Canzoniero, L.M.T.; Ciavardelli, D.; Rizzarelli, E.; Sensi, S.L. Effects of Dietary Supplementation of Carnosine on Mitochondrial Dysfunction, Amyloid Pathology, and Cognitive Deficits in 3xTg-AD Mice. *PLoS ONE* **2011**, *6*, e17971. [CrossRef]
10. Boldyrev, A.A.; Aldini, G.; Derave, W. Physiology and Pathophysiology of Carnosine. *Physiol. Rev.* **2013**, *93*, 1803–1845. [CrossRef] [PubMed]
11. Aldini, G.; de Courten, B.; Regazzoni, L.; Gilardoni, E.; Ferrario, G.; Baron, G.; Altomare, A.; D'Amato, A.; Vistoli, G.; Carini, M. Understanding the antioxidant and carbonyl sequestering activity of carnosine: Direct and indirect mechanisms. *Free Radic. Res.* **2021**, *55*, 321–330. [CrossRef] [PubMed]
12. Holliday, R.; McFarland, G.A. A role for carnosine in cellular maintenance. *Biochemistry (Mosc.)* **2000**, *65*, 843–848. [PubMed]
13. Villari, V.; Attanasio, F.; Micali, N. Control of the Structural Stability of α-Crystallin under Thermal and Chemical Stress: The Role of Carnosine. *J. Phys. Chem. B* **2014**, *118*, 13770–13776. [CrossRef]
14. Nagai, K.; Suda, T. Immunoregulative effects of carnosine and beta-alanine. *Nihon Seirigaku Zasshi* **1986**, *48*, 564–571.
15. Hipkiss, A.R.; Baye, E.; de Courten, B. Carnosine and the processes of ageing. *Maturitas* **2016**, *93*, 28–33. [CrossRef] [PubMed]
16. Turner, M.D.; Sale, C.; Garner, A.C.; Hipkiss, A.R. Anti-cancer actions of carnosine and the restoration of normal cellular homeostasis. *Biochim. Biophys. Acta—Mol. Cell Res.* **2021**, *1868*, 119117. [CrossRef]

17. Li, X.; Yang, K.; Gao, S.; Zhao, J.; Liu, G.; Chen, Y.; Lin, H.; Zhao, W.; Hu, Z.; Xu, N. Carnosine Stimulates Macrophage-Mediated Clearance of Senescent Skin Cells Through Activation of the AKT2 Signaling Pathway by CD36 and RAGE. *Front. Pharmacol.* **2020**, *11*, 593832. [CrossRef]

18. Mikuła-Pietrasik, J.; Książek, K. L-Carnosine Prevents the Pro-cancerogenic Activity of Senescent Peritoneal Mesothelium Towards Ovarian Cancer Cells. *Anticancer Res.* **2016**, *36*, 665–671.

19. Wu, C.C.; Lai, P.Y.; Hsieh, S.; Cheng, C.C.; Hsieh, S.L. Suppression of Carnosine on Adhesion and Extravasation of Human Colorectal Cancer Cells. *Anticancer Res.* **2019**, *39*, 6135–6144. [CrossRef]

20. Prakash, M.D.; Fraser, S.; Boer, J.C.; Plebanski, M.; de Courten, B.; Apostolopoulos, V. Anti-Cancer Effects of Carnosine—A Dipeptide Molecule. *Molecules* **2021**, *26*, 1644. [CrossRef]

21. Hwang, B.; Shin, S.S.; Song, J.H.; Choi, Y.H.; Kim, W.J.; Moon, S.K. Carnosine exerts antitumor activity against bladder cancers in vitro and in vivo via suppression of angiogenesis. *J. Nutr. Biochem.* **2019**, *74*, 108230. [CrossRef]

22. Hipkiss, A.R.; Gaunitz, F. Inhibition of tumour cell growth by carnosine: Some possible mechanisms. *Amino Acids* **2014**, *46*, 327–337. [CrossRef]

23. Dolan, E.; Swinton, P.A.; de Salles Painelli, V.; Hemingway, B.S.; Mazzolani, B.; Smaira, F.I.; Saunders, B.; Artioli, G.G.; Gualano, B. A Systematic Risk Assessment and Meta-Analysis on the Use of Oral β-Alanine Supplementation. *Adv. Nutr.* **2019**, *10*, 452–463. [CrossRef]

24. Saunders, B.; Elliott-Sale, K.; Artioli, G.G.; Swinton, P.A.; Dolan, E.; Roschel, H.; Sale, C.; Gualano, B. β-alanine supplementation to improve exercise capacity and performance: A systematic review and meta-analysis. *Br. J. Sports Med.* **2017**, *51*, 658–669. [CrossRef]

25. Trexler, E.T.; Smith-Ryan, A.E.; Stout, J.R.; Hoffman, J.R.; Wilborn, C.D.; Sale, C.; Kreider, R.B.; Jäger, R.; Earnest, C.P.; Bannock, L.; et al. International society of sports nutrition position stand: Beta-Alanine. *J. Int. Soc. Sports Nutr.* **2015**, *12*, 30. [CrossRef]

26. Turck, D.; Bohn, T.; Castenmiller, J.; De Henauw, S.; Hirsch-Ernst, K.I.; Maciuk, A.; Mangelsdorf, I.; McArdle, H.J.; Naska, A.; Pelaez, C.; et al. Safety of zinc l-carnosine as a Novel food pursuant to Regulation (EU) 2015/2283 and the bioavailability of zinc from this source in the context of Directive 2002/46/EC on food supplements. *EFSA J.* **2022**, *20*, 7332. [CrossRef]

27. Wu, G. *Amino Acids*; CRC Press: Boca Raton, FL, USA, 2013. [CrossRef]

28. Artioli, G.G.; Sale, C.; Jones, R.L. Carnosine in health and disease. *Eur. J. Sport Sci.* **2019**, *19*, 30–39. [CrossRef] [PubMed]

29. Spelnikov, D.; Harris, R.C. A kinetic model of carnosine synthesis in human skeletal muscle. *Amino Acids* **2019**, *51*, 115–121. [CrossRef] [PubMed]

30. Asatoor, A.M.; Bandoh, J.K.; Lant, A.F.; Milne, M.D.; Navab, F. Intestinal absorption of carnosine and its constituent amino acids in man. *Gut* **1970**, *11*, 250–254. [CrossRef]

31. Chmielewska, K.; Dzierzbicka, K.; Inkielewicz-Stępniak, I.; Przybyłowska, M. Therapeutic Potential of Carnosine and Its Derivatives in the Treatment of Human Diseases. *Chem. Res. Toxicol.* **2020**, *33*, 1561–1578. [CrossRef]

32. Gaunitz, F.; Hipkiss, A.R. Carnosine and cancer: A perspective. *Amino Acids* **2012**, *43*, 135–142. [CrossRef]

33. Jin, C.-L.; Yang, L.-X.; Wu, X.-H.; Li, Q.; Ding, M.-P.; Fan, Y.-Y.; Zhang, W.-P.; Luo, J.-H.; Chen, Z. Effects of carnosine on amygdaloid-kindled seizures in Sprague–Dawley rats. *Neuroscience* **2005**, *135*, 939–947. [CrossRef] [PubMed]

34. Rocha, S.M.; Pires, J.; Esteves, M.; GraÃ§a, B.; Bernardino, L. Histamine: A new immunomodulatory player in the neuron-glia crosstalk. *Front. Cell. Neurosci.* **2014**, *8*, 120. [CrossRef] [PubMed]

35. Charoensin, S.; Laopaiboon, B.; Boonkum, W.; Phetcharaburanin, J.; Villareal, M.O.; Isoda, H.; Duangjinda, M. Thai Native Chicken as a Potential Functional Meat Source Rich in Anserine, Anserine/Carnosine, and Antioxidant Substances. *Animals* **2021**, *11*, 902. [CrossRef]

36. Jones, G.; Smith, M.; Harris, R. Imidazole dipeptide content of dietary sources commonly consumed within the British diet. *Proc Nutr. Soc.* **2011**, *70*, E363. [CrossRef]

37. Sarkar, P.; Kar, K.; Mondal, M.C.; Chakraborty, I.; Kar, M. Elevated level of carbonyl compounds correlates with insulin resistance in type 2 diabetes. *Ann. Acad. Med. Singapore* **2010**, *39*, 909–912. [CrossRef] [PubMed]

38. Freund, M.A.; Chen, B.; Decker, E.A. The Inhibition of Advanced Glycation End Products by Carnosine and Other Natural Dipeptides to Reduce Diabetic and Age-Related Complications. *Compr. Rev. Food Sci. Food Saf.* **2018**, *17*, 1367–1378. [CrossRef] [PubMed]

39. de Courten, B.; Jakubova, M.; de Courten, M.P.; Kukurova, I.J.; Vallova, S.; Krumpolec, P.; Valkovic, L.; Kurdiova, T.; Garzon, D.; Barbaresi, S.; et al. Effects of carnosine supplementation on glucose metabolism: Pilot clinical trial. *Obesity* **2016**, *24*, 1027–1034. [CrossRef]

40. Siriwattanasit, N.; Satirapoj, B.; Supasyndh, O. Effect of Oral carnosine supplementation on urinary TGF-β in diabetic nephropathy: A randomized controlled trial. *BMC Nephrol.* **2021**, *22*, 236. [CrossRef]

41. Karkabounas, S.; Papadopoulos, N.; Anastasiadou, C.; Gubili, C.; Peschos, D.; Daskalou, T.; Fikioris, N.; Simos, Y.V.; Kontargiris, E.; Gianakopoulos, X.; et al. Effects of α-Lipoic Acid, Carnosine, and Thiamine Supplementation in Obese Patients with Type 2 Diabetes Mellitus: A Randomized, Double-Blind Study. *J. Med. Food* **2018**, *21*, 1197–1203. [CrossRef]

42. Nealon, R.S.; Sukala, W.R. The Effect of 28 Days of Beta-alanine Supplementation on Exercise Capacity and Insulin Sensitivity in Individuals with Type 2 Diabetes Mellitus: A Randomised, Double-blind and Placebo-controlled Pilot Trial. *Sport Nutr. Ther.* **2016**, *1*, 1–7. [CrossRef]

43. Elbarbary, N.S.; Ismail, E.A.R.; El-Naggar, A.R.; Hamouda, M.H.; El-Hamamsy, M. The effect of 12 weeks carnosine supplementation on renal functional integrity and oxidative stress in pediatric patients with diabetic nephropathy: A randomized placebo-controlled trial. *Pediatr. Diabetes* **2018**, *19*, 470–477. [CrossRef]
44. Houjeghani, S.; Kheirouri, S.; Faraji, E.; Jafarabadi, M.A. l -Carnosine supplementation attenuated fasting glucose, triglycerides, advanced glycation end products, and tumor necrosis factor—α levels in patients with type 2 diabetes: A double-blind placebo-controlled randomized clinical trial. *Nutr. Res.* **2018**, *49*, 96–106. [CrossRef]
45. Matthews, J.J.; Dolan, E.; Swinton, P.A.; Santos, L.; Artioli, G.G.; Turner, M.D.; Elliott-Sale, K.J.; Sale, C. Effect of Carnosine or β -Alanine Supplementation on Markers of Glycemic Control and Insulin Resistance in Humans and Animals: A Systematic Review and Meta-analysis. *Adv. Nutr.* **2021**, *12*, 2216–2231. [CrossRef]
46. Sureshkumar, K.; Durairaj, M.; Srinivasan, K.; Goh, K.W.; Undela, K.; Mahalingam, V.T.; Ardianto, C.; Ming, L.C.; Ganesan, R.M. Effect of L-Carnosine in Patients with Age-Related Diseases: A Systematic Review and Meta-Analysis. *Front. Biosci.* **2023**, *28*, 18. [CrossRef]
47. Menon, K.; Marquina, C.; Liew, D.; Mousa, A.; Courten, B. Histidine-containing dipeptides reduce central obesity and improve glycaemic outcomes: A systematic review and meta-analysis of randomized controlled trials. *Obes. Rev.* **2020**, *21*, e12975. [CrossRef] [PubMed]
48. Spaas, J.; Van Noten, P.; Keytsman, C.; Nieste, I.; Blancquaert, L.; Derave, W.; Eijnde, B.O. Carnosine and skeletal muscle dysfunction in a rodent multiple sclerosis model. *Amino Acids* **2021**, *53*, 1749–1761. [CrossRef]
49. Takeuchi, K.; Toyohara, H.; Sakaguchi, M. A hyperosmotic stress-induced mRNA of carp cell encodes Na+- and Cl−-dependent high affinity taurine transporter1The sequence reported in this paper has been deposited in the DDBJ/EMBL/GenBank database with accession no. AB006986.1. *Biochim. Biophys. Acta—Biomembr.* **2000**, *1464*, 219–230. [CrossRef] [PubMed]
50. Lenney, J.F.; George, R.P.; Weiss, A.M.; Kucera, C.M.; Chan, P.W.H.; Rinzler, G.S. Human serum carnosinase: Characterization, distinction from cellular carnosinase, and activation by cadmium. *Clin. Chim. Acta* **1982**, *123*, 221–231. [CrossRef] [PubMed]
51. Manoharan, S.; Guillemin, G.J.; Abiramasundari, R.S.; Essa, M.M.; Akbar, M.; Akbar, M.D. The Role of Reactive Oxygen Species in the Pathogenesis of Alzheimer's Disease, Parkinson's Disease, and Huntington's Disease: A Mini Review. *Oxid. Med. Cell. Longev.* **2016**, *2016*, 8590578. [CrossRef]
52. Schön, M.; Mousa, A.; Berk, M.; Chia, W.L.; Ukropec, J.; Majid, A.; Ukropcová, B.; de Courten, B. The Potential of Carnosine in Brain-Related Disorders: A Comprehensive Review of Current Evidence. *Nutrients* **2019**, *11*, 1196. [CrossRef] [PubMed]
53. Rokicki, J.; Li, L.; Imabayashi, E.; Kaneko, J.; Hisatsune, T.; Matsuda, H. Daily Carnosine and Anserine Supplementation Alters Verbal Episodic Memory and Resting State Network Connectivity in Healthy Elderly Adults. *Front. Aging Neurosci.* **2015**, *7*, 219. [CrossRef]
54. Szcześniak, D.; Budzeń, S.; Kopeć, W.; Rymaszewska, J. Anserine and carnosine supplementation in the elderly: Effects on cognitive functioning and physical capacity. *Arch Gerontol. Geriatr.* **2014**, *59*, 485–490. [CrossRef]
55. Katakura, Y.; Totsuka, M.; Imabayashi, E.; Matsuda, H.; Hisatsune, T. Anserine/Carnosine Supplementation Suppresses the Expression of the Inflammatory Chemokine CCL24 in Peripheral Blood Mononuclear Cells from Elderly People. *Nutrients* **2017**, *9*, 1199. [CrossRef]
56. Hisatsune, T.; Kaneko, J.; Kurashige, H.; Cao, Y.; Satsu, H.; Totsuka, M.; Katakura, Y.; Imabayashi, E.; Matsuda, H. Effect of Anserine/Carnosine Supplementation on Verbal Episodic Memory in Elderly People. *J. Alzheimer's Dis.* **2015**, *50*, 149–159. [CrossRef] [PubMed]
57. Baraniuk, J.N.; El-Amin, S.; Corey, R.; Rayhan, R.; Timbol, C. Carnosine Treatment for Gulf War Illness: A Randomized Controlled Trial. *Glob. J. Health Sci.* **2013**, *5*, 69–81. [CrossRef] [PubMed]
58. Solis, M.Y.; Cooper, S.; Hobson, R.M.; Artioli, G.G.; Otaduy, M.C.; Roschel, H.; Robertson, J.; Martin, D.; Painelli, V.S.; Harris, R.C.; et al. Effects of Beta-Alanine Supplementation on Brain Homocarnosine/Carnosine Signal and Cognitive Function: An Exploratory Study. *PLoS ONE* **2015**, *10*, e0123857. [CrossRef] [PubMed]
59. Masuoka, N.; Yoshimine, C.; Hori, M.; Tanaka, M.; Asada, T.; Abe, K.; Hisatsune, T. Effects of Anserine/Carnosine Supplementation on Mild Cognitive Impairment with APOE4. *Nutrients* **2019**, *11*, 1626. [CrossRef]
60. Boldyrev, A.; Fedorova, T.; Stepanova, M.; Dobrotvorskaya, I.; Kozlova, E.; Boldanova, N.; Bagyeva, G.; Ivanova-Smolenskaya, I.; Illarioshkin, S. Carnisone Increases Efficiency of DOPA Therapy of Parkinson's Disease: A Pilot Study. *Rejuvenation Res.* **2008**, *11*, 821–827. [CrossRef]
61. Cornelli, U. Treatment of Alzheimer's Disease with a Cholinesterase Inhibitor Combined with Antioxidants. *Neurodegener. Dis.* **2010**, *7*, 193–202. [CrossRef]
62. Checkoway, H.; Lundin, J.I.; Kelada, S.N. Neurodegenerative diseases. *IARC Sci. Publ.* **2011**, *163*, 407–419.
63. Fonteh, A.N.; Harrington, R.J.; Tsai, A.; Liao, P.; Harrington, M.G. Free amino acid and dipeptide changes in the body fluids from Alzheimer's disease subjects. *Amino Acids* **2007**, *32*, 213–224. [CrossRef] [PubMed]
64. Molina, J.A.; Jiménez-Jiménez, F.J.; Gomez, P.; Vargas, C.; Navarro, J.A.; Ortí-Pareja, M.; Gasalla, T.; Benito-León, J.; Bermejo, F.; Arenas, J. Decreased cerebrospinal fluid levels of neutral and basic amino acids in patients with Parkinson's disease. *J. Neurol. Sci.* **1997**, *150*, 123–127. [CrossRef]
65. Millan, M.J.; Fone, K.; Steckler, T.; Horan, W.P. Negative symptoms of schizophrenia: Clinical characteristics, pathophysiological substrates, experimental models and prospects for improved treatment. *Eur. Neuropsychopharmacol.* **2014**, *24*, 645–692. [CrossRef]

66. Beck, K.; Javitt, D.C.; Howes, O.D. Targeting glutamate to treat schizophrenia: Lessons from recent clinical studies. *Psychopharmacology (Berl.)* **2016**, *233*, 2425–2428. [CrossRef] [PubMed]
67. Ghajar, A.; Khoaie-Ardakani, M.-R.; Shahmoradi, Z.; Alavi, A.-R.; Afarideh, M.; Shalbafan, M.-R.; Ghazizadeh-Hashemi, M.; Akhondzadeh, S. L-carnosine as an add-on to risperidone for treatment of negative symptoms in patients with stable schizophrenia: A double-blind, randomized placebo-controlled trial. *Psychiatry Res.* **2018**, *262*, 94–101. [CrossRef]
68. Chengappa, K.N.R.; Turkin, S.R.; DeSanti, S.; Bowie, C.R.; Brar, J.S.; Schlicht, P.J.; Murphy, S.L.; Hetrick, M.L.; Bilder, R.; Fleet, D. A preliminary, randomized, double-blind, placebo-controlled trial of l-carnosine to improve cognition in schizophrenia. *Schizophr. Res.* **2012**, *142*, 145–152. [CrossRef]
69. Woo, H.-I.; Chun, M.-R.; Yang, J.-S.; Lim, S.-W.; Kim, M.-J.; Kim, S.-W.; Myung, W.-J.; Kim, D.-K.; Lee, S.-Y. Plasma Amino Acid Profiling in Major Depressive Disorder Treated With Selective Serotonin Reuptake Inhibitors. *CNS Neurosci. Ther.* **2015**, *21*, 417–424. [CrossRef]
70. Ali-Sisto, T.; Tolmunen, T.; Kraav, S.-L.; Mäntyselkä, P.; Valkonen-Korhonen, M.; Honkalampi, K.; Ruusunen, A.; Velagapudi, V.; Lehto, S.M. Serum levels of carnosine may be associated with the duration of MDD episodes. *J. Affect Disord.* **2023**, *320*, 647–655. [CrossRef] [PubMed]
71. Sauer, A.K.; Stanton, J.E.; Hans, S.; Grabrucker, A.M. Autism Spectrum Disorders: Etiology and Pathology. In *Autism Spectrum Disorders*; Exon Publications: Brisbane, Australia, 2021. [CrossRef]
72. Anyachor, C.P.; Dooka, D.B.; Orish, C.N.; Amadi, C.N.; Bocca, B.; Ruggieri, F.; Senofonte, M.; Frazzoli, C.; Orisakwe, O.E. Mechanistic considerations and biomarkers level in nickel-induced neurodegenerative diseases: An updated systematic review. *IBRO Neurosci. Reports* **2022**, *13*, 136–146. [CrossRef]
73. Błażewicz, A.; Grabrucker, A.M. Metal Profiles in Autism Spectrum Disorders: A Crosstalk between Toxic and Essential Metals. *Int. J. Mol. Sci.* **2022**, *24*, 308. [CrossRef] [PubMed]
74. Abraham, D.A.; Undela, K.; Narasimhan, U.; Rajanandh, M.G. Effect of L-Carnosine in children with autism spectrum disorders: A systematic review and meta-analysis of randomised controlled trials. *Amino Acids* **2021**, *53*, 575–585. [CrossRef] [PubMed]
75. Bala, K.A.; Doğan, M.; Mutluer, T.; Kaba, S.; Aslan, O.; Balahoroğlu, R.; Çokluk, E.; Üstyol, L.; Kocaman, S. Plasma amino acid profile in autism spectrum disorder (ASD). *Eur. Rev. Med. Pharmacol. Sci.* **2016**, *20*, 923–929.
76. Ming, X.; Stein, T.P.; Barnes, V.; Rhodes, N.; Guo, L. Metabolic Perturbance in Autism Spectrum Disorders: A Metabolomics Study. *J. Proteome Res.* **2012**, *11*, 5856–5862. [CrossRef] [PubMed]
77. Chez, M.G.; Buchanan, C.P.; Aimonovitch, M.C.; Becker, M.; Schaefer, K.; Black, C.; Komen, J. Double-Blind, Placebo-Controlled Study of L-Carnosine Supplementation in Children with Autistic Spectrum Disorders. *J. Child Neurol.* **2002**, *17*, 833–837. [CrossRef] [PubMed]
78. Mehrazad-Saber, Z.; Kheirouri, S.; Noorazar, S.G. Effects of L-Carnosine Supplementation on Sleep Disorders and Disease Severity in Autistic Children: A Randomized, Controlled Clinical Trial. *Basic Clin. Pharmacol. Toxicol.* **2018**, *123*, 72–77. [CrossRef]
79. Hajizadeh-Zaker, R.; Ghajar, A.; Mesgarpour, B.; Afarideh, M.; Mohammadi, M.R.; Akhondzadeh, S. l-Carnosine As an Adjunctive Therapy to Risperidone in Children with Autistic Disorder: A Randomized, Double-Blind, Placebo-Controlled Trial. *J. Child Adolesc. Psychopharmacol.* **2018**, *28*, 74–81. [CrossRef] [PubMed]
80. Ghajar, A.; Aghajan-Nashtaei, F.; Afarideh, M.; Mohammadi, M.R.; Akhondzadeh, S. l-Carnosine as Adjunctive Therapy in Children and Adolescents with Attention-Deficit/Hyperactivity Disorder: A Randomized, Double-Blind, Placebo-Controlled Clinical Trial. *J. Child Adolesc. Psychopharmacol.* **2018**, *28*, 331–338. [CrossRef]
81. Esposito, D.; Belli, A.; Ferri, R.; Bruni, O. Sleeping without Prescription: Management of Sleep Disorders in Children with Autism with Non-Pharmacological Interventions and over-the-Counter Treatments. *Brain Sci.* **2020**, *10*, 441. [CrossRef]
82. Siafis, S.; Çıray, O.; Wu, H.; Schneider-Thoma, J.; Bighelli, I.; Krause, M.; Rodolico, A.; Ceraso, A.; Deste, G.; Huhn, M.; et al. Pharmacological and dietary-supplement treatments for autism spectrum disorder: A systematic review and network meta-analysis. *Mol. Autism.* **2022**, *13*, 10. [CrossRef]
83. Ann Abraham, D.; Narasimhan, U.; Christy, S.; Muhasaparur Ganesan, R. Effect of l-Carnosine as adjunctive therapy in the management of children with autism spectrum disorder: A randomized controlled study. *Amino Acids* **2020**, *52*, 1521–1528. [CrossRef] [PubMed]
84. Lindfield, R.; Vishwanath, K.; Ngounou, F.; Khanna, R. The challenges in improving outcome of cataract surgery in low and middle income countries. *Indian J. Ophthalmol.* **2012**, *60*, 464. [CrossRef]
85. Wang, A.M.; Ma, C.; Xie, Z.H.; Shen, F. Use of carnosine as a natural anti-senescence drug for human beings. *Biochemistry (Mosc).* **2000**, *65*, 869–871.
86. Babizhayev, M.A.; Deyev, A.I.; Yermakova, V.N.; Semiletov, Y.A.; Davydova, N.G.; Kurysheva, N.I.; Zhukotskii, A.V.; Goldman, I.M. N-Acetylcarnosine, a natural histidine-containing dipeptide, as a potent ophthalmic drug in treatment of human cataracts. *Peptides* **2001**, *22*, 979–994. [CrossRef]
87. Babizhayev, M.; Kasus-Jacobi, A. State of the Art Clinical Efficacy and Safety Evaluation of N-acetylcarnosine Dipeptide Ophthalmic Prodrug. Principles for the Delivery, Self-Bioactivation, Molecular Targets and Interaction with a Highly Evolved Histidyl-Hydrazide Structure in the Treatm. *Curr. Clin. Pharmacol.* **2009**, *4*, 4–37. [CrossRef] [PubMed]
88. Babizhayev, M.A.; Burke, L.; Micans, P.; Richer, S.P. N-acetylcarnosine sustained drug delivery eye drops to control the signs of ageless vision: Glare sensitivity, cataract amelioration and quality of vision currently available treatment for the challenging 50,000-patient population. *Clin. Interv. Aging* **2009**, *4*, 31–50. [CrossRef] [PubMed]

89. Dubois, V.D.P.; Bastawrous, A. *N*-acetylcarnosine (NAC) drops for age-related cataract. *Cochrane Database Syst. Rev.* **2017**, *2*, CD009493. [CrossRef]

90. Yang, Q.-J.; Zhao, J.-R.; Hao, J.; Li, B.; Huo, Y.; Han, Y.-L.; Wan, L.-L.; Li, J.; Huang, J.; Lu, J.; et al. Serum and urine metabolomics study reveals a distinct diagnostic model for cancer cachexia. *J. Cachexia Sarcopenia Muscle.* **2018**, *9*, 71–85. [CrossRef]

91. Stuerenburg, H.J. The roles of carnosine in aging of skeletal muscle and in neuromuscular diseases. *Biochemistry (Mosc.)* **2000**, *65*, 862–865.

92. Hipkiss, A.R. Would Carnosine or a Carnivorous Diet Help Suppress Aging and Associated Pathologies? *Ann. N. Y. Acad. Sci.* **2006**, *1067*, 369–374. [CrossRef]

93. Hipkiss, A.R. Glycation, ageing and carnosine: Are carnivorous diets beneficial? *Mech. Ageing Dev.* **2005**, *126*, 1034–1039. [CrossRef]

94. del Favero, S.; Roschel, H.; Solis, M.Y.; Hayashi, A.P.; Artioli, G.G.; Otaduy, M.C.; Benatti, F.B.; Harris, R.C.; Wise, J.A.; Leite, C.C.; et al. Beta-alanine (Carnosyn^{TM}) supplementation in elderly subjects (60–80 years): Effects on muscle carnosine content and physical capacity. *Amino Acids* **2012**, *43*, 49–56. [CrossRef] [PubMed]

95. McCormack, W.P.; Stout, J.R.; Emerson, N.S.; Scanlon, T.C.; Warren, A.M.; Wells, A.J.; Gonzalez, A.M.; Mangine, G.T.; Robinson, E.H., 4th; Fragala, M.S.; et al. Oral nutritional supplement fortified with beta-alanine improves physical working capacity in older adults: A randomized, placebo-controlled study. *Exp. Gerontol.* **2013**, *48*, 933–939. [CrossRef] [PubMed]

96. Furst, T.; Massaro, A.; Miller, C.; Williams, B.T.; LaMacchia, Z.M.; Horvath, P.J. β-Alanine supplementation increased physical performance and improved executive function following endurance exercise in middle aged individuals. *J. Int. Soc. Sports Nutr.* **2018**, *15*, 32. [CrossRef]

97. Dubois-Deruy, E.; Peugnet, V.; Turkieh, A.; Pinet, F. Oxidative Stress in Cardiovascular Diseases. *Antioxidants.* **2020**, *9*, 864. [CrossRef]

98. Roberts, P.R.; Zaloga, G.P. Cardiovascular effects of carnosine. *Biochemistry (Mosc.)* **2000**, *65*, 856–861.

99. Stvolinsky, S.L.; Dobrota, D. Anti-ischemic activity of carnosine. *Biochemistry (Mosc.)* **2000**, *65*, 849–855. [PubMed]

100. Lombardi, C.; Carubelli, V.; Lazzarini, V.; Vizzardi, E.; Bordonali, T.; Ciccarese, C.; Castrini, A.I.; Cas, A.D.; Nodari, S.; Metra, M. Effects of oral administration of orodispersible levo-carnosine on quality of life and exercise performance in patients with chronic heart failure. *Nutrition* **2015**, *31*, 72–78. [CrossRef]

101. Yoshikawa, F.; Nakajima, T.; Hanada, M.; Hirata, K.; Masuyama, T.; Aikawa, R. Beneficial effect of polaprezinc on cardiac function post-myocardial infarction: A prospective and randomized clinical trial. *Medicine* **2019**, *98*, e14637. [CrossRef]

102. Hall, A.G.; King, J.C. Zinc Fortification: Current Trends and Strategies. *Nutrients* **2022**, *14*, 3895. [CrossRef]

103. Li, M.; Sun, Z.; Zhang, H.; Liu, Z. Recent advances on polaprezinc for medical use (Review). *Exp. Ther. Med.* **2021**, *22*, 1445. [CrossRef] [PubMed]

104. Tang, W.; Liu, H.; Ooi, T.C.; Rajab, N.F.; Cao, H.; Sharif, R. Zinc carnosine: Frontiers advances of supplement for cancer therapy. *Biomed Pharmacother.* **2022**, *151*, 113157. [CrossRef]

105. Hewlings, S.J.; Medeiros, D.M. *Nutrition: Real People, Real Choices*; Pearson Prentice Hall: Upper Saddle River, NJ, USA, 2008; ISBN-13 978-0757579035, ISBN-10 0757579035.

106. Sakagami, M.; Ikeda, M.; Tomita, H.; Ikui, A.; Aiba, T.; Takeda, N.; Inokuchi, A.; Kurono, Y.; Nakashima, M.; Shibasaki, Y.; et al. A zinc-containing compound, Polaprezinc, is effective for patients with taste disorders: Randomized, double-blind, placebo-controlled, multi-center study. *Acta Otolaryngol.* **2009**, *129*, 1115–1120. [CrossRef]

107. Fujii, H.; Hirose, C.; Ishihara, M.; Iihara, H.; Imai, H.; Tanaka, Y.; Matsuhashi, N.; Takahashi, T.; Yamaguchi, K.; Yoshida, K.; et al. Improvement of Dysgeusia by Polaprezinc, a Zinc-L-carnosine, in Outpatients Receiving Cancer Chemotherapy. *Anticancer. Res.* **2018**, *38*, 6367–6373. [CrossRef]

108. Doi, H.; Fujiwara, M.; Suzuki, H.; Niwa, Y.; Nakayama, M.; Shikata, T.; Odawara, S.; Takada, Y.; Kimura, T.; Kamikonya, N.; et al. Polaprezinc reduces the severity of radiation-induced mucositis in head and neck cancer patients. *Mol. Clin. Oncol.* **2015**, *3*, 381–386. [CrossRef]

109. Hayashi, H.; Kobayashi, R.; Suzuki, A.; Ishihara, M.; Nakamura, N.; Kitagawa, J.; Kanemura, N.; Kasahara, S.; Kitaichi, K.; Hara, T.; et al. Polaprezinc prevents oral mucositis in patients treated with high-dose chemotherapy followed by hematopoietic stem cell transplantation. *Anticancer Res.* **2014**, *34*, 7271–7277.

110. Hayashi, H.; Kobayashi, R.; Suzuki, A.; Yamada, Y.; Ishida, M.; Shakui, T.; Kitagawa, J.; Hayashi, H.; Sugiyama, T.; Takeuchi, H.; et al. Preparation and clinical evaluation of a novel lozenge containing polaprezinc, a zinc-L-carnosine, for prevention of oral mucositis in patients with hematological cancer who received high-dose chemotherapy. *Med. Oncol.* **2016**, *33*, 91. [CrossRef] [PubMed]

111. Suzuki, A.; Kobayashi, R.; Shakui, T.; Kubota, Y.; Fukita, M.; Kuze, B.; Aoki, M.; Sugiyama, T.; Mizuta, K.; Itoh, Y. Effect of polaprezinc on oral mucositis, irradiation period, and time to discharge in patients with head and neck cancer. *Head Neck* **2016**, *38*, 1387–1392. [CrossRef] [PubMed]

112. Mahmood, A.; FitzGerald, A.J.; Marchbank, T.; Ntatsaki, E.; Murray, D.; Ghosh, S.; Playford, R.J. Zinc carnosine, a health food supplement that stabilises small bowel integrity and stimulates gut repair processes. *Gut* **2007**, *56*, 168–175. [CrossRef]

113. Watanabe, T.; Ishihara, M.; Matsuura, K.; Mizuta, K.; Itoh, Y. Polaprezinc prevents oral mucositis associated with radiochemotherapy in patients with head and neck cancer. *Int. J. Cancer* **2010**, *127*, 1984–1990. [CrossRef]

114. Yanase, K.; Funaguchi, N.; Iihara, H.; Yamada, M.; Kaito, D.; Endo, J.; Ito, F.; Ohno, Y.; Tanaka, H.; Itoh, Y.; et al. Prevention of radiation esophagitis by polaprezinc (zinc L-carnosine) in patients with non-small cell lung cancer who received chemoradiotherapy. *Int. J. Clin. Exp. Med.* **2015**, *8*, 16215–16222. [PubMed]

115. Tan, B.; Luo, H.-Q.; Xu, H.; Lv, N.-H.; Shi, R.-H.; Luo, H.-S.; Li, J.-S.; Ren, J.-L.; Zou, Y.-Y.; Li, Y.-Q.; et al. Polaprezinc combined with clarithromycin-based triple therapy for Helicobacter pylori-associated gastritis: A prospective, multicenter, randomized clinical trial. *PLoS ONE* **2017**, *12*, e0175625. [CrossRef] [PubMed]

116. Takaoka, T.; Sarukura, N.; Ueda, C.; Kitamura, Y.; Kalubi, B.; Toda, N.; Abe, K.; Yamamoto, S.; Takeda, N. Effects of zinc supplementation on serum zinc concentration and ratio of apo/holo-activities of angiotensin converting enzyme in patients with taste impairment. *Auris Nasus Larynx* **2010**, *37*, 190–194. [CrossRef]

117. Masayuki, F.; Norihiko, K.; Keita, T.; Miwa, I.; Masayuki, I.; Toshihiko, I.; Hiromi, F.; Chikaaki, M.; Norio, N. Efficacy and safety of polaprezinc as a preventive drug for radiation-induced stomatitis. *Nihon Igaku Hoshasen Gakkai Zasshi* **2002**, *62*, 144–150.

118. Kitagawa, J.; Kobayashi, R.; Nagata, Y.; Kasahara, S.; Ono, T.; Sawada, M.; Ohata, K.; Kato-Hayashi, H.; Hayashi, H.; Shimizu, M.; et al. Polaprezinc for prevention of oral mucositis in patients receiving chemotherapy followed by hematopoietic stem cell transplantation: A multi-institutional randomized controlled trial. *Int. J. Cancer* **2021**, *148*, 1462–1469. [CrossRef] [PubMed]

119. Jung, D.H.; Park, J.C.; Lee, Y.C.; Lee, S.K.; Shin, S.K.; Chung, H.; Park, J.J.; Kim, J.-H.; Youn, Y.H.; Park, H. Comparison of the Efficacy of Polaprezinc Plus Proton Pump Inhibitor and Rebamipide Plus Proton Pump Inhibitor Treatments for Endoscopic Submucosal Dissection-induced Ulcers. *J. Clin. Gastroenterol.* **2021**, *55*, 233–238. [CrossRef] [PubMed]

120. Hewlings, S.; Kalman, D. A Review of Zinc-L-Carnosine and Its Positive Effects on Oral Mucositis, Taste Disorders, and Gastrointestinal Disorders. *Nutrients* **2020**, *12*, 665. [CrossRef]

121. FDA (Food and Drug Administration). NDI 134—Zinc Carnosine—Original NDI Notification. Published Online 2007. Available online: https://www.regulations.gov/document/FDA-2003-S-0732-0044 (accessed on 1 February 2023).

nutrients

MDPI

Review

The Beneficial Effects of Soybean Proteins and Peptides on Chronic Diseases

Sumei Hu [†], Caiyu Liu [†] and Xinqi Liu *

Beijing Advanced Innovation Center for Food Nutrition and Human Health, Beijing Engineering and Technology Research Center of Food Additives, National Soybean Processing Industry Technology Innovation Center, Beijing Technology and Business University, Beijing 100048, China; husumei@btbu.edu.cn (S.H.); 2130021016@st.btbu.edu.cn (C.L.)
* Correspondence: liuxinqi@btbu.edu.cn
† These authors contributed equally to this work.

Abstract: With lifestyle changes, chronic diseases have become a public health problem worldwide, causing a huge burden on the global economy. Risk factors associated with chronic diseases mainly include abdominal obesity, insulin resistance, hypertension, dyslipidemia, elevated triglycerides, cancer, and other characteristics. Plant-sourced proteins have received more and more attention in the treatment and prevention of chronic diseases in recent years. Soybean is a low-cost, high-quality protein resource that contains 40% protein. Soybean peptides have been widely studied in the regulation of chronic diseases. In this review, the structure, function, absorption, and metabolism of soybean peptides are introduced briefly. The regulatory effects of soybean peptides on a few main chronic diseases were also reviewed, including obesity, diabetes mellitus, cardiovascular diseases (CVD), and cancer. We also addressed the shortcomings of functional research on soybean proteins and peptides in chronic diseases and the possible directions in the future.

Keywords: chronic diseases; soybean peptides; diabetes mellitus; obesity; cardiovascular diseases; cancer

Citation: Hu, S.; Liu, C.; Liu, X. The Beneficial Effects of Soybean Proteins and Peptides on Chronic Diseases. *Nutrients* **2023**, *15*, 1811. https://doi.org/10.3390/nu15081811

Academic Editor: Sara Baldassano

Received: 19 March 2023
Revised: 4 April 2023
Accepted: 5 April 2023
Published: 7 April 2023

1. Introduction

With the development of urbanization and the increase in sedentary habits, chronic diseases have become a worldwide public health problem. Chronic diseases are non-infectious diseases, but not some specific diseases, with complex etiology, slow development, and long duration [1]. Chronic diseases may be caused by lifestyle, environment, diet, or genetic factors. There were 28 million people who died of chronic diseases worldwide in 1990, and this number increased to 36 million in 2008 and 39 million in 2016 [1]. More than two-thirds of the deaths worldwide are believed to be caused by chronic diseases [1]. In 2019, seven of the top ten causes of death were due to non-communicable diseases, according to the World Health Organization (WHO) [2]. At present, chronic diseases mainly include obesity, diabetes, cardiovascular diseases (CVD), and cancer, and they are key causes of premature death in humans [3]. The pathogenesis of chronic diseases is relatively complex, such as imbalances in the protease network, which can lead to malfunctions in the cellular signal network [4].

The expression of matrix metalloproteinases (MMPs) and the imbalance of the phosphoryladylinositol 3-kinase (PI3K)/AKT/major target of rapamycin (mTOR) signaling pathway may both have an impact on the development of chronic diseases such as CVD, type 2 diabetes (T2D), and cancer [4,5].

As metabolic diseases, chronic diseases are associated with metabolic syndrome (MetS). MetS is not a disease itself but a comprehensive concept, representing factors that increase the risks of individual diseases (as shown in Figure 1) [6]. It has been reported that several components of MetS led to a significant increase in the risk of chronic diseases and that the risk factors for chronic diseases and the definition of MetS partially overlap [7].

Therefore, MetS may be a risk factor leading to the development of chronic diseases. Several diseases associated with both conditions have been identified, including obesity, T2D, CVD, and cancer [7].

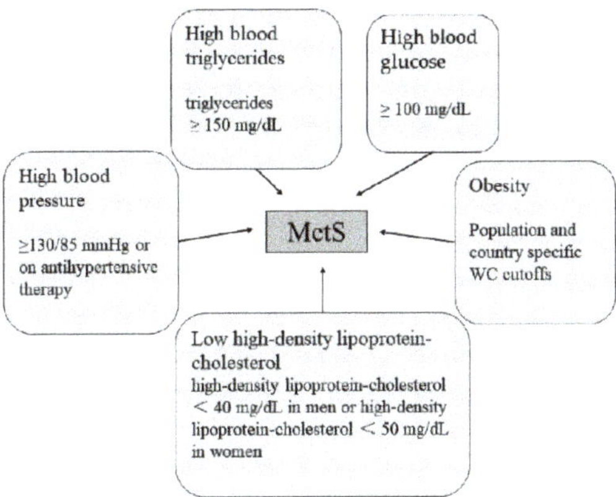

Figure 1. Main diagnosis of MetS. Three of the above five conditions are considered MetS [8].

Among these conditions (Figure 1), obesity is a major cause of MetS. Obesity may increase the susceptibility to insulin resistance, thus causing MetS. In 2005, the International Diabetes Federation recognized obesity as a necessary factor in the diagnosis of MetS [9], while Dr. Reaven objected to this and believed that insulin resistance might be the main cause of MetS [10]. MetS is an important risk factor for T2D and CVD (see Figure 2). With the increase in the number of patients with MetS, the number of patients with T2D and CVD also increased significantly [8].

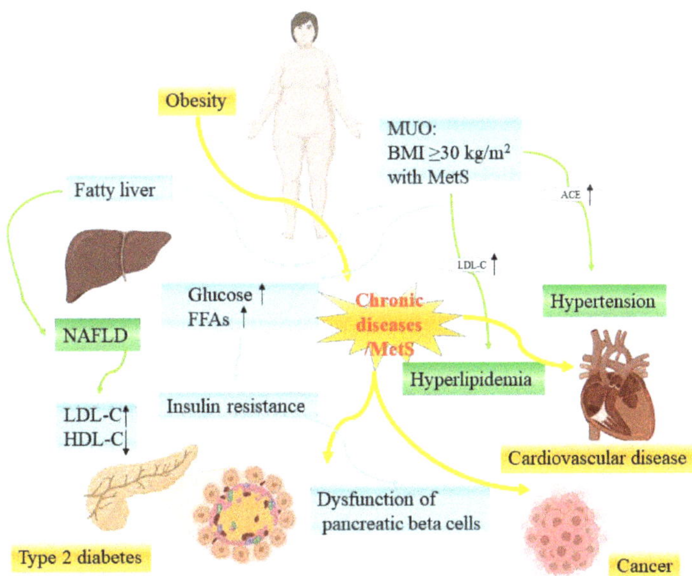

Figure 2. The relationship between chronic diseases and MetS-related diseases (some picture elements are from the BioRender). MUO—metabolic unhealthy obesity; NAFLD—nonalcoholic fatty liver disease; LDL-C—low-density lipoprotein cholesterol; HDL-C—high-density lipoprotein cholesterol; ACE—angiotensin-converting enzyme; FFAs—free fatty acids.

Obesity includes metabolic healthy obesity (MHO) and metabolic unhealthy obesity (MUO) [11], while MHO is unstable and transient, and most patients with MHO will transition to MUO stage with the accumulation of fat [12]. Morgan Mongraw Chaffin and colleagues found that MHO would develop into MUO when it exceeded a certain baseline (odds ratio [OR]: 1.60; 95% confidence interval [CI]: 1.14 to 2.25), and the risk of CVD for those with MUO increased significantly [13]. Due to the fact that obesity is the cause of heart diseases and the relationship between obesity and CVD is mediated by MetS, MetS can be regarded as a sign of obesity accumulation at the exposure threshold [13].

On the other direction, obesity can cause MetS to develop into diabetes, while nonalcoholic fatty liver disease (NAFLD), the most common metabolic liver disease, is a continuum between them [14]. In the United States, about 30% of adults have NAFLD, and 20% of them are developed by individuals with obesity [15]. In a meta-analysis of 24 studies involving 35,599 type 2 diabetic patients, the prevalence of NAFLD in ordinary diabetes patients was 59.67% but increased to 77.87% in diabetes patients with obesity [16]. MetS leads to an increase in glucose levels in the body and excessive production of free fatty acids [17], and the degree of dysfunction of pancreatic beta cells was related to the severity of MetS [18]. When pancreatic beta cells exceed a certain metabolic capacity for a long time, their quality and function will be reduced, and their metabolic function will be damaged [17].

At present, the treatments for chronic diseases include physical exercise and diet therapy, as well as drugs for related symptoms. Bioactive peptides derived from food proteins have been recognized by the industry as improving health because of their low costs and low side effects. A variety of bioactive peptides from different foods have been reported for their bioactivities, including anti-hypertension, anti-diabetes, and anti-cancer activities [19–21]. Studies have shown that increased intake of plant proteins is associated with decreased risk of obesity, CVD, diabetes, cancer, and other symptoms [21]. Soybean peptides, as one of the popular bioactive ingredients derived from soybean proteins, have been utilized in many health aspects, such as anti-obesity, anti-diabetes, anti-CVD, anti-cancer, and antioxidant activities [22].

As a traditional plant, soybean has been planted in China for nearly 5000 years [23]. The United States introduced soybeans in 1965 [24] and has now become the world's highest soybean production country, with the production volume reaching 45% of the world's total output [25]. Later on, the cultivation of soybeans gradually developed in other countries, and it has become a popular cash crop in the world. Proteins are the most abundant nutrient in soybean, accounting for about 40% of all nutrients, and they are a very important plant source of dietary proteins [26]. Soybean proteins contain all twenty types of amino acids, including nine essential amino acids [27,28]. Its nutritional value is equivalent to that of animal protein, and therefore, it is considered to be a full-value protein [22].

Soybean peptides are derived by the hydrolysis of soybean proteins using different proteases, and they are mixtures of oligopeptides with 3–6 amino acids and molecular weights of 300–700 Da [29]. The physiological activities of soybean peptides are determined by the size of their relative molecular weights and the sequences of amino acids [30]. Their amino acid composition and proportion are the same as those in soybean proteins, but they are easier to absorb and more stable [31]. Soy proteins and peptides have been shown to be safe and non-toxic in the past, which is important for their further utilization [32].

In this paper, we reviewed the structure, function, absorption, and metabolic characteristics of soybean peptides. We also discussed their potential effects on the regulation and improvement of chronic diseases.

2. Structure and Metabolism of Soybean Peptides

Bioactive peptides are sequences consisting of 2–20 amino acids that can regulate or improve physiological functions and thus prevent or treat chronic diseases [33]. Enzymolysis is an effective method to produce functional peptides, but different durations of the hydrolysis process and different enzymes used in the process have a great impact on their functions and intensity [33]. Soybean peptides are also generated with the enzymolysis approach (Figure 3). The separation and identification of protein hydrolysates can help us understand the relationship between the structures and functions of some peptides. This is very important for improving the bioavailability of bioactive substances in the future.

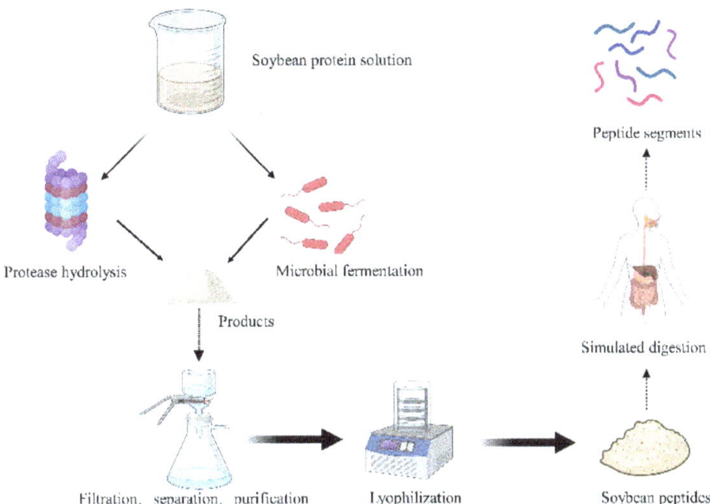

Figure 3. Preparation process of soybean peptide segments (some picture elements are from BioRender).

2.1. Structure and Function

The structures of bioactive peptides are very important for their biofunctions. Understanding the structures of the peptides, including disulfide bond position, amino acid composition and sequence, molecular weight, hydrophobicity, and other structural characteristics, is very important for the design of new peptides and the improvement of their efficacy, bioavailability, physical, and chemical properties [34]. For instance, peptide segments with three or more disulfide bonds have higher stability [35].

Peptides with specific functions generally have certain structural characteristics. On the other hand, specific structures may contribute to specific functions. For instance, peptides with proline or hydroxyproline at the C or N ends have good angiotensin-converting enzyme inhibitory (ACE-I) activity [34]. The ability of peptides to bind to ACE [36] and the antioxidant activity of soybean protein hydrolysates [37] depend on the presence of hydrophobic amino acids at the C-terminus. The presence of three aromatic amino acids (Trp, Tyr, and Pre), hydrophilic and basic amino acids (His, Lys, and Pro), and hydrophobic amino acids (Leu, Phe, and Val) in the polypeptides enhances their antioxidant capacity [38,39]. Soybean β-conglycinin is one of the most abundant proteins in soybean, accounting for 24.7–45.3% (w/w) of total protein components [40]. Compared with normal soybean β-conglycinin, deglycation enhanced its antioxidant performance, and thus deglycosylation may be an innovative strategy to improve its performance [41]. The presence of Glu, His, Asp, Met, and Val can significantly enhance the antioxidant capacity of soybean peptides, even for those with large molecular weights [42]. Although it is generally believed that peptides with low molecular weights have strong antioxidant properties, the functional effectiveness of peptides is also related to the processing methods. For example, higher substrate concentration in the digestive process generates more small soybean peptides, which enhance the free radical scavenging activity of, α, α-diphenyl-β-picrylhydrazyl (DPPH)—an important indicator for the evaluation of antioxidant activity [43]. These studies indicate that the structures of soybean peptides play an important role in their functions. Lingrong Wen et al. identified 46 peptides with immunomodulatory activity, and most of them contained Try, Glu, and hydrophobic amino acid residues (Pro, Gly, Phe, Val,

Leu) [29]. The binding of hydrophobic amino acids with Cys, Glu, Tyr, Asp, Trp, and Gln in the sequence is also important for immune regulatory activity [44]. Soybean peptides are mixtures of small peptides with different molecular weights. There is still a lot to do for the isolation and identification of small peptides with different functions.

The functions of soybean peptides have been studied in many aspects, including reducing blood fat [45], acting as an antioxidant [46], being anti-cancer [47], intestinal flora [48], being immunomodulatory [49], being anti-inflammatory [50], being anti-hypertensive [51], being anti-diabetic [52], and other physiological activities. However, there are still relatively few studies focusing on the effect of structures on the functions of peptides. Some peptides derived from soybean proteins have displayed many functions, yet there is not enough evidence to conclude that specific functions are associated with specific structures. Daliri and colleagues believed that peptides with multiple biological activities are better than those with a single biological activity. This is because peptides with multiple activities can play multiple beneficial roles at the same time [53]. Therefore, how to efficiently derive stable soybean peptides with specific structures and multiple functions merits further research.

2.2. Absorption and Metabolism

Bioactive peptides play physiological roles beyond their nutritional values. However, most bioactive peptides are in an encrypted state when they exist in the parent protein, where they cannot perform their functions [54]. Short peptides containing 2–20 amino acids have to be released from the parent proteins through enzymatic hydrolysis to activate their bioactivities [55]. Protein hydrolysates or short peptides have higher biological activities than complete proteins and/or amino acid mixtures [55]. After being digested in the digestive tract, some bioactive peptides can be absorbed through the intestinal tract to enter the blood circulation completely and play a role when they reach the corresponding target organs, while the others have local effects in the gastrointestinal tract [55].

Intestinal epithelial cells are a major obstacle to the absorption of any food ingredient. The activity of proteases on the surface of intestinal epithelial cells covered by microvilli may be the key factor that affects the stability and integrity of peptides as well as their operation and biological activities [56]. Differentiated Caco-2 cells have the morphology and function of mature intestinal epithelial cells and express brush border peptidase and transporters. They are useful in vitro models and can be used to study the stability, absorption, and transport of peptides [57]. To determine the effective utilization of a bioactive peptide, the differentiated Caco-2 cell lines were also used as an intestinal model to investigate the absorption of the peptides [58]. Although the Caco-2 cell model has been used as the best model in vitro for studying the absorption of different compounds for 35 years [57], its use in studying the absorption of food bioactive peptides is relatively new. One study examined the absorption of soybean β-conglycinin on Caco-2 cells after in vitro digestion mimicking the gastrointestinal tract and showed that 22 of 25 different peptide segments from the apical chamber samples were detected at the basolateral side of the transwell. This indicates that they could be absorbed by Caco-2 cells in vitro [41]. Gilda Aiello et al. found that three soybean peptides (IAVPGEVA, IAVPTGVA, and LPYP) were partially absorbed by Caco-2 cells in vitro and improved cholesterol metabolism in HepG2 cells by inhibiting the activity of 3-hydroxy-3-methylglutamate CoA reductase (HMGCoAR) [59].

In addition to absorption, Caco-2 cells are also used to evaluate the stability of bioactive peptides [60]. This is because Caco-2 cells can express brushborder peptidases and transporters, which can affect the stability and transport of peptide segments [60]. The transport of the peptide KPVAAP was detected for up to 60 min in the apical and basolateral sides of the transwell, indicating that it can be stably absorbed by Caco-2 cells [60]. After the soybean peptide segment, WGAPSL is digested in the gastrointestinal tract. The degradation of WGAPSL on both sides of apical and basal samples during the transportation of Caco-2 cells is determined. This indicates that WGAPSL can pass through the intestinal peptidase and mucus layer and be completely absorbed by the human body. This shows that WGAPSL has good stability after being digested in the gastrointestinal tract [61]. Due to the hydrolysis and absorption of the gastrointestinal tract, the biological activity and absorption stability of bioactive peptides in vivo may be different from those in vitro, while peptides need to be in active form to exert their biological activity in vivo. Therefore, we should pay attention to the biological stability and metabolic changes after the peptides are transported to the blood.

As peptidases exist widely in the body, including in the liver, kidney, blood, and other tissues, the mode, rate, and degree of how soybean peptides metabolize may be different in different target organs [62]. After entering the body, the peptide bonds within bioactive peptides are cut by the endopeptidases to form oligopeptides, and then the *N*-terminals or *C*-terminals are hydrolyzed by the exopeptidase (carboxypeptidase, aminopeptidase) into amino acids [62]. In addition, due to the high molecular weight, charged functional groups, and low lipophilicity of some peptide segments, they are easily blocked by the intestinal epithelial barrier. This results in a decrease in their bioavailability [63]. To improve their stability and bioavailability, various biochemical methods have been adopted, such as the substitution of unnatural amino acids and D-amino acids, cyclization, chemical modification (*N*-terminal and *C*-terminal), main chain modification, and nanoparticle formulation [64]. Understanding how bioactive peptides are metabolized and degraded by endogenous proteases is very important for functional food or drug design and the improvement of the metabolic stability of peptides [65].

The absorption and metabolism of soybean peptides are crucial to their biological effectiveness. It is very significant to investigate their absorption rate in vitro and the pathways they may have an impact on in vivo.

3. The Effects of Soybean Peptides on Chronic Diseases

Researchers are interested in exploring peptides and protein hydrolysates as active ingredients to prevent or treat chronic diseases. As a potent bioactive with an abundant source, soybean peptides have attracted a lot of attention, and the functions of soybean peptides have been widely investigated (Table 1). The functions of soybean proteins and peptides on chronic diseases, including anti-obesity, anti-diabetes, anti-CVD diseases, and anti-cancer activities, are of interest in this review (Table 1 and Figure 4).

Table 1. The functions of soybean proteins and peptides in relation to chronic diseases.

Function	Bioactive Substances of Soybean Peptides	Detection Model (Females or Males)	Main Results	References
Anti-obesity effect	β-conglycinin	C57BL/6 mice (males)	Weight decreased.	[66]
	β-conglycinin	C57BL/6 mice (males)	FGF21 increased.	[67]
	β-conglycinin	Obese rats (males)	Abdominal fat and lipid contents decreased.	[68]
	β-conglycinin	Rats (males)	Serum cholesterol decreased from 146 mg/dL to 124 mg/dL, and liver triglycerides decreased from 214 mg to 163 mg.	[69]
	Soybean protein isolates	Obese rats (females)	AST level decreased from 222.5 U/L to 103.4 U/L, and ALT level decreased from 71.9 U/L to 56.2 U/L.	[70]
	Soybean proteins	Obese OLETF rats (males)	Serum cholesterol decreased to 142 mg/dL.	[71]
	Soybean proteins	C57BL/6J mice (males)	*Firmicutes* to *Bacteriodetes* increased; Serum triglycerides decreased.	[72]
Anti-diabetes effect	Glu-Ala-Lys and Gly-Ser-Arg		The inhibitory effect of α-glucosidase activity was 45.89%.	[73]
	Soybean protein isolates and soybean peptides	Human (both)	Plasma insulin response significantly increased after 30 min of SPI consumption.	[74]
	Soybean proteins	Patients with diabetes (both)	Fasting blood glucose decreased by 1.68% after 2 months.	[75]
	VHVV	H9c2 cells and ICR mice (males)	Cell viability increased; Cell apoptosis decreased; Postprandial blood glucose level decreased.	[76]

Table 1. *Cont.*

Function	Bioactive Substances of Soybean Peptides	Detection Model (Females or Males)	Main Results	References
Anti-CVD	VAWWMY / Soystatin	Rats (males)	Serum and liver cholesterol levels were reduced to 0.03%.	[77]
	IAVPTGVA, IAVPGEVA and LPYP	HepG2 cells	Catalytic activity of HMGCoAR and the level of LDL decreased.	[78]
	Soybean protein hydrolysates	Caco-2 cells	The solubility of dietary cholesterol micelles decreased.	[79]
	YVVNPDNDEN and YVVNPDNNEN	HepG2 cells	After 24 h, the relative expression of LDL-C and PCSK9 decreased by about 20%.	[80]
	ALEPDHRVESEGGL and SLVNNDDDRDSYRLQS-GDAL	Caco-2 cells	Blood lipids decreased.	[45]
	VHVV	Hypertensive rats (males)	ACE activity and inflammatory factors decreased.	[51]
	Small molecule peptides	Hypertensive rats (males)	The inhibition rate of ACE activity was about 60% and the concentration of angiotensin II decreased.	[81]
	Polypeptide content of soybean meal		ACE activity decreased.	[82]
Anti-cancer effect	Lunasin	NSCLC cell line H661	Decreased proliferation of cancer cells.	[83]
	Lunasin	Human breast cancer cells	Decreased proliferation of cancer cells.	[84]
	Lunasin	Colorectal cancer HCT-116 cells	After treatment with 10 µM lunasin for 72 h, cell growth decreased by 12.9%.	[85]
	Germinated soybean peptides	Human colon cancer cell lines	After treatment with 10 mg/mL soybean peptide segments for 24 h, cell viability decreased by 82–66%.	[86]
	Black soybean peptides Leu/Ile-Val-Pro-Lys	HepG2 cells MCF-7cells HeLa cells	With high cytotoxicity, the IC_{50} values are 0.22, 0.15, and 0.32 µM, respectively.	[47]

Note: FGF21: fibroblast growth factor 21 gene; AST: aspartate aminotransferase; ALT: alanine aminotransferase; SPI: soybean protein isolates; HMGCoAR: 3-hydroxy-3-methylglutamate CoA reductase; LDL-C: low-density lipoprotein cholesterol; PCSK9: protein convertase subtilisin/kexin type 9; ACE: angiotensin-converting enzyme; IC_{50}: half maximal inhibitory concentration.

3.1. The Effect of Soybean Peptides on Obesity

The rapid increase in the prevalence of obesity has become a major public health issue globally. According to the WHO, when a person's BMI is $\geq$ 30 kg/m^2, it is considered obesity, and BMI $\geq$ 25 kg/m^2 is considered overweight [87,88]. In 2016, more than 39% of adults worldwide were overweight and about 13% were obese [89]. Abdominal obesity is closely related to chronic metabolic diseases such as T2D and CVD [87]. There are many treatments for obesity, such as drugs, surgery, and diets, with diet being the easiest and cheapest way to lose weight with no side effects. Protein has been widely used as a diet strategy to lose weight because of its high satiety effect [90]. It has long been shown that soybean protein has an anti-obesity effect, even better than whey and casein [91]. A random cross-balance experiment showed that fermented soybean had a better regulation effect on appetite regulating hormones (Acyl-ghrelin, insulin, and arginine) in obese girls than non-fermented soybean [92]. It showed a higher insulin-stimulating effect, which may be because the fermentation process can accelerate the degradation rate of protein and increase the bioavailability of short peptides [92]. Current studies show that soybean protein components can play a certain anti-obesity role.

β-conglycinin accounts for about 20% of the total soybean proteins, making it an important component for the beneficial effects of soybean proteins. Studies have found that β-conglycinin can reduce serum triglyceride and cholesterol levels, and thus it may have an anti-obesity effect [69]. The diet containing soybean protein reduced weight and fat tissue accumulation in C57BL/6 mice, which showed that β-conglycinin played an anti-obesity role [66]. After a single intake of β-conglycinin, both fibroblast growth factor 21 gene (FGF21) expression in the liver and FGF21 in the circulating body of the mice increased significantly [67]. In addition, β-conglycinin feeding for up to 9 weeks kept FGF21 levels in the liver and circulating FGF21 at a certain level, thereby reducing weight gain associated with a high-fat diet and thus ameliorating obesity [67]. Similarly, conglycinin peptide also reduced the liver lipase activity in obese rats, thereby reducing the abdominal fat accumulation and lipid contents. This indicates that it may have an anti-obesity effect [68]. As shown above, β-conglycinin, as the main component of soybean proteins, may be a potential compound for the treatment of obesity. In the future, the anti-obesity effect of β-conglycinin and the underlying mechanism will need further investigation. It may be a good choice for patients with obesity to lose weight.

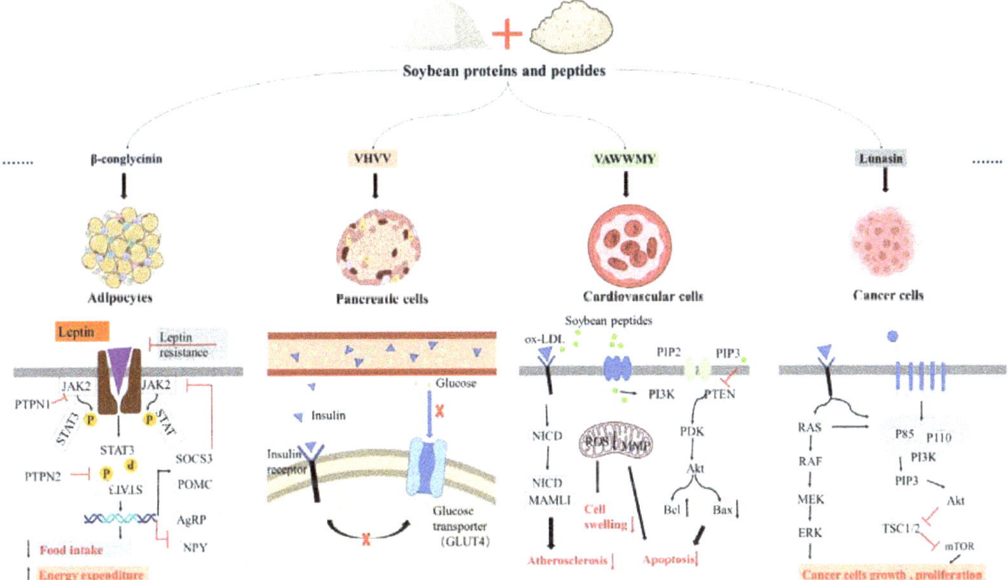

Figure 4. Partial potential mechanisms of the various activities of soybean proteins and peptides on chronic diseases (some picture elements are from the BioRender). JAK—Janus kinase; STAT—Signal transducer and activator of transcription; PTP—protein tyrosine phosphatases; SOCS—suppressor of cytokine signaling; POMC—proopiomelanocortin; AgRP—Agouti-related peptide; NPY—neuropeptide Y; PIP2—phosphatidylinositol-4,5-biphosphate; PIP3—phosphatidylinositol-3,4,5-triphosphate; ox-LDL—oxidized low-density lipoprotein; PI3K—phosphoinositide 3-kinase; PTEN—phosphatase and tensin homolog; ROS—reactive oxygen species; MMP—matrix metalloproteinase; NICD—Notch intracellular domain; MAML1—Mastermind-like proteins; MEK—mitogen-activated protein kinase; ERK—extracellular signal-regulated kinases; mTOR—mammalian target of rapamycin.

Soybean proteins were also involved in the regulation of the gastrointestinal microbiome and bile acid homeostasis [21,71,72,93]. The diverse microbiota plays a key role in the development of obesity, and their interactions (via signaling molecules/communication) can be maintained through diet/supplements [94]. The maintenance of the microbiota through dietary strategies may be of great importance in the treatment of chronic diseases. Consumption of soybean proteins improved the intestinal microbiota, increased the diversity of intestinal microbes, and improved the transmission of bile acid metabolism signals [71]. Muhammad Umair Ijaz et al. found that in high-fat diet-fed C57BL/6J mice, soybean protein supplementation increased the ratio of Firmicutes to Bacteriodetes, improved the composition of intestinal microorganisms, and reduced the accumulation of serum triglycerides [72]. The changes in intestinal microorganisms have a big impact on the risk of chronic diseases. Soybean proteins can improve the microbial diversity of the gastrointestinal tract, regulate fat synthesis, and thus have an anti-obesity effect. Soybean proteins can also reduce adipocyte hypertrophy, the concentrations of free fatty acids, and the accumulation of triglycerides in the liver after high-fat diet intake [21]. Reza Hakkak and colleagues reported that in obese Zucker rats, soybean protein isolate (SPI) feeding for 8 weeks reduced the denaturation of fat in the liver and decreased the levels of aspartate aminotransferase (AST) and alanine aminotransferase (ALT) in serum, which were beneficial to bile acid homeostasis, and the effect was even better than that of casein [70]. Soybean protein consumption also reduced the production of fat in the liver of obese OLETF rats and reduced their cholesterol levels [71]. In another study, SPI reduced the weight and improved the body fat of rats significantly, and this may be related to the reduction of perirenal fat [93].

Previous studies have focused on the effect of soybean proteins on serum hormone levels, cholesterol metabolism, and gut microbiota, through which soybean proteins showed their anti-obesity effect [69,72]. However, there are relatively few studies investigating the anti-obesity effect of soybean peptides and the underlying mechanisms. As soybean peptides have been released from the parent soybean protein, they have a big potential to show anti-obesity effects. Further research is needed to unravel the anti-obesity effect of soybean peptides.

3.2. The Effect of Soybean Peptides on Diabetes Mellitus

At present, diabetes has become a public health problem worldwide, being the seventh leading cause of death globally [95,96] and causing a huge burden for the families of patients and the world economy. According to the WHO and the International Diabetes Federation, the prevalence of diabetes is increasing year by year. Diabetes is a systemic metabolic disease caused by abnormal blood glucose levels, and it is one of the fastest-growing and most common chronic diseases in the world [97]. There are two main types of diabetes, type 1 diabetes (T1D) and T2D [97]. T1D is mainly an early-onset autoimmune disease caused by genetic factors, which occurs in childhood mainly due to the reduction of pancreatic beta cells [97]. T2D is a late-onset non-autoimmune disease caused by environmental factors and characterized by beta cell dysfunction of the pancreas and insulin resistance [97]. The incidence of T2D is as high as 90%, and it is associated with a high mortality rate and high medical costs, as well as various complications, including retinopathy, kidney disease, microvascular complications, and nerve damage [97]. It is of great importance to find effective treatments for diabetes. Of many options, diet intervention is easy and cheap. Food-derived bioactive substances could be a good strategy to control blood glucose as a diet intervention strategy.

Soybean proteins are also high-quality proteins, but with different amino acid patterns, so they stimulate insulin secretion to different degrees than caseins [68]. Not only soybean proteins, but bioactives with amino acid patterns similar to soybean proteins also improved insulin sensitivity [68]. Soybean proteins are known as a hypoglycemic functional food as they contain specific amino acids, including Leu, Arg, Ala, Phe, Ile, Lys, and Met, which can stimulate insulin secretion and act as a trypsin inhibitor [98].

The relationship between the consumption of legumes and soybean products and the incidence of T2D was analyzed by Jun Tang and colleagues, and the results showed that specific soybean components, including soybean isoflavones and soybean proteins, were negatively related to the risk of T2D [99]. An increase of 10 g of soybean protein per day can reduce the risk of T2D by 9% [99]. The intake of soybean proteins was negatively correlated with the risk of diabetes in women in a dose-dependent manner but not in men [100]. Judit Konya et al. administered soybean proteins with or without soybean isoflavones to 60 diabetic patients aged 45–80 years and found that soybean proteins without isoflavones had a certain intrinsic activity in controlling blood glucose. This demonstrates the hypoglycemic effect of soybean proteins [75]. The structure, activity, and mechanism of plant compounds with therapeutic and ameliorative effects on T2D have been reviewed, and it showed that soybean proteins and bioactives play a certain role in the regulation of blood glucose [101]. Soybean proteins are rich in glycine and arginine, and their amino acid pattern is conducive to insulin sensitivity and glucose utilization [100]. These studies showed that the structural pattern of soybean proteins and their special amino acids can improve the sensitivity of pancreatic β cells and stimulate insulin secretion. Therefore, some soybean proteins and their derived bioactives may become effective substances for regulating blood glucose.

Dipeptidyl peptidase-IV (DPP-IV), α-glucosidase, and α-amylase are key enzymes that directly regulate blood glucose, and inhibition of these enzymes is an effective strategy for the treatment of T2D [102]. Gonzalez-Montoya et al. derived soybean peptides with different molecular weights after in vitro simulated gastrointestinal digestion. They found that peptides at 5–10 kDa inhibited DPP-IV, α-amylase, and α-glucoside, and further separation of these peptides yielded four components, three of which contained most polypeptides with encrypted dipeptide and tripeptide amino acid sequences, whose structure may be the main reason for their diabetes-inhibiting effects [52]. Another study showed that the soybean protein hydrolysates inhibited α-glucosidase activity and the peptides with molecular weight < 5 kDa and amino acid sequences Glu-Ala-Lys and Gly-Ser-Arg showed the strongest inhibitory effect [73]. Val-His-Val-Val (VHVV) is also a short peptide isolated from soybean protein hydrolysates. VHVV could restore the viability of H9c2 cells at high glucose conditions, and 10 μg/mL VHVV reduced the number of H9c2 cells undergoing apoptosis and postprandial blood glucose level in diabetic mice and improved the morphological structure and number of pancreatic cells [76]. Hatsumi Ueoka et al. observed an increase in plasma insulin levels after oral administration of both soybean protein isolate solution and soybean peptide solution, with even higher plasma insulin levels in the soybean peptide group at 30 min. This demonstrates a greater insulin secretion-stimulating effect of the peptide due to easier digestion and absorption [74].

Both soybean proteins and peptides have good amino acid sequences and absorption characteristics. However, most research still focuses on the study of the phenotype of diabetic mice, and the underlying mechanisms remain unknown. More in-depth research is still needed. In addition, more refined screening, separation, and purification of peptides with anti-diabetes effects will shed light on the regulation and treatment of diabetes.

3.3. The Effect of Soybean Peptides on CVD

CVD is a type of disease involving the heart and blood vessels, including hyperlipidemia, hypercholesterolemia, hypertension, atherosclerosis, and other major diseases. The risk of CVD is closely related to insulin resistance and obesity [103]. At present, CVD accounts for 46.2% of global non-communicable disease deaths, which is one of the main causes of premature death [104]. Nearly 17.9 million people die from CVD each year, and the number of deaths is estimated to increase to 23.6 million by 2030 [105].

The effect of soybean proteins on the improvement and prevention of CVD has received much attention. As early as 1999, the United States Food and Drug Administration (FDA) approved the food label containing soybean proteins to prevent CVD [106]. The FDA approved the health statement that 25 g of soybean proteins per day can reduce the risk of CVD [106]. A large-scale meta-analysis showed that soybean intake was negatively correlated with CVD risk [107]. This may be due to the fact that the soybean bioactive peptides could reduce the total cholesterol levels in the body [108]. Another meta-analysis showed that a daily intake of 25 g of soybean proteins can reduce low-density lipoprotein cholesterol (LDL-C) levels in adults by 3–4% [109].

Both hyperlipidemia and hypercholesterolemia are major risk factors for CVD [110]. Soybean peptides have been reported to lower cholesterol levels. In 2010, scientists first found a new peptide, VAWWMY, with a cholesterol-lowering effect in soybean glycine, named "soystatin," which has the same binding capacity with bile acid as cholesterol-lowering drugs, although soystatin is the only low-cholesterol peptide isolated from soybean [77]. Three soybean globulin glycin peptides (IAVPTGVA, IAVPGEVA, and LPYP) downregulated the catalytic activity of HMGCoAR, activated the LDLR-SREBP2 pathway, and improved the ability to absorb LDL in vitro, which in turn regulated the cholesterol metabolism of HepG2 cells [78]. In another study, digested soybean protein hydrolysates reduced the solubility of dietary cholesterol micelles by 37.6% and the absorbability by 18.99%, respectively, in Caco-2 cells [79].

Abnormality in blood lipids, especially the elevation of plasma LDL-C levels, is a major risk factor for CVD [80]. In addition to pharmacological methods, researchers are paying more and more attention to nutritional intervention strategies to prevent chronic diseases. In a previous study, two soybean peptides, YVVNPDNDEN and YVVNPDNNEN, both reduced the levels of LDL-C by inhibiting HMG-CoAR activity, while the latter one downregulated the protein level of protein convertase subtilisin/kexin type 9 (PCSK9), a key regulator of LDL-R [80]. In another study, oral administration of two other soybean peptides, ALEPDHRVESEGGL and SLVNNDDDRDSYRLQSGDAL, up-regulated trans-intestinal cholesterol excretion (TICE), inhibited the expression of cytochrome P450 family members (CYP7A1 and CYP8B1), reduced bile acid synthesis, and increased cholesterol excretion in the liver, and thus both peptides have blood lipid-lowering effects [45]. The blood lipid regulatory effect of soybean peptides may play an important role in the treatment of CVD.

Hypertension is also a major risk factor for CVD. Effective treatments for hypertension can reduce the burden of the population with CVD related to high blood pressure [111]. The soybean peptide VHVV inhibited the ACE activity in hypertensive rats and activated the SIRT1-PGC1α/Nrf2 pathway, which reduced the production of renal inflammatory factors and the apoptosis of renal cells. This suggests that VHVV can improve hypertensive renal damage [51]. Another study found that soybean peptides hydrolyzed by alkaline protease and neutral protease had the highest inhibitory effect on ACE activity in hypertensive rats, with an inhibition rate of 71.2% [81]. Ultrasonic fermentation increased the polypeptide content of soybean meal by 36.2%, and thus the ACE inhibitory activity of the soybean meal was increased in vitro by up to 70.05% [82]. In spontaneously hypertensive rats, feeding with soybean oligopeptide at 4.50 g/kg for 30 days significantly reduced both systolic and diastolic blood pressure as well as the quality and concentration of angiotensin II [81]. The investigation of the blood pressure-lowering effect of soybean peptides has mostly focused on the ACE inhibitory activity, and only a very few studies have seen the direct effects of soybean peptides on blood pressure. More studies will be needed to investigate the direct blood pressure-lowering effect in vivo and the effect of anti-hypertension on CVD.

There have been relatively few studies on the effects of soybean peptides on CVD. Relatively few peptides were screened out. Further studies are still needed to screen more potent peptides for CVD-regulating effects and to investigate the underlying mechanisms of the anti-CVD activity of the peptides.

3.4. The Effect of Soybean Peptides on Cancer

Cancer has become the second-leading cause of death in the United States [112]. More than 600,000 people in the United States died of cancer in 2021 [112]. Traditional cancer treatments such as drug therapy and chemotherapy are expensive and can cause adverse reactions or complications [113]. In recent years, some anti-tumor peptides have been reported, including soybean peptides [114]. As a relatively inexpensive method, soybean peptides will play an important role in the prevention and remission of cancer development [114].

Lunasin is a bioactive peptide with 43 amino acids and a molecular weight of 5.5 kDa, originally isolated from soybean [83]. It has chemopreventive and therapeutic effects [83]. Studies have shown that Lunasin can effectively inhibit the proliferation of the non-small cell lung cancer (NSCLC) cell line H661 by inhibiting the G1/S phase of the cell cycle and altering the expression of related protein kinase complex components [83]. Thus, the expression level of p27Kip1 and the phosphorylation level of Akt at S473 are altered, and finally, the anti-cancer effect is achieved [83]. Lunasin had a significant inhibitory effect on cell proliferation of human breast cancer cells, and the inhibitory rates of Lunasin extracted from transgenic soybean and wild-type soybean were 43 and 23.8%, respectively [84]. Lunasin at the concentrations of 40 and 80 μM significantly increased the apoptosis of colorectal cancer HCT-116 cells by reducing the level of the DNA repair enzyme (PARP) protein (a marker of cell apoptosis) and increasing the expression of caspase-3 protein and playing a certain role in inhibiting the tumorigenesis by prolonging the G1 phase [85]. As discussed above, Lunasin can inhibit cell proliferation or increase apoptosis in various cancer cells. It could be a potent substance for the prevention and treatment of cancer.

The size of soybean peptides also has a certain influence on the inhibition of cancer cells. Gonzalez-Montoya and colleagues treated three human colon cancer cell lines (Caco-2, HT-29, and HWT-116) with peptides of different lengths obtained after simulating gastrointestinal digestion in vitro. They showed that the inhibitory effects of germinated soybean peptides of different lengths on the proliferation of cancer cells were different [86]. However, more research on the underlying mechanisms of the most active peptides and their potential protective effect on colon health in animal models with colon cancer is still needed. Soybean peptides with different molecular weights inhibited cancer cell proliferation in human blood, breast, and prostate at different degrees, with the 10–50 kDa peptide from the N98-4445A soybean strain inhibiting CCRF-CEM blood cancer cells by 68% [115]. Peptides with different molecular weights from black bean also inhibited the human hepatoma cell line (HepG2), cervical cancer cell line (HeLa), and lung cancer cell line (MCF-7) at different rates, and the inhibitory effects of those with molecular weight < 4 kDa on the growth of HepG2, HeLa, and MCF-7 cancer cells were 2.28-, 1.96-, and 5.91-fold, respectively [47]. The maximum inhibition rate of these peptide segments on the growth of HeLa cancer cells can reach 6.44-fold, which may be related to their hydrophobic interaction and hydrogen bonds with target proteins such as Bcl-2, caspase-7, and caspase-3 [47].

Cancer is a chronic disease that cannot be completely cured at present. The anti-cancer research on soybean peptides is helpful in inhibiting cancer cells. However, only a few studies are focusing on the anti-cancer properties of soybean peptides. More studies will be needed to screen and investigate the anti-cancer effect of more soybean bioactives in animal models with different cancers.

4. Conclusions

Chronic diseases are comprehensive diseases with complex pathogenetic mechanisms. With lifestyle changes, chronic diseases have become the main cause of human death in the world. Finding bioactive substances that can regulate and treat chronic diseases has become the unanimous desire of researchers. Bioactive peptides have been more and more recognized for their activities in improving health and preventing or treating chronic diseases. In addition to providing nutrition, food protein peptides can also provide more functions through changes in specific biochemical pathways. Soybean-derived peptides have received a lot of attention for their potent activities of anti-obesity, anti-diabetes, CVD regulation, and anti-cancer, which are very important for the prevention and treatment of chronic diseases (see Figure 4). After digestion, most peptides can be completely absorbed by intestinal cells and transported to the corresponding target organs and cells.

This paper reviewed the bioactivity-related structures of soybean proteins and peptides. It briefly introduced absorption and metabolism in the body and broadly reviewed their functions related to chronic diseases, including anti-obesity, anti-diabetes, CVD regulation, and anti-cancer activity. According to previous reports, soybean proteins and peptides are potent ingredients that may have a major impact on chronic diseases. It is worth further investigation for more potential bioactivities and the underlying mechanisms for these functions.

5. Prospect

In the past 20 years, soybean proteins and peptides have attracted extensive attention for their variety of functions. Although it has fewer side effects, it is not as effective as drug therapy. Functional studies on soybean peptides are extensive and inaccurate.

Single-functional soybean peptides have been widely studied, while multifunctional peptides are still a challenging topic. It is necessary to extend the research from single-functional peptides to multifunctional peptides in the future. This change should focus on the method of protein hydrolysis, as protein hydrolysates are complex mixed peptides, and only some of these peptides have bioactive functions.

Although soybean peptides have multiple biological activities, their sensitivity to gastrointestinal proteases and peptidases may lead to a loss of activity before reaching the target organs. This aspect needs to be taken into account, and in vivo or clinical validation will be needed before they can be utilized as a treatment strategy. The application of advanced multi-omics technology and bioinformatics may be of great significance.

The main disadvantage of bioactive peptides is that they are easily degraded by proteases and could be quickly cleared by the kidney, resulting in low bioavailability and poor transmembrane absorption [64]. Small peptides are more easily absorbed than large peptides because they can cross the intestinal barrier more easily and reach their target organs [116]. Soybean peptides have great potential for improving human health. However, there is still a lack of clinical data. In terms of commercialization, how to improve the production technology and their bioavailability while ensuring their quality needs further study.

Generally, food-derived bioactive peptides are characterized by poor absorption, distribution, metabolism, excretion, and toxicity (ADME-T). Currently, ADME characteristics of bioactive peptides have been evaluated in vitro, in vivo, and in silico using various tools [117]. However, how to improve their undesirable characteristics through structural modification or other aspects is still under investigation. The research on improving the bioavailability of soybean peptides will bring great potential and challenges to the development of functional foods or drugs.

Finally, long-term or excessive consumption of soybean peptides may cause allergic reactions and other side effects or reduce digestive capacity in a small group of people. Further research should also investigate the dose and duration for the effective utilization of soybean peptides and avoid potential side effects for the small population. By addressing most of these concerns, soybean peptides will play a significant role in the prevention and treatment of chronic diseases.

Author Contributions: S.H. supervised, wrote the grant and co-wrote the paper. C.L. collected the references and co-wrote the paper. X.L. directed the paper. All authors have read and agreed to the published version of the manuscript.

Funding: This research was funded by the National Natural Science Foundation of China (31900831) to S.H. and supported by the Cultivation Project of Double First-Class Disciplines of Food Science and Engineering, Beijing Technology and Business University (No. BTBUYXTD202207).

Institutional Review Board Statement: Not applicable.

Informed Consent Statement: Not applicable.

Data Availability Statement: No new data were created or analyzed in this study.

Conflicts of Interest: The authors declare no conflict of interest.

References

1. Anderson, E.; Durstine, J.L. Physical activity, exercise, and chronic diseases: A brief review. *Sports Med. Health Sci.* **2019**, *1*, 3–10. [CrossRef]
2. Zulhendri, F.; Ravalia, M.; Kripal, K.; Chandrasekaran, K.; Fearnley, J.; Perera, C.O. Propolis in Metabolic Syndrome and Its Associated Chronic Diseases: A Narrative Review. *Antioxidants* **2021**, *10*, 348. [CrossRef]
3. Gil, A.; Ortega, R.M. Introduction and Executive Summary of the Supplement, Role of Milk and Dairy Products in Health and Prevention of Noncommunicable Chronic Diseases: A Series of Systematic Reviews. *Adv. Nutr.* **2019**, *10*, S67–S73. [CrossRef] [PubMed]
4. Kumar, L.; Bisen, M.; Khan, A.; Kumar, P.; Patel, S.K.S. Role of Matrix Metalloproteinases in Musculoskeletal Diseases. *Biomedicines* **2022**, *10*, 2477. [CrossRef]
5. Very, N.; Vercoutter-Edouart, A.S.; Lefebvre, T.; Hardiville, S.; El Yazidi-Belkoura, I. Cross-Dysregulation of O-GlcNAcylation and PI3K/AKT/mTOR Axis in Human Chronic Diseases. *Front. Endocrinol.* **2018**, *9*, 602. [CrossRef] [PubMed]
6. Fahed, G.; Aoun, L.; Bou Zerdan, M.; Allam, S.; Bou Zerdan, M.; Bouferraa, Y.; Assi, H.I. Metabolic Syndrome: Updates on Pathophysiology and Management in 2021. *Int. J. Mol. Sci.* **2022**, *23*, 786. [CrossRef]

7.   Rao, D.; Dai, S.; Lagace, C.; Krewski, D. Metabolic syndrome and chronic disease. *Chronic Dis. Inj. Can.* **2014**, *34*, 36–45. [CrossRef] [PubMed]

8.   Xu, H.; Li, X.; Adams, H.; Kubena, K.; Guo, S. Etiology of Metabolic Syndrome and Dietary Intervention. *Int. J. Mol. Sci.* **2018**, *20*, 128. [CrossRef]

9.   Alberti, K.G.; Eckel, R.H.; Grundy, S.M.; Zimmet, P.Z.; Cleeman, J.I.; Donato, K.A.; Fruchart, J.C.; James, W.P.; Loria, C.M.; Smith, S.C.; et al. Harmonizing the metabolic syndrome: A joint interim statement of the International Diabetes Federation Task Force on Epidemiology and Prevention; National Heart, Lung, and Blood Institute; American Heart Association; World Heart Federation; International Atherosclerosis Society; and International Association for the Study of Obesity. *Circulation* **2009**, *120*, 1640–1645. [CrossRef]

10. Reaven, G.M. The metabolic syndrome: Is this diagnosis necessary? *Am. J. Clin. Nutr.* **2006**, *83*, 1237–1247. [CrossRef]

11. Primeau, V.; Coderre, L.; Karelis, A.; Brochu, M.; Lavoie, M.; Messier, V.; Sladek, R.; Rabasa-Lhoret, R. Characterizing the profile of obese patients who are metabolically healthy. *Int. J. Obes.* **2011**, *35*, 971–981. [CrossRef] [PubMed]

12. Kim, N.H.; Seo, J.A.; Cho, H.; Seo, J.H.; Yu, J.H.; Yoo, H.J.; Kim, S.G.; Choi, K.M.; Baik, S.H.; Choi, D.S.; et al. Risk of the Development of Diabetes and Cardiovascular Disease in Metabolically Healthy Obese People: The Korean Genome and Epidemiology Study. *Medicine* **2016**, *95*, e3384. [CrossRef] [PubMed]

13. Mongraw-Chaffin, M.; Foster, M.C.; Anderson, C.A.M.; Burke, G.L.; Haq, N.; Kalyani, R.R.; Ouyang, P.; Sibley, C.T.; Tracy, R.; Woodward, M.; et al. Metabolically Healthy Obesity, Transition to Metabolic Syndrome, and Cardiovascular Risk. *J. Am. Coll. Cardiol.* **2018**, *71*, 1857–1865. [CrossRef]

14. Godoy-Matos, A.F.; Silva Junior, W.S.; Valerio, C.M. NAFLD as a continuum: From obesity to metabolic syndrome and diabetes. *Diabetol. Metab. Syndr.* **2020**, *12*, 60. [CrossRef] [PubMed]

15. Zhu, L.; Baker, S.S.; Gill, C.; Liu, W.; Alkhouri, R.; Baker, R.D.; Gill, S.R. Characterization of gut microbiomes in nonalcoholic steatohepatitis (NASH) patients: A connection between endogenous alcohol and NASH. *Hepatology* **2013**, *57*, 601–609. [CrossRef] [PubMed]

16. Dai, W.; Ye, L.; Liu, A.; Wen, S.W.; Deng, J.; Wu, X.; Lai, Z. Prevalence of nonalcoholic fatty liver disease in patients with type 2 diabetes mellitus. *Medicine* **2017**, *96*, e8179. [CrossRef]

17. Gupta, D.; Jetton, T.L.; LaRock, K.; Monga, N.; Satish, B.; Lausier, J.; Peshavaria, M.; Leahy, J.L. Temporal characterization of beta cell-adaptive and -maladaptive mechanisms during chronic high-fat feeding in C57BL/6NTac mice. *J. Biol. Chem.* **2017**, *292*, 12449–12459. [CrossRef]

18. Hudish, L.I.; Reusch, J.E.; Sussel, L. beta Cell dysfunction during progression of metabolic syndrome to type 2 diabetes. *J. Clin. Investig.* **2019**, *129*, 4001–4008. [CrossRef]

19. Mann, B.; Athira, S.; Sharma, R.; Kumar, R.; Sarkar, P. Bioactive Peptides from Whey Proteins. In *Whey Proteins*; Academic Press: Cambridge, MA, USA, 2019; pp. 519–547.

20. Cicero, A.F.G.; Fogacci, F.; Colletti, A. Potential role of bioactive peptides in prevention and treatment of chronic diseases: A narrative review. *Br. J. Pharmacol.* **2017**, *174*, 1378–1394. [CrossRef]

21. Tovar, A.R.; Torres, N. The role of dietary protein on lipotoxicity. *Biochim. Biophys. Acta* **2010**, *1801*, 367–371. [CrossRef]

22. Chatterjee, C.; Gleddie, S.; Xiao, C.W. Soybean Bioactive Peptides and Their Functional Properties. *Nutrients* **2018**, *10*, 1211. [CrossRef]

23. Hymowitz, T. On the domestication of the soybean. *Econ. Bot.* **1970**, *24*, 408–421. [CrossRef]

24. Hymowitz, T.; Harlan, J.R. Introduction of soybean to North America by Samuel Bowen in 1765. *Econ. Bot.* **1983**, *37*, 371–379. [CrossRef]

25. Qin, P.; Wang, T.; Luo, Y. A review on plant-based proteins from soybean: Health benefits and soy product development. *J. Agric. Food Res.* **2022**, *7*, 100265. [CrossRef]

26. Wen, Y.; Liu, A.; Meng, C.; Li, Z.; He, P. Quantification of lectin in soybeans and soy products by liquid chromatography-tandem mass spectrometry. *J. Chromatogr. B Analyt. Technol. Biomed. Life Sci.* **2021**, *1185*, 122987. [CrossRef]

27. Fukushima, D. Recent progress of soybean protein foods: Chemistry, technology, and nutrition. *Food Rev. Int.* **1991**, *7*, 323–351. [CrossRef]

28. Zhang, Y.; Zhou, F.; Zhao, M.; Ning, Z.; Sun-Waterhouse, D.; Sun, B. Soy peptide aggregates formed during hydrolysis reduced protein extraction without decreasing their nutritional value. *Food Funct.* **2017**, *8*, 4384–4395. [CrossRef]

29. Wen, L.; Jiang, Y.; Zhou, X.; Bi, H.; Yang, B. Structure identification of soybean peptides and their immunomodulatory activity. *Food Chem.* **2021**, *359*, 129970. [CrossRef]

30. Lin, Q.; Xu, Q.; Bai, J.; Wu, W.; Hong, H.; Wu, J. Transport of soybean protein-derived antihypertensive peptide LSW across Caco-2 monolayers. *J. Funct. Foods* **2017**, *39*, 96–102. [CrossRef]

31. Singh, B.P.; Vij, S. In vitro stability of bioactive peptides derived from fermented soy milk against heat treatment, pH and gastrointestinal enzymes. *LWT* **2018**, *91*, 303–307. [CrossRef]

32. Park, S.Y.; Lee, J.-S.; Baek, H.-H.; Lee, H.G. Purification and Characterization of Antioxidant Peptides from Soy Protein Hydrolysate. *J. Food Biochem.* **2010**, *34*, 120–132. [CrossRef]

33. Karami, Z.; Akbari-Adergani, B. Bioactive food derived peptides: A review on correlation between structure of bioactive peptides and their functional properties. *J. Food Sci. Technol.* **2019**, *56*, 535–547. [CrossRef]

34. Ganguly, A.; Sharma, K.; Majumder, K. Food-derived bioactive peptides and their role in ameliorating hypertension and associated cardiovascular diseases. *Adv. Food Nutr. Res.* **2019**, *89*, 165–207. [CrossRef]
35. Wang, G.; Chen, Y.; Li, L.; Tang, W.; Wright, J.M. First-line renin–angiotensin system inhibitors vs. other first-line antihypertensive drug classes in hypertensive patients with type 2 diabetes mellitus. *J. Hum. Hypertens.* **2018**, *32*, 494–506. [CrossRef]
36. Iwaniak, A.; Minkiewicz, P.; Darewicz, M. Food-originating ACE inhibitors, including antihypertensive peptides, as preventive food components in blood pressure reduction. *Compr. Rev. Food Sci. Food Saf.* **2014**, *13*, 114–134. [CrossRef]
37. Ngoh, Y.-Y.; Gan, C.-Y. Enzyme-assisted extraction and identification of antioxidative and α-amylase inhibitory peptides from Pinto beans (*Phaseolus vulgaris* cv. Pinto). *Food Chem.* **2016**, *190*, 331–337. [CrossRef]
38. Girgih, A.T.; He, R.; Malomo, S.; Offengenden, M.; Wu, J.; Aluko, R.E. Structural and functional characterization of hemp seed (*Cannabis sativa* L.) protein-derived antioxidant and antihypertensive peptides. *J. Funct. Foods* **2014**, *6*, 384–394. [CrossRef]
39. Najafian, L.; Babji, A.S. Isolation, purification and identification of three novel antioxidative peptides from patin (*Pangasius sutchi*) myofibrillar protein hydrolysates. *LWT-Food Sci. Technol.* **2015**, *60*, 452–461. [CrossRef]
40. Martinez-Villaluenga, C.; Bringe, N.A.; Berhow, M.A.; de Mejia, E.G. beta-Conglycinin Embeds Active Peptides That Inhibit Lipid Accumulation in 3T3-L1 Adipocytes in Vitro. *J. Agric. Food Chem.* **2008**, *56*, 10533–10543. [CrossRef]
41. Amigo-Benavent, M.; Clemente, A.; Caira, S.; Stiuso, P.; Ferranti, P.; del Castillo, M.D. Use of phytochemomics to evaluate the bioavailability and bioactivity of antioxidant peptides of soybean beta-conglycinin. *Electrophoresis* **2014**, *35*, 1582–1589. [CrossRef]
42. Chen, J.; Cui, C.; Zhao, H.; Wang, H.; Zhao, M.; Wang, W.; Dong, K. The effect of high solid concentrations on enzymatic hydrolysis of soya bean protein isolate and antioxidant activity of the resulting hydrolysates. *Int. J. Food Sci. Technol.* **2018**, *53*, 954–961. [CrossRef]
43. Kedare, S.B.; Singh, R.P. Genesis and development of DPPH method of antioxidant assay. *J. Food Sci. Technol.* **2011**, *48*, 412–422. [CrossRef]
44. Chalamaiah, M.; Yu, W.; Wu, J. Immunomodulatory and anticancer protein hydrolysates (peptides) from food proteins: A review. *Food Chem.* **2018**, *245*, 205–222. [CrossRef] [PubMed]
45. Lee, H.; Shin, E.; Kang, H.; Youn, H.; Youn, B. Soybean-Derived Peptides Attenuate Hyperlipidemia by Regulating Trans-Intestinal Cholesterol Excretion and Bile Acid Synthesis. *Nutrients* **2021**, *14*, 95. [CrossRef] [PubMed]
46. Cavaliere, C.; Montone, A.M.I.; Aita, S.E.; Capparelli, R.; Cerrato, A.; Cuomo, P.; Lagana, A.; Montone, C.M.; Piovesana, S.; Capriotti, A.L. Production and Characterization of Medium-Sized and Short Antioxidant Peptides from Soy Flour-Simulated Gastrointestinal Hydrolysate. *Antioxidants* **2021**, *10*, 734. [CrossRef] [PubMed]
47. Chen, Z.; Li, W.; Santhanam, R.K.; Wang, C.; Gao, X.; Chen, Y.; Wang, C.; Xu, L.; Chen, H. Bioactive peptide with antioxidant and anticancer activities from black soybean [*Glycine max* (L.) Merr.] byproduct: Isolation, identification and molecular docking study. *Eur. Food Res. Technol.* **2018**, *245*, 677–689. [CrossRef]
48. Li, W.; Li, H.; Zhang, Y.; He, L.; Zhang, C.; Liu, X. Different effects of soybean protein and its derived peptides on the growth and metabolism of *Bifidobacterium animalis* subsp. *animalis* JCM 1190. *Food Funct.* **2021**, *12*, 5731–5744. [CrossRef]
49. Fang, J.; Lu, J.; Zhang, Y.; Wang, J.; Wang, S.; Fan, H.; Zhang, J.; Dai, W.; Gao, J.; Yu, H. Structural properties, antioxidant and immune activities of low molecular weight peptides from soybean dregs (Okara). *Food Chem. X* **2021**, *12*, 100175. [CrossRef]
50. Zhao, F.; Liu, W.; Yu, Y.; Liu, X.; Yin, H.; Liu, L.; Yi, G. Effect of small molecular weight soybean protein-derived peptide supplementation on attenuating burn injury-induced inflammation and accelerating wound healing in a rat model. *RSC Adv.* **2019**, *9*, 1247–1259. [CrossRef]
51. Tsai, B.C.-K.; Kuo, W.-W.; Day, C.H.; Hsieh, D.J.-Y.; Kuo, C.-H.; Daddam, J.; Chen, R.-J.; Padma, V.V.; Wang, G.; Huang, C.-Y. The soybean bioactive peptide VHVV alleviates hypertension-induced renal damage in hypertensive rats via the SIRT1-PGC1α/Nrf2 pathway. *J. Funct. Foods* **2020**, *75*, 104255. [CrossRef]
52. Gonzalez-Montoya, M.; Hernandez-Ledesma, B.; Mora-Escobedo, R.; Martinez-Villaluenga, C. Bioactive Peptides from Germinated Soybean with Anti-Diabetic Potential by Inhibition of Dipeptidyl Peptidase-IV, alpha-Amylase, and alpha-Glucosidase Enzymes. *Int. J. Mol. Sci.* **2018**, *19*, 2883. [CrossRef] [PubMed]
53. Daliri, E.B.; Oh, D.H.; Lee, B.H. Bioactive Peptides. *Foods* **2017**, *6*, 32. [CrossRef] [PubMed]
54. Rutherfurd-Markwick, K.J. Food proteins as a source of bioactive peptides with diverse functions. *Br. J. Nutr.* **2012**, *108*, S149–S157. [CrossRef] [PubMed]
55. Erdmann, K.; Cheung, B.W.; Schroder, H. The possible roles of food-derived bioactive peptides in reducing the risk of cardiovascular disease. *J. Nutr. Biochem.* **2008**, *19*, 643–654. [CrossRef] [PubMed]
56. Aiello, G.; Ferruzza, S.; Ranaldi, G.; Sambuy, Y.; Arnoldi, A.; Vistoli, G.; Lammi, C. Behavior of three hypocholesterolemic peptides from soy protein in an intestinal model based on differentiated Caco-2 cell. *J. Funct. Foods* **2018**, *45*, 363–370. [CrossRef]
57. Lammi, C.; Aiello, G.; Boschin, G.; Arnoldi, A. Multifunctional peptides for the prevention of cardiovascular disease: A new concept in the area of bioactive food-derived peptides. *J. Funct. Foods* **2019**, *55*, 135–145. [CrossRef]
58. Lammi, C.; Aiello, G.; Vistoli, G.; Zanoni, C.; Arnoldi, A.; Sambuy, Y.; Ferruzza, S.; Ranaldi, G. A multidisciplinary investigation on the bioavailability and activity of peptides from lupin protein. *J. Funct. Foods* **2016**, *24*, 297–306. [CrossRef]
59. Lammi, C.; Bollati, C.; Ferruzza, S.; Ranaldi, G.; Sambuy, Y.; Arnoldi, A. Soybean- and Lupin-Derived Peptides Inhibit DPP-IV Activity on In Situ Human Intestinal Caco-2 Cells and Ex Vivo Human Serum. *Nutrients* **2018**, *10*, 1082. [CrossRef]
60. Gallego, M.; Grootaert, C.; Mora, L.; Aristoy, M.C.; Van Camp, J.; Toldrá, F. Transepithelial transport of dry-cured ham peptides with ACE inhibitory activity through a Caco-2 cell monolayer. *J. Funct. Foods* **2016**, *21*, 388–395. [CrossRef]

61. Zhang, H.; Duan, Y.; Feng, Y.; Wang, J. Transepithelial Transport Characteristics of the Cholesterol- Lowing Soybean Peptide, WGAPSL, in Caco-2 Cell Monolayers. *Molecules* **2019**, *24*, 2843. [CrossRef]

62. Yao, J.F.; Yang, H.; Zhao, Y.Z.; Xue, M. Metabolism of Peptide Drugs and Strategies to Improve their Metabolic Stability. *Curr. Drug Metab.* **2018**, *19*, 892–901. [CrossRef]

63. Lau, J.L.; Dunn, M.K. Therapeutic peptides: Historical perspectives, current development trends, and future directions. *Bioorg. Med. Chem.* **2018**, *26*, 2700–2707. [CrossRef] [PubMed]

64. Erak, M.; Bellmann-Sickert, K.; Els-Heindl, S.; Beck-Sickinger, A.G. Peptide chemistry toolbox—Transforming natural peptides into peptide therapeutics. *Bioorg. Med. Chem.* **2018**, *26*, 2759–2765. [CrossRef] [PubMed]

65. Brian Chia, C.S. A Review on the Metabolism of 25 Peptide Drugs. *Int. J. Pept. Res. Ther.* **2021**, *27*, 1397–1418. [CrossRef]

66. Watanabe, K.; Igarashi, M.; Li, X.; Nakatani, A.; Miyamoto, J.; Inaba, Y.; Sutou, A.; Saito, T.; Sato, T.; Tachibana, N.; et al. Dietary soybean protein ameliorates high-fat diet-induced obesity by modifying the gut microbiota-dependent biotransformation of bile acids. *PLoS ONE* **2018**, *13*, e0202083. [CrossRef]

67. Hashidume, T.; Kato, A.; Tanaka, T.; Miyoshi, S.; Itoh, N.; Nakata, R.; Inoue, H.; Oikawa, A.; Nakai, Y.; Shimizu, M.; et al. Single ingestion of soy beta-conglycinin induces increased postprandial circulating FGF21 levels exerting beneficial health effects. *Sci. Rep.* **2016**, *6*, 28183. [CrossRef] [PubMed]

68. Wanezaki, S.; Saito, S.; Inoue, N.; Tachibana, N.; Shirouchi, B.; Sato, M.; Yanagita, T.; Nagao, K. Soy beta-Conglycinin Peptide Attenuates Obesity and Lipid Abnormalities in Obese Model OLETF Rats. *J. Oleo Sci.* **2020**, *69*, 495–502. [CrossRef]

69. Wanezaki, S.; Tachibana, N.; Nagata, M.; Saito, S.; Nagao, K.; Yanagita, T.; Kohno, M. Soy β-conglycinin improves obesity-induced metabolic abnormalities in a rat model of nonalcoholic fatty liver disease. *Obes. Res. Clin. Pract.* **2015**, *9*, 168–174. [CrossRef]

70. Hakkak, R.; Gauss, C.H.; Bell, A.; Korourian, S. Short-Term Soy Protein Isolate Feeding Prevents Liver Steatosis and Reduces Serum ALT and AST Levels in Obese Female Zucker Rats. *Biomedicines* **2018**, *6*, 55. [CrossRef]

71. Panasevich, M.R.; Schuster, C.M.; Phillips, K.E.; Meers, G.M.; Chintapalli, S.V.; Wankhade, U.D.; Shankar, K.; Butteiger, D.N.; Krul, E.S.; Thyfault, J.P.; et al. Soy compared with milk protein in a Western diet changes fecal microbiota and decreases hepatic steatosis in obese OLETF rats. *J. Nutr. Biochem.* **2017**, *46*, 125–136. [CrossRef]

72. Ijaz, M.U.; Ahmed, M.I.; Zou, X.; Hussain, M.; Zhang, M.; Zhao, F.; Xu, X.; Zhou, G.; Li, C. Beef, Casein, and Soy Proteins Differentially Affect Lipid Metabolism, Triglycerides Accumulation and Gut Microbiota of High-Fat Diet-Fed C57BL/6J Mice. *Front. Microbiol.* **2018**, *9*, 2200. [CrossRef] [PubMed]

73. Jiang, M.; Yan, H.; He, R.; Ma, Y. Purification and a molecular docking study of α-glucosidase-inhibitory peptides from a soybean protein hydrolysate with ultrasonic pretreatment. *Eur. Food Res. Technol.* **2018**, *244*, 1995–2005. [CrossRef]

74. Ueoka, H.; Fukuba, Y.; Yamaoka Endo, M.; Kobayashi, T.; Hamada, H.; Kashima, H. Effects of soy protein isolate and soy peptide preload on gastric emptying rate and postprandial glycemic control in healthy humans. *J. Physiol. Anthropol.* **2022**, *41*, 25. [CrossRef] [PubMed]

75. Konya, J.; Sathyapalan, T.; Kilpatrick, E.S.; Atkin, S.L. The Effects of Soy Protein and Cocoa with or Without Isoflavones on Glycemic Control in Type 2 Diabetes. A Double-Blind, Randomized, Placebo-Controlled Study. *Front. Endocrinol.* **2019**, *10*, 296. [CrossRef] [PubMed]

76. Marthandam Asokan, S.; Wang, T.; Su, W.T.; Lin, W.T. Short Tetra-peptide from soy-protein hydrolysate attenuates hyperglycemia associated damages in H9c2 cells and ICR mice. *J. Food Biochem.* **2018**, *42*, e12638. [CrossRef]

77. Nagaoka, S.; Nakamura, A.; Shibata, H.; Kanamaru, Y. Soystatin (VAWWMY), a novel bile acid-binding peptide, decreased micellar solubility and inhibited cholesterol absorption in rats. *Biosci. Biotechnol. Biochem.* **2010**, *74*, 1738–1741. [CrossRef]

78. Lammi, C.; Zanoni, C.; Arnoldi, A. IAVPGEVA, IAVPTGVA, and LPYP, three peptides from soy glycinin, modulate cholesterol metabolism in HepG2 cells through the activation of the LDLR-SREBP2 pathway. *J. Funct. Foods* **2015**, *14*, 469–478. [CrossRef]

79. Jia, H.; Tian, L.; Zhang, B.; Fan, X.; Zhao, D. The soluble fraction of soy protein peptic hydrolysate reduces cholesterol micellar solubility and uptake. *Int. J. Food Sci. Technol.* **2019**, *54*, 2123–2131. [CrossRef]

80. Macchi, C.; Greco, M.F.; Ferri, N.; Magni, P.; Arnoldi, A.; Corsini, A.; Sirtori, C.R.; Ruscica, M.; Lammi, C. Impact of Soy beta-Conglycinin Peptides on PCSK9 Protein Expression in HepG2 Cells. *Nutrients* **2021**, *14*, 193. [CrossRef]

81. Li, W.; Zhang, J.; Ying, X.; Wang, Y.; Zhang, L.; Li, H.; Liu, X. Effect of soybean oligopeptides on blood pressure and plasma angiotensin in spontaneously hypertensive rats. *Shipin Kexue/Food Sci.* **2019**, *40*, 152–158. [CrossRef]

82. Ruan, S.; Luo, J.; Li, Y.; Wang, Y.; Huang, S.; Lu, F.; Ma, H. Ultrasound-assisted liquid-state fermentation of soybean meal with Bacillus subtilis: Effects on peptides content, ACE inhibitory activity and biomass. *Process Biochem.* **2020**, *91*, 73–82. [CrossRef]

83. Mcconnell, E.J.; Devapatla, B.; Yaddanapudi, K.; Davis, K.R. The soybean-derived peptide lunasin inhibits non-small cell lung cancer cell proliferation by suppressing phosphorylation of the retinoblastoma protein. *Oncotarget* **2015**, *6*, 4649–4662. [CrossRef] [PubMed]

84. Hao, Y.; Fan, X.; Guo, H.; Yao, Y.; Ren, G.; Lv, X.; Yang, X. Overexpression of the bioactive lunasin peptide in soybean and evaluation of its anti-inflammatory and anti-cancer activities in vitro. *J. Biosci. Bioeng.* **2020**, *129*, 395–404. [CrossRef] [PubMed]

85. Fernández-Tomé, S.; Xu, F.; Han, Y.; Hernández-Ledesma, B.; Xiao, H. Inhibitory Effects of Peptide Lunasin in Colorectal Cancer HCT-116 Cells and Their Tumorsphere-Derived Subpopulation. *Int. J. Mol. Sci.* **2020**, *21*, 537. [CrossRef]

86. Gonzalez-Montoya, M.; Hernandez-Ledesma, B.; Silvan, J.M.; Mora-Escobedo, R.; Martinez-Villaluenga, C. Peptides derived from in vitro gastrointestinal digestion of germinated soybean proteins inhibit human colon cancer cells proliferation and inflammation. *Food Chem.* **2018**, *242*, 75–82. [CrossRef]

87. Kopelman, P.G. Obesity as a medical problem. *Nature* **2000**, *404*, 635–643. [CrossRef] [PubMed]
88. Smith, K.B.; Smith, M.S. Obesity Statistics. *Prim. Care* **2016**, *43*, 121–135. [CrossRef]
89. Alghnam, S.; Alessy, S.A.; Bosaad, M.; Alzahrani, S.; Al Alwan, I.I.; Alqarni, A.; Alshammari, R.; Al Dubayee, M.; Alfadhel, M. The Association between Obesity and Chronic Conditions: Results from a Large Electronic Health Records System in Saudi Arabia. *Int. J. Environ. Res. Public Health* **2021**, *18*, 2361. [CrossRef]
90. Belza, A.; Ritz, C.; Sorensen, M.Q.; Holst, J.J.; Rehfeld, J.F.; Astrup, A. Contribution of gastroenteropancreatic appetite hormones to protein-induced satiety. *Am. J. Clin. Nutr.* **2013**, *97*, 980–989. [CrossRef]
91. Aoyama, T.; Fukui, K.; Nakamori, T.; Hashimoto, Y.; Yamamoto, T.; Takamatsu, K.; Sugano, M. Effect of soy and milk whey protein isolates and their hydrolysates on weight reduction in genetically obese mice. *Biosci. Biotechnol. Biochem.* **2000**, *64*, 2594–2600. [CrossRef]
92. Noer, E.R.; Dewi, L.; Kuo, C.H. Fermented soybean enhances post-meal response in appetite-regulating hormones among Indonesian girls with obesity. *Obes. Res. Clin. Pract.* **2021**, *15*, 339–344. [CrossRef] [PubMed]
93. Astawan, M.; Mardhiyyah, Y.S.; Wijaya, C.H. Potential of Bioactive Components in Tempe for the Treatment of Obesity. *J. Gizi Dan Pangan* **2018**, *13*, 79–86. [CrossRef]
94. Christopoulou, M.E.; Papakonstantinou, E.; Stolz, D. Matrix Metalloproteinases in Chronic Obstructive Pulmonary Disease. *Int. J. Mol. Sci.* **2023**, *24*, 3786. [CrossRef]
95. Cardenas-Diaz, F.L.; Leavens, K.F.; Kishore, S.; Osorio-Quintero, C.; Chen, Y.J.; Stanger, B.Z.; Wang, P.; French, D.; Gadue, P. A Dual Reporter EndoC-betaH1 Human beta-Cell Line for Efficient Quantification of Calcium Flux and Insulin Secretion. *Endocrinology* **2020**, *161*, bqaa005. [CrossRef] [PubMed]
96. Zheng, Y.; Ley, S.H.; Hu, F.B. Global aetiology and epidemiology of type 2 diabetes mellitus and its complications. *Nat. Rev. Endocrinol.* **2018**, *14*, 88–98. [CrossRef] [PubMed]
97. Cole, J.B.; Florez, J.C. Genetics of diabetes mellitus and diabetes complications. *Nat. Rev. Endocrinol.* **2020**, *16*, 377–390. [CrossRef]
98. Kanetro, B. Amino acid profile of soybean (Glicine max) sprout protein for determining insulin stimulation amino acids. *Int. Food Res. J.* **2018**, *25*, 2497–2502.
99. Tang, J.; Wan, Y.; Zhao, M.; Zhong, H.; Zheng, J.S.; Feng, F. Legume and soy intake and risk of type 2 diabetes: A systematic review and meta-analysis of prospective cohort studies. *Am. J. Clin. Nutr.* **2020**, *111*, 677–688. [CrossRef]
100. Woo, H.W.; Kim, M.K.; Lee, Y.H.; Shin, D.H.; Shin, M.H.; Choi, B.Y. Sex-specific associations of habitual intake of soy protein and isoflavones with risk of type 2 diabetes. *Clin. Nutr.* **2021**, *40*, 127–136. [CrossRef]
101. Chang, C.L.; Lin, Y.; Bartolome, A.P.; Chen, Y.-C.; Chiu, S.-C.; Yang, W.-C. Herbal therapies for type 2 diabetes mellitus: Chemistry, biology, and potential application of selected plants and compounds. *Evid. -Based Complement. Altern. Med.* **2013**, *2013*, 378657. [CrossRef]
102. Nagaoka, S.; Takeuchi, A.; Banno, A. Plant-derived peptides improving lipid and glucose metabolism. *Peptides* **2021**, *142*, 170577. [CrossRef]
103. Katagiri, R.; Sawada, N.; Goto, A.; Yamaji, T.; Iwasaki, M.; Noda, M.; Iso, H.; Tsugane, S.; Japan Public Health Center-based Prospective Study Group. Association of soy and fermented soy product intake with total and cause specific mortality: Prospective cohort study. *BMJ* **2020**, *368*, m34. [CrossRef] [PubMed]
104. World Health Organization. *Global Status Report on Noncommunicable Diseases 2014*; World Health Organization: Geneva, Switzerland, 2014.
105. Roth, G.A.; Huffman, M.D.; Moran, A.E.; Feigin, V.; Mensah, G.A.; Naghavi, M.; Murray, C.J. Global and regional patterns in cardiovascular mortality from 1990 to 2013. *Circulation* **2015**, *132*, 1667–1678. [CrossRef]
106. Food and Drug Administration. Food labeling: Health claims; soy protein and coronary heart disease. Food and Drug Administration, HHS. Final rule. *Fed Regist* **1999**, *64*, 57700–57733.
107. Yan, Z.; Zhang, X.; Li, C.; Jiao, S.; Dong, W. Association between consumption of soy and risk of cardiovascular disease: A meta-analysis of observational studies. *Eur. J. Prev. Cardiol.* **2017**, *24*, 735–747. [CrossRef] [PubMed]
108. Zampelas, A. The Effects of Soy and its Components on Risk Factors and End Points of Cardiovascular Diseases. *Nutrients* **2019**, *11*, 2621. [CrossRef] [PubMed]
109. Blanco Mejia, S.; Messina, M.; Li, S.S.; Viguiliouk, E.; Chiavaroli, L.; Khan, T.A.; Srichaikul, K.; Mirrahimi, A.; Sievenpiper, J.L.; Kris-Etherton, P.; et al. A Meta-Analysis of 46 Studies Identified by the FDA Demonstrates that Soy Protein Decreases Circulating LDL and Total Cholesterol Concentrations in Adults. *J. Nutr.* **2019**, *149*, 968–981. [CrossRef]
110. Nagaoka, S. Structure-function properties of hypolipidemic peptides. *J. Food Biochem.* **2019**, *43*, e12539. [CrossRef] [PubMed]
111. Fuchs, F.D.; Whelton, P.K. High Blood Pressure and Cardiovascular Disease. *Hypertension* **2020**, *75*, 285–292. [CrossRef]
112. Siegel, R.L.; Miller, K.D.; Fuchs, H.E.; Jemal, A. Cancer Statistics, 2021. *CA Cancer J. Clin.* **2021**, *71*, 7–33. [CrossRef]
113. Toric, J.; Markovic, A.K.; Brala, C.J.; Barbaric, M. Anticancer effects of olive oil polyphenols and their combinations with anticancer drugs. *Acta Pharm.* **2019**, *69*, 461–482. [CrossRef] [PubMed]
114. Ortiz-Martinez, M.; Winkler, R.; García-Lara, S. Preventive and therapeutic potential of peptides from cereals against cancer. *J. Proteom.* **2014**, *111*, 165–183. [CrossRef] [PubMed]
115. Rayaprolu, S.J.; Hettiarachchy, N.S.; Horax, R.; Phillips, G.K.; Mahendran, M.; Chen, P. Soybean peptide fractions inhibit human blood, breast and prostate cancer cell proliferation. *J. Food Sci. Technol.* **2017**, *54*, 38–44. [CrossRef] [PubMed]

116. Roberts, P.R.; Burney, J.D.; Black, K.W.; Zaloga, G.P. Effect of chain length on absorption of biologically active peptides from the gastrointestinal tract. *Digestion* **1999**, *60*, 332–337. [CrossRef] [PubMed]
117. Di, L. Strategic approaches to optimizing peptide ADME properties. *AAPS J.* **2015**, *17*, 134–143. [CrossRef] [PubMed]

Article

The Protective Effects of Corn Oligopeptides on Acute Alcoholic Liver Disease by Inhibiting the Activation of Kupffer Cells NF-κB/AMPK Signal Pathway

Ying Wei [1,*,†], Mingliang Li [2,†], Zhiyuan Feng [3], Di Zhang [4], Meiling Sun [1], Yong Wang [5] and Xiangning Chen [1,*]

[1]   Department of Food Science and Engineering, Beijing University of Agriculture, Beijing 102206, China
[2]   School of Food Science and Technology, Jiangnan University, Wuxi 214122, China
[3]   Key Laboratory of Industrial Fermentation Microbiology, Ministry of Education, Tianjin University of Science and Technology, Tianjin 300457, China
[4]   School of Food and Biological Engineering, Jiangsu University, Zhenjiang 212013, China
[5]   Academy of National Food and Strategic Reserves Administration, Beijing 100037, China
*   Correspondence: proudwy@126.com (Y.W.); cxn@bua.edu.cn (X.C.)
†   These authors contributed equally to this work.

Abstract: Alcohol can cause injury and lead to an inflammatory response in the liver. The NF-κB/AMPK signaling pathway plays a vital role in regulating intracellular inflammatory cytokine levels. In this study, corn oligopeptides (CPs), as the research objects, were obtained from corn gluten meal, and their regulation of the activation of the Kupffer cell NF-κB/AMPK signal pathway induced by LPS was investigated. Results showed that ALT, AST, and inflammatory cytokines in mice serum after the administration of CPs at 0.2, 0.4, and 0.8 g/kg of body weight displayed a distinct ($p < 0.05$) reduction. On the other hand, the CPs also inhibited the expression of recognized receptor CD14 and TLR4, down-regulated P-JNK, P-ERK, and P-p-38, and thus inhibited inflammatory cytokine levels in Kupffer cells (KCs). Furthermore, four kinds of dipeptides with a leucine residue at the C-terminus that might exhibit down-regulated inflammatory cytokines in the NF-κB/AMPK signaling pathway functions were detected using HPLC-MS/MS. These results indicated that CPs have a potential application value in acute alcoholic liver disease.

Keywords: corn oligopeptide; ALD; Kupffer cell; inflammatory cytokines; NF-κB/AMPK

Citation: Wei, Y.; Li, M.; Feng, Z.; Zhang, D.; Sun, M.; Wang, Y.; Chen, X. The Protective Effects of Corn Oligopeptides on Acute Alcoholic Liver Disease by Inhibiting the Activation of Kupffer Cells NF-κB/AMPK Signal Pathway. *Nutrients* **2022**, *14*, 4194. https://doi.org/10.3390/nu14194194

Academic Editors: Chiara Rosso and Antoni Sureda

Received: 29 August 2022
Accepted: 27 September 2022
Published: 8 October 2022

1. Introduction

Alcoholic liver disease (ALD) is the hepatic manifestation of alcohol overconsumption. ALD could be generally aroused by excessive and acute ingestion of alcohol [1,2]. According to disease courses and severity, there are many forms of ALD, including fatty liver, alcoholic hepatitis, and chronic hepatitis with hepatic fibrosis or cirrhosis [3,4]. It has been the major cause of liver diseases in western countries. In the United States, the predicted number of ALD patients is over two million [5]. In China, according to general statistics, there are about 300 million drinkers. Varying degrees of alcoholic fatty liver can be observed in about 80% of the people in this group, and 20% have suffered from severe alcoholic liver disease [6]. With the increasing population of heavy drinkers, especially in northern regions, ALD has become the second-largest liver disease in China, following the viral hepatitis A and hepatitis B [7].

ALD has also given rise to a high mortality rate because of ineffective treatments. The Veterans Administration Cooperative Studies have shown that the mortality rate of patients with alcoholic hepatitis and cirrhosis exceeds 60% after 4 years [4,8,9]. Until now, no satisfactory effects have been obtained by the treatments for ALD. This is mainly attributed to the severe side effects of the drugs used in current treatments [10]. Thus, ALD has become a major cause of morbidity and mortality worldwide [11,12]. Various mechanisms have been proposed as the consequences of ALD, such as inflammatory cytokines, mitochondrial injury, and oxidative stress. Thus, the development of treatment for ALD has mainly focused on drugs targeting these mechanisms, aiming to find safer treatments with fewer side effects [13,14].

Food-derived oligopeptides are functional components derived from food protein and exist in dietary protein with a specific amino acid sequence. After the protein is degraded, these functional peptides are released and exhibit superior biological activities in the process, including antioxidant, anti-inflammatory, immune regulating, antibacterial, etc. Recently, several bioactive peptides extracted from natural products were studied, such as ganodermalucidum peptides [15] and cassia seed peptides [16]. The protective impacts of these peptides against hepatic damage induced by D-galactosamine or acetaminophen have been proven [4]. Thus, the ALD-protective ingredients extracted from plants and animals have attracted attention from researchers in academia and industry. Numerous studies have confirmed that food-derived oligopeptides can be used as potential dietary therapeutics and new functional raw materials for enhancing health [17].

Corn gluten meal (CGM) is a byproduct in the starch processing industry, and it includes roughly 60% (*w/w*) protein [18]. Corn oligopeptides (CPs) are low molecular weight peptides decomposed from CGM using enzymolysis [18,19]. In previous studies, multiple functions of CPs have been studied and verified, such as the alleviation of fatigue, resistance to the peroxidative reaction of lipids, suppression of angiotensin I-converting enzyme, and promotion of ethanolic metabolism [1,19,20]. However, few studies have detailed the effects and relevant mechanism of CPs on ALD, although existing results have shown that they would be beneficial for treating ALD due to their antioxidant activity and ethanolic metabolism promotion [3].

This study investigated the effects of CPs on alcoholic hepatic damage in an animal model of mice. Biochemical markers assessed hepatic damage and treatment effects in mice serum. In addition, the primary culture of KCs and relevant cytokines were applied to explore the potential molecular mechanism of protective effects from CPs on alcoholic hepatic damage. This work aims to provide a reference for functional activity research of anti-inflammatories in ADL, nutritional food development, and the clinical applications of CPs.

2. Materials and Methods

2.1. Preparation of CPs

CPs were prepared from CGM [21]. The CPs involved in this study were provided by CF Haishi Biotechnology Ltd. Co. (Beijing, China). The CPs were prepared with the following protocols: the CGM was suspended in distilled water after being ground with a 60-mesh sieve (1:10, *w/w*). The suspension was then hydrolyzed at pH 11.0 and 90 °C for 1 h. The suspension was neutralized and centrifuged to recover the insoluble protein precipitate. The insoluble protein precipitate was resuspended and subjected to the procedures above. The wet corn protein isolate (CPI) was thus obtained.

The wet CPI was resuspended with a concentration of 6% (*w/w*) and subjected to a two-step enzymatic hydrolysis. In the first step, the enzymatic hydrolysis was performed with crude alkaline proteinases at pH 8.5 and 55 °C for 3 h. The second step was performed with crude neutral proteinases at pH 7.0 and 45 °C for 2 h (Angel Yeast, Hubei, China). The obtained hydrolysates were centrifuged to remove the insoluble impurities. The supernatant was filtered successively through 10 and 1 kDa MWCO ceramic membranes. The step of nanofiltration was carried out to remove the mineral salt. The salt-removed solution was concentrated by cry concentration under vacuum at 70 °C, with an evaporation rate of 500 kg/h. When the solution concentration was about 30 Baumé degrees, it was decolored with 12% active carbon at 75 °C for 1 h. The carbon was then removed by normal filtration after de-coloration. Most of the water was removed by spray drying with a pressure of 20 MPa. The CP powder was obtained, and it was applied in the following experiments.

2.2. Identification of Corn Oligopeptides

CPs have been analyzed to determine the chemical components, amino acid compositions, and molecular weight distribution. The crude protein, moisture, and ash content were determined according to the methods specified by the Association of Official Analytical Chemists. The amino acid composition was determined using an amino acid analyzer (L-8900, Hitachi, Tokyo, Japan). The molecular weight distribution of the corn oligopeptide was established using HPLC (LC-20AD, Shimadzu, Kyoto, Japan) according to the previously reported method. A total amino acid analysis was conducted with an amino acid analyzer (835–50, Hitachi, Tokyo, Japan) [4,22]. The amino acid sequences of CPs were detected by HPLC-MS/MS (8060, Shimadzu Corporation, Japan) with reference to Wei's method [23], and their main structures were revealed. Contents of the peptides in CPs were estimated using LC−MS/MS in the multiple reaction monitoring (MRM) mode by using the same RP-HPLC condition as described above. The synthetic peptides were used for the optimization of the MRM condition using LabSolution Ver. 5.80 software.

2.3. Animal Models

Male Kunming mice were used for this study (approval number: 2016-0006). The mice were aged 6 to 12 weeks and weighed 20 ± 2 g. All the mice were maintained in an environmentally controlled room at 22 ± 1 °C, with a 12 h light/dark cycle (light from 7:00 to 19:00). The treatment and maintenance of animals were conducted according to the Principle of Laboratory Animal Care (NIH Publication No. 85-23, revised 1985) and the Peking University Animal Research Committee guidelines.

All mice were fed a normal AIN-93M rodent diet (Vital River Limited Company, Beijing, China), and the main protein source was casein. The animals were randomly assigned to 5 groups: a normal control group (defined as the standard group; $n = 5$), an alcohol control group (defined as the control group; $n = 10$), and 3 CPs intervention groups with different doses (designated as the LCP, MCP, and HCP groups; $n = 10$).

The mice in the control and experimental groups were administered with 50% ethanol on day 7 in addition to the standard diet with a dose of 12 mL/kg of body weight. The standard group was administered with saline solution in the same manner. In the experimental groups, mice were pretreated with CPs 1 h before the ethanol administration, and the dose was 0.2, 0.4, and 0.8 g/kg of body weight (respectively designated as CP-0.2, CP-0.4, and CP-0.8) (the amount of corn peptides to be gavaged was based on the optimal recommended daily dosage for humans ($\leq$4.5 g/day)). The mice in the control group were administrated with ethanol, without any treatment. All the mice fasted for 12 h after ethanol treatment. Then, all the mice were anesthetized with pentobarbital. Blood was taken from the mouse's heart, and serum was obtained from the blood by centrifugation at $3000 \times g$ for 20 min at room temperature. The liver was carefully removed and immediately frozen in liquid nitrogen and stored in a $-80°$ freezer until use.

2.4. Evaluation of Hepatic Biomarkers

2.4.1. Enzyme Activities

The activities of alanine aminotransferase (ALT) and aspartate aminotransferase (AST) in mice serum were analyzed by the Mouse Aspartate Aminotransferase ELISA Kit (Cusabio Biotech Co., Ltd. Lot: CSB-E12649m, Wuhan, China) and Mouse Alanine Aminotransferase ELISA Kit (Cusabio Biotech Co., Ltd. Lot: CSB-E16539m, Wuhan, China) following the manufacturer's instructions.

2.4.2. ELISA for TNF-α, IL-1, and IL-6 in Mice Serum

Mouse serum samples were analyzed for TNF-α, IL-1, and IL-6 levels with ELISA following the manufacturer's instructions. The involved ELISA kit was obtained from Jiancheng Bioengineering Institute (Nanjing, China).

2.4.3. Real-Time Quantitative PCR

Total RNA was extracted from the liver with the SV Total RNA Isolation System. cDNA was synthesized from 1 μg of RNA by the First Strand cDNA Synthesis Kit for RT-PCR (AMV). Real-time PCR was performed for TNF-a, IL-1, IL-6, and the housekeeping gene, encoding glyceraldehyde-3-phosphate-dehydrogenase (GAPDH). It was carried out with ABI PRISM 7000 Sequence Detection systems (Applied Biosystems, USA). The reaction mixture was composed of Absolute TM QPCR SYBR Green Mixes (12.5 μL), forward and reverse primers (5 μM and 1 μL each), nuclease-free water (8 μL), and a cDNA sample (2.5 μL).

All primers were synthesized by Invitrogen (Invitrogen, Hong Kong, China). The GAPDH gene was used as an internal control. The real-time PCR primer sequences for these genes are shown in Table 1. The PCR conditions were as follows: 30 s at 95 °C for 1 cycle; 5 s at 95 °C; 31 s at 60 °C for 45 cycles; 15 s at 95 °C; 1 min at 60 °C; and 15 s at 95 °C. Results were analyzed with ABI sequence Detection System software (Applied Biosystems, Foster, CA, USA).

Table 1. PCR primers of mice used in this study.

Gene	Primer Sequences (5′-3′)	Product Length
TNF-α	forward: CATCTTCTCAAAATTCGAGTGACAA reverse: TGGGAGTAGACAAGGTACAACCC	447 bp
IL-1	forward: CTTCATCTTTGAAGAAGAGCCC reverse: CTCTGCAGACTCAAACTCCAC	418 bp
IL-6	forward: TTCACAAGTCCGGACAGGAG reverse: TGGTCTTGGTCCTTAGCCAC 3	488 bp
GAPDH	forward: GAAGGTGAAGGTCGGAGTCA reverse: TTCACACCCATGACGAACAT	402 bp

2.4.4. Isolation and Culture of Murine Kupffer Cells (KCs)

KCs were isolated according to the method described previously [24]. Briefly, each liver was first perfused with calcium- and magnesium-free D-Hank's solution until the liver became completely blanched. After gently mashing the liver, the digestion was allowed to proceed in this solution for 10 min at 37 °C, with stirring. It was then centrifuged at $300\times g$ at 4 °C for 5 min, and the pellet was washed three times with 40 mL cold HBSS containing 10 mmol/L HEPES and 10 mg/mL DNase (HBSS + HEPES/DNase). The final pellet was resuspended in 40 mL HBSS + HEPES/DNase and centrifuged at $100\times g$ at 4 °C for 1 min. The resulting supernatant (containing most of the hepatic non-parenchymal cells) was layered on a sterile Percoll gradient (15 mL 25% Percoll over 15 mL 50% Percoll), which was then centrifuged at $900\times g$ for 20 min. The lower zone, including the interface zone, was collected and resuspended in 40 mL cold HBSS + HEPES/DNase and centrifuged at $900\times g$ for 5 min. The pellet was resuspended, washed twice with HBSS + HEPES/DNase, and then resuspended in RPMI 1640 medium with 10 mmol/L HEPES. The final cell pellets were resuspended in the appropriate volume of RPMI 1640 medium supplemented with 10% endotoxin-free FCS, 100 U/mL penicillin, 100 U/mL streptomycins, 15 mM L-glutamine, and 10 mM HEPES to achieve a final concentration of 1×10^6 viable cells/mL. The cell suspensions were seeded in 90 mm culture dishes and incubated at 37 °C in 5% CO_2 air for 3 h to allow the adhesion of KCs. Non-adherent cells were removed by vigorous washing with Hank's solution. Over 90% of the adherent cells were identified to be KCs by positive peroxide staining. The adherent cells were then trypsinized for cell detachment. They were sub-cultured at 2×10^6 cells/mL in 35 mm dishes for 24 h to recover from isolation and adherence for in vitro stimulation. The cells were exposed to *E. coli* LPS (60 ng/mL); at the same time, the CPs with a concentration of 0.1, 0.5, and 1 mg/mL were added respectively and incubated at 37 °C for 1 h or 12 h. Kupffer cell viability with CPs was assessed using the MTT, as described previously [25].

2.4.5. Extraction of Proteins

After 1 h incubation, the cells (2×10^6) were washed with ice-cold phosphate-buffered saline (PBS) and were lysed by adding ice-cold SDS sample buffer containing 62.5 mM Tris-HCL (pH 6.8), 2% SDS, 10% glycerol, 50 mM DTT, and 0.1% bromphenol blue. The cell extract was collected into a microfuge tube and was sonicated for 10 to 15 s to shear DNA and reduce sample viscosity. The sample was followed by heating to 95 °C to 100 °C for 5 min, and it was spun for 20 min at 4 °C at 15,000 g in a microfuge. The supernatant was decanted, and a small aliquot was removed for protein assessment. The rest of the sample was aliquoted and frozen at −70 °C until the application for western blotting.

2.5. Western Blot Analyses

Protein samples were resolved by SDS-PAGE and transferred to nitrocellulose membranes (Roche Diagnostics, Rotkreuz, Switzerland). These nitrocellulose membranes were blocked with 5% skimmed milk powder/Tris-buffered saline with 0.1% Tween20 (*w/v*) at 4 °C overnight. After that, the membranes were incubated with the primary antibodies:inhibitor of κB kinase-α (IκB-α; Santa Cruz Biotechnology, Delaware Ave Santa Cruz, CA, USA; dilution ratio; 1:1000), TLR4, CD14 (Santa Cruz Biotechnology, USA; dilution ratio; 1:200), p-p38, p-JNK, p-ERK, total p38, total JNK, total ERK (Cell Signaling Technology, Boston, CA, USA; all at 1:200), and β-actin (Cell Signaling Technology, Boston, CA, USA; 1:500); the samples were incubated overnight at 4 °C. Then, membranes were treated with HRP-conjugated anti-rabbit IgG (H + L) as the second antibody (Promega, Madison, WI, USA; dilution ratio; 1:2000). Chemiluminescent HRP Substrate examined the immunoblotting (Cat. NO: WBKLS0100; ImmobilonTMWestern, MA, USA) according to the manufacturer's instructions. Membranes were exposed by a FUJIFILM Luminescent Image Analyzer LAS-1000 (Macintosh TM, USA). The intensities of the resulting bands were quantified by Quantity One software on an AGS-800 densitometer (BioRad, Hercules, CA, USA).

2.6. Statistical Analyses

The data were expressed as mean ± SD. ANOVA and multiple comparisons were applied to calculate the statistical difference between groups. The significance level was set at 95%. All statistical calculations were performed with the SPSS 21.0 software for windows.

3. Results

3.1. The Chemical Composition of CPs

Table 2 displays the specific chemical compositions obtained by the analysis and indicates that corn oligopeptide had a high protein (83.60%) and peptide (79.23%) content. Other components in the powder had 3.78% water and 3.90% ash. The results of Table 1 show that corn oligopeptide contained considerable amounts of branched-chain amino acids (21.44%), including valine (2.72%), leucine (16.73%), and isoleucine (1.99%). Rich branched-chain amino acid compositions play an essential role in the regulation of human liver physiological functions.

Table 2. Chemical composition of CPs.

	Corn Oligopetides
Moisture (%)	3.78 ± 0.11
Ash (%)	3.90 ± 0.13
Protein (%)	83.60 ± 1.21
Peptide (%)	79.23 ± 1.19
Amino acid composition (%)	
Ala	8.17
Pro	6.52
Val	2.72
Met	1.97

Table 2. *Cont.*

	Corn Oligopetides
Ile	1.99
Leu	16.73
Phe	4.33
Trp	0.23
Asp	4.73
Ser	4.14
Glu	22.72
His	1.03
Gly	1.18
Arg	1.35
Thr	2.22
Cys	2.17
Tyr	4.29
Lys	0.25

Table 3 and Figure 1 show that 96.51% of the peptides in the CPs were below a molecular weight of 1000 Da and that the average molecular weight in the CP mixture was 349 Da. The average molecular weight of amino acids was 137 Da, and the mean peptide length was approximately 2.5 residues. Peptide compositions of CPs mostly were represented by dipeptides or tripeptides, which can be absorbed and transported more efficiently than either amino acids or intact proteins.

Figure 2 shows that the main peptide sequences in the CPs are pEL, LL, VL, and TL, and their proportions in the CPs are 0.72%, 0.27%, 0.17%, and 0.14%, respectively. Previous studies have shown that VL and LL exhibit protect liver effects [1,26] and TL exhibits inflammatory level inhibitory effects [27,28]. Therefore, CPS may have anti-inflammatory and protect liver functions due to the presence of these peptides.

Table 3. Molecular weight distributions of corn oligopeptides.

Molecular Weight (u)	Over 10,000	3000–10,000	1000–3000	150–1000	Below 150
Distributions (%)	0.0000	0.1513	3.3431	77.5960	18.9096

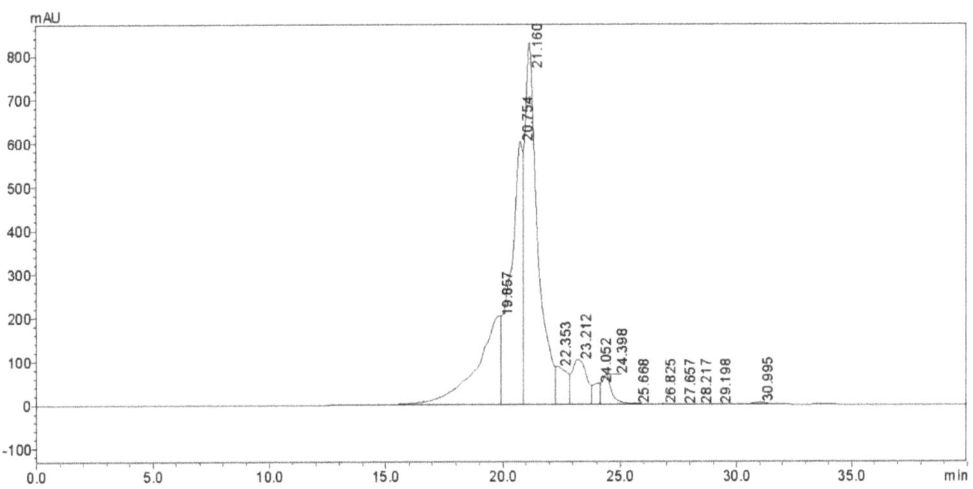

Figure 1. Corn oligopeptide chromatogram (220 nm).

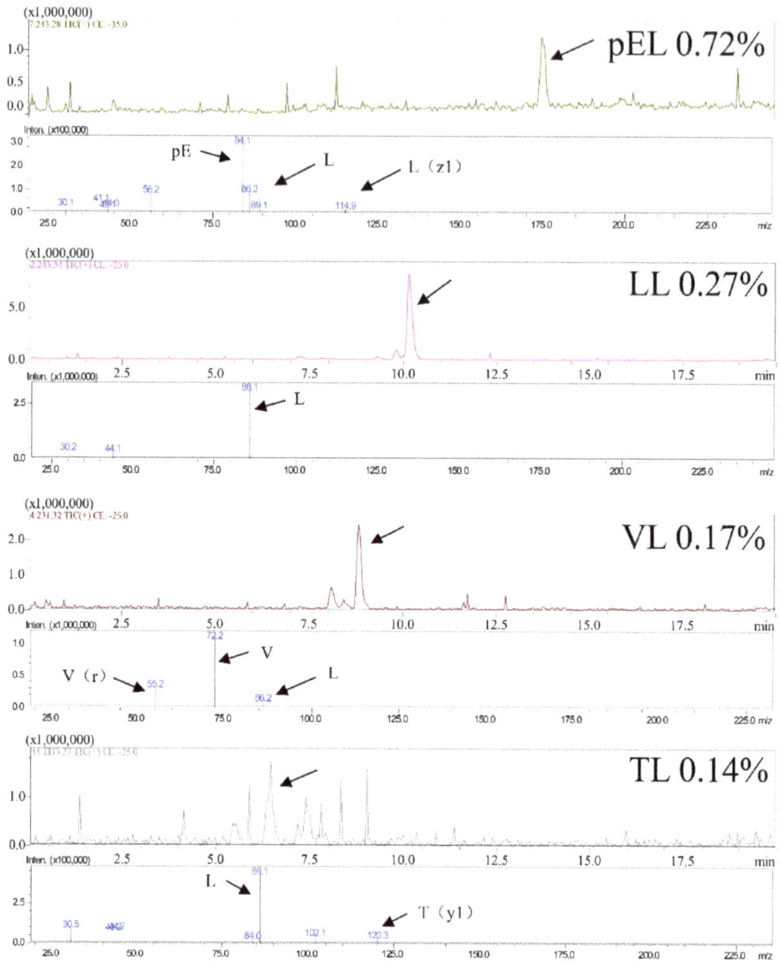

Figure 2. CPs main peptides sequences.

3.2. The Inhibition Effects of CPs on AST and ALT in Mice Serum

AST and ALT exist in organs, including the liver, heart, skeletal muscle, and kidneys. The concentrations of AST and ALT in serum were maintained within a certain level. When there was liver cell damage, the levels of AST and ALT increased correspondingly. Therefore, in clinical practice, liver damage will be reflected by testing AST and ALT. In this work, the ALT and AST in the standard group were less than 70 IU/L and 100 IU/L (Figure 3A,B). With the treatment of ethanol, the levels of AST and ALT were increased to more than 600 IU/L and 180 IU/L. If mice were pretreated with CPs (0.2, 0.4, and 0.8 g/kg of body weight), after ethanol treatment, the levels of AST and ALT were reduced with the increased concentration of CPs (Figure 3A,B). In the group of CP-0.8, the levels of both AST and ALT were close to that of the standard group without ethanol treatment, with no significant damage to liver cells. This indicated that CPs had an excellent protective effect on the liver.

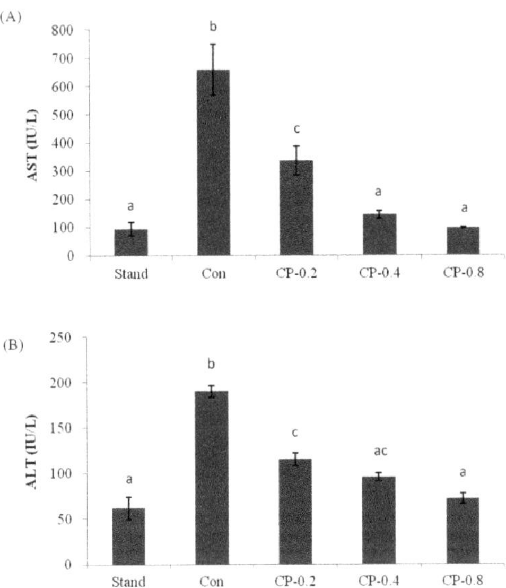

Figure 3. The effects of CPs on transaminase activities in mice serum at 12 h after ethanol treatment. (**A**) The level of AST. (**B**) The level of ALT. Values are expressed as means $\pm$ SEM (n = 5–10). Values with different letters are significantly different ($p < 0.05$).

3.3. The Inhibition Effects of CPs on Inflammatory Factors in Mice Serum

IL-1, IL-6, and TNF-α are pro-inflammatory cytokines in the inflammatory response. The pathogenic role of pro-inflammatory factor signaling in ALD is attributable to the activation of the inflammatory response. In this work, after the treatment of ethanol, the level of IL-1, IL-6, and TNF-α were all dramatically increased compared with standard groups without ethanol (Figure 4A–C). However, with the pretreatment of CPs, the factors maintained a relatively low concentration. The level was decreased with the increased CPs concentration.

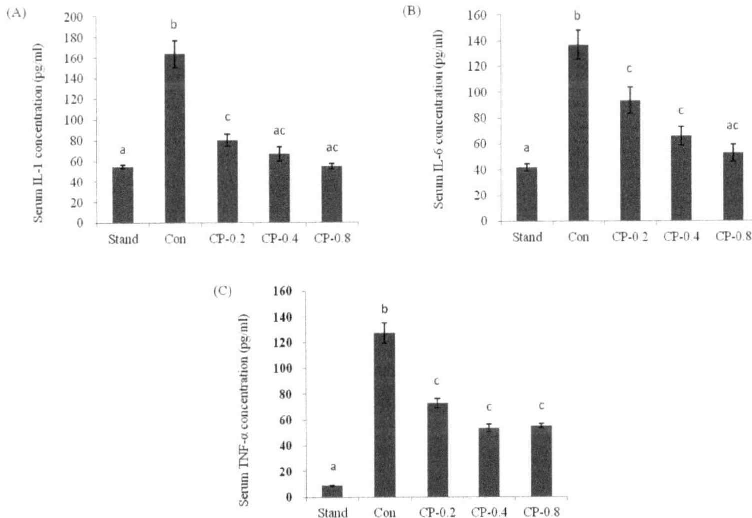

Figure 4. The effects of CPs on the levels of IL-1 (**A**), IL-6 (**B**), and TNF-α (**C**) in mice serum at 12 h after ethanol treatment. Values are expressed as mean $\pm$ SEM (n = 5–10). Values with different letters are significantly different at $p < 0.05$.

The mRNA expression levels of pro-inflammatory cytokines IL-1, IL-6, and TNF-α in the liver were analyzed after ethanol treatment with or without CPs pretreatment. Real-time PCR was performed to quantify the mRNA expression level. This is expressed as the ratio to the housekeeping gene encoding GAPDH. In this study, the mRNA expression levels of IL-1, IL-6, and TNF-α were upregulated after ethanol treatment, compared to that of the standard group without any treatment (Figure 5A–C). With the pretreatment of CPs (CP-0.8), the IL-1, IL-6, and TNF-α were maintained at relatively low levels compared to the control group after ethanol treatment (Figure 5A–C). These results indicated that CPs played an essential role in regulating the mRNA expression levels of inflammatory cytokines.

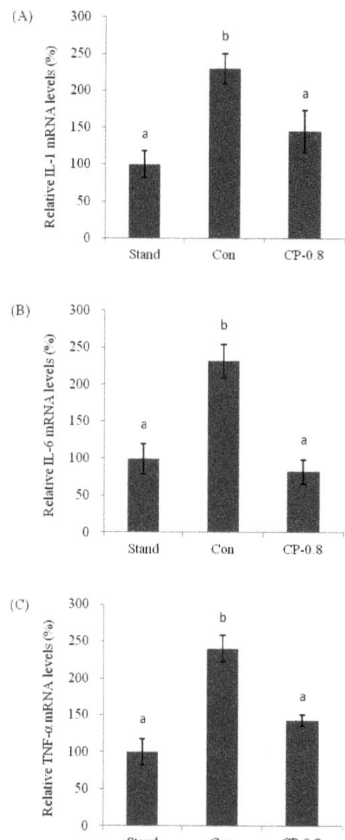

Figure 5. The effects of CPs on the mRNA expression of IL-1 (**A**), IL-6 (**B**), and TNF-α (**C**) in mice liver at 12 h after ethanol treatment. Values are expressed as means ± SEM (n = 5–10). Values with different letters are significantly different at $p < 0.05$.

3.4. The Molecular Mechanism of Effects from CPs on LPS-Induced Kuffer Cells (KCs)

The LPS-induced signaling is typical and critical in the initiation and process of ALD. LPS can be recognized by cell differentiation antigen (CD14) and toll-like receptor 4 (TLR4). The downstream signaling pathways are activated and end in activating transcription factors, including nuclear factor (NF)-κB [13,29].

The cell viability of Kupffer cells was determined after treatment with different concentrations of CPs (0.1, 0.5, 1, 2, 5, 10 mg/mL). The results are shown in Figure 6. Compared with the control, the cell viability after treatment with 1 mg/mL did not show significant statistical difference for 12 h, while the treatment at the concentration of 2–10 mg/mL slightly decreased the cell viability; thus, the optimal concentration of CPS is below 2 mg/mL.

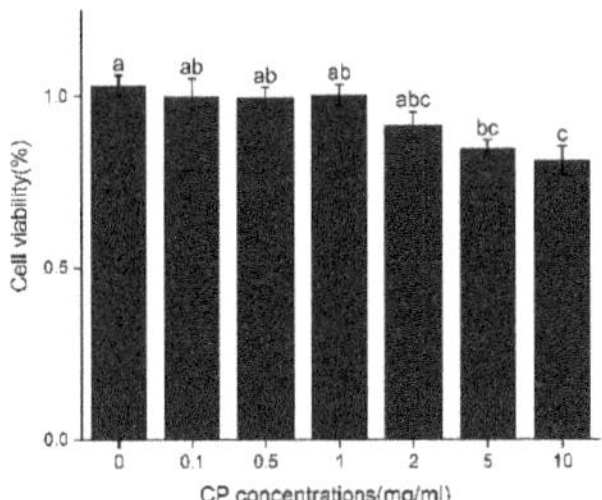

Figure 6. Effect of CPs concentration on cell viability in Kupffer cells after 12 h treatment. Different letters represented the significant difference at $p < 0.05$.

The western blot analysis was used to compare the expression of CD14 and TLR4 in three groups: standard group without any treatment, LPS treated group, and both LPS and CPs treated group (1 mg/mL, Figure 7A). In the LPS treated group, the expression level of CD14 was upregulated to be 150% and 180%, respectively, compared to the expression of β-actin (Figure 7B,C). However, in the presence of CPs with a concentration of 1 mg/mL, the expression of CD14 and TLR4 in the LPS treated group was down-regulated. The expression level was approximately equal to or a little more than that in the standard group (Figure 7B,C).

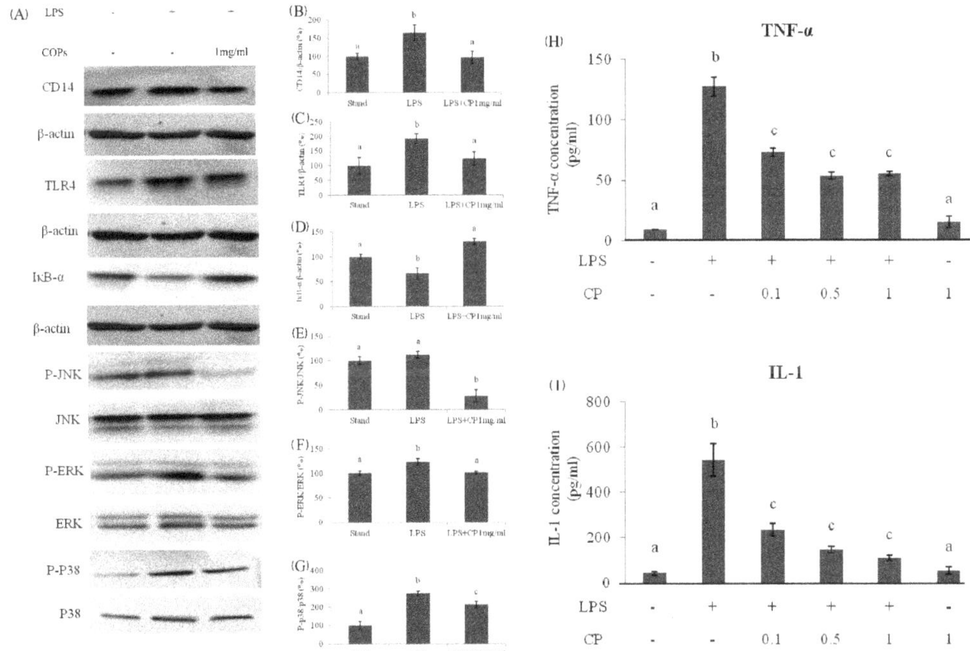

Figure 7. The effects of protein expression at 1 h after LPS administration in Kupffer cells: (**A**) Bands of proteins. The values of CD14 (**B**) and TLR4 (**C**) were normalized by the value of the total protein. (**D**) The effects of CPs on IkB-α degradation at 1 h after LPS administration in Kupffer cells. The expression levels of phosphorylated and total JNK, ERK, and p-38 protein were detected by western blotting. The value of JNK (**E**), ERK (**F**), and p-38 (**G**) were normalized by the value of the total protein. The effects of CPs on the TNF-α (**H**) and IL-1 (**I**) expression level in supernatants of KCs at 12 h after LPS treatment. Different letters represented the significant difference at $p < 0.05$.

In the LPS treated group, the NF-κB was activated, and JNK, ERK, and p-38 protein phosphorylation was enhanced. This led to an inflammatory response and the upregulated expression of inflammatory cytokines, such as TNF-α and IL-1. It also led to additional liver cell damage. In the western blot analysis, after adding both CPs and LPS, the expression level of IkB-α was increased, and its inhibition effect on NF-κB was enhanced (Figure 2D) As a result, the phosphorylation of JNK, ERK, and p-38 protein was down-regulated compared to the LPS treated control group (Figure 7E–G). After adding CPs and LPS, the expression of TNF-α and IL-1 were significantly down-regulated, although the level was still higher than that in the standard group and the group treated only with CPs (Figure 7H,I). Therefore, CPs as a functional food have the potential activity to prevent up-regulation of inflammation levels induced by alcohol.

4. Discussion

The pathogenesis of ALD involves many factors, such as genetics and nutrition, in addition to many injurious factors such as oxidative stress, bacterial lipopolysaccharides (LPS) and cytokines [30,31]. The development of treatment of ALD has been limited since the 1970s [11,32], and most of the treatments have been associated with side effects. As a result, more and more research is beginning to explore new treatments that combine the pathogenesis of ALD from natural active ingredients with the ability to protect the liver from alcohol damage.

Maize is a popular food for people all over the world. Studies have proven the safety of CPs. They also have various functions, such as antioxidant, anti-inflammatory, anti-hypertensive, immune boosting, and anti-fatigue [4,17]. Some researchers have found a protective effect of CPs against early alcoholic liver injury in mice and rats [33,34]. They focused mainly on AST, ALT, and SOD activity and MDA levels in serum. In addition, abnormal lipid metabolism could be improved [4]. Our results found that maize peptides contain a large number of branched-chain amino acids, including valine, leucine, and isoleucine. The study showed that oral BCAA supplements improved the manifestations of recurrent hepatic encephalopathy in patients with cirrhosis, without effects on mortality, nutrition, or adverse events [35]. Tedesco, Laura et al. found that branched-chain amino acid supplementation could be used to reduce the risk of cirrhosis, improves mitochondrial functional integrity against EtOH toxicity, and preserves liver integrity in mammals [36].

In recent years, an increasing number of studies have focused on the inflammatory mechanisms (innate immune mechanisms) of ALD. It is believed that activation of KCs is a major trigger of hepatotoxicity and liver injury [37]. KCs are macrophages in the liver and account for 15% of all liver cells. lPS is phagocytosed by KCs, which then produce TNF-α and other inflammatory mediators. Degeneration and necrosis of hepatocytes is promoted [38,39]. The LPS-Kupffer cell signaling modality has been implicated in the protective effect of CPs against ALD. Previous studies have demonstrated that CPs are an effective model for screening drugs for ALD treatment [40,41]. The researchers demonstrated that the protective effect of the Chinese herbal formula Qing Gan Hou Pu Fang was achieved by regulating related molecules in a similar signaling pathway to LPS-KCs.

In previous studies, ethanol ingestion led to damage to the gut barrier. Gut barrier dysfunction results in an elevation of circling bacterial endotoxins, which plays a key role in triggering LPS as recognized by the ethanol-induced expression of the CD14/TLR4 receptor. This leads to the production of pro-inflammatory cytokines. In our study, we found that liver tissue inflammatory factors were abundantly expressed. We speculated that this was caused by elevated LPS [42]. Therefore, we examined the role of LPS in inducing inflammation onset in KCs cells. The results showed that the presence of CPs tended to reduce the expression of CD14/TLR4 on the cell membrane of KCs. The translocation of NF-κB was significantly inhibited in the nucleus of KCs. Then, the phosphorylation of JNK, ERK, and p-38 protein was down-regulated. The expression of inflammatory cytokines such as TNF-α and IL-1 could be reduced. Thus, the inflammatory response could be inhibited. The results were consistent with the concentration variation of TNF-α and IL-1 in serum and the level of ALT and AST. Therefore, it was concluded that CPs could inhibit intracellular inflammation by improving LPS-induced KCs cell activation and reducing intracellular NF-κB and MAPK cell pathway activation to protect mice from hepatocyte injury. However, it is worth noting that, although the CPs have natural ingredients, more clinical studies should be carried out to prove the clinical performance of CPs on ALD. KCs signal pathway-based evaluation methods could also be applied to evaluate other natural ingredients in ALD treatments.

5. Conclusions

In this study, the role of natural CP in protecting the liver from alcohol damage by inhibiting LPS-induced activation of KCs cells was successfully demonstrated. It proves to be potential food therapy for patients with ALD. The primary culture was involved in evaluating the effect mechanism from CPs on KCs. The KCs-related signal pathway could also be applied to evaluate other treatments for ALD. In addition, the KCs are closely related to liver transplants. Further studies should be performed to develop treatments based on the functions of CPs and other natural ingredients.

Author Contributions: Project administration & Supervision, Y.W. (Ying Wei); Writing-review & editing, M.L. and Z.F.; Investigation, D.Z. and X.C.; Editing, M.S.; Methodology, Y.W. (Yong Wang) All authors have read and agreed to the published version of the manuscript.

Funding: This research was funded by Young Teachers Research Innovation Capacity Enhancement Program of Beijing Universityof Agricultural [NO:QJKC2022045].

Institutional Review Board Statement: This experiment was approved by the Beijing Branch of China National137 Committee (License number: SCXK [Beijing] 2018–0010).

Informed Consent Statement: Not applicated.

Data Availability Statement: Not applicated.

Acknowledgments: The authors acknowledge the Young Teachers Research Innovation Capacity Enhancement Program of Beijing University of Agriculture (grant No. QJKC2022045).

Conflicts of Interest: The authors declare no conflict of interest.

References

1. Ma, Z.; Hou, T.; Shi, W.; Liu, W.; He, H. Inhibition of Hepatocyte Apoptosis: An Important Mechanism of Corn Peptides Attenuating Liver Injury Induced by Ethanol. *Int. J. Mol. Sci.* **2015**, *16*, 22062–22080. [CrossRef] [PubMed]
2. O'Shea, R.S.; Dasarathy, S.; McCullough, A.J. Alcoholic liver disease. *Hepatology* **2010**, *51*, 307–328. [CrossRef] [PubMed]
3. Yu, Y.; Wang, L.; Wang, Y.; Lin, D.; Liu, J. Hepatoprotective Effect of Albumin Peptides from Corn Germ Meal on Chronic Alcohol-Induced Liver Injury in Mice. *J. Food Sci.* **2017**, *82*, 2997–3004. [CrossRef] [PubMed]
4. Zhang, F.; Zhang, J.; Li, Y. Corn oligopeptides protect against early alcoholic liver injury in rats. *Food Chem. Toxicol.* **2012**, *50*, 2149–2154. [CrossRef]
5. Mathurin, P.; Bataller, R. Trends in the management and burden of alcoholic liver disease. *J. Hepatol.* **2015**, *62* (Suppl. 1), S38–S46. [CrossRef]
6. Fan, J.G.; Farrell, G.C. Epidemiology of non-alcoholic fatty liver disease in China. *J. Hepatol.* **2009**, *50*, 204–210. [CrossRef]
7. Fan, J.G. Epidemiology of alcoholic and nonalcoholic fatty liver disease in China. *J. Gastroenterol. Hepatol.* **2013**, *28* (Suppl. 1), 11–17. [CrossRef]
8. Ong, J.P.; Pitts, A.; Younossi, Z.M. Increased overall mortality and liver-related mortality in non-alcoholic fatty liver disease. *J. Hepatol.* **2008**, *49*, 608–612. [CrossRef]
9. Targher, G.; Arcaro, G. Non-alcoholic fatty liver disease and increased risk of cardiovascular disease. *Atherosclerosis* **2007**, *191*, 235–240. [CrossRef]
10. Lieber, C.S. Alcoholic liver disease: New insights in pathogenesis lead to new treatments. *J. Hepatol.* **2000**, *32*, 113–128. [CrossRef]
11. Hao, F.; Cubero, F.J.; Ramadori, P.; Liao, L.; Haas, U.; Lambertz, D.; Nevzorova, Y.A. Inhibition of Caspase-8 does not protect from alcohol-induced liver apoptosis but alleviates alcoholic hepatic steatosis in mice. *Cell Death Dis.* **2017**, *8*, e3152. [CrossRef] [PubMed]
12. Vaduganathan, M.; van Meijgaard, J.; Mehra, M.R.; Joseph, J.; O'Donnell, C.J.; Warraich, H.J. Prescription Fill Patterns for Commonly Used Drugs During the COVID-19 Pandemic in the United States. *JAMA* **2020**, *323*, 2524–2526. [CrossRef] [PubMed]

13. Miller, A.M.; Horiguchi, N.; Jeong, W.I.; Radaeva, S.; Gao, B. Molecular mechanisms of alcoholic liver disease: Innate immunity and cytokines. *Alcohol. Clin. Exp. Res.* **2011**, *35*, 787–793. [CrossRef]
14. Seitz, H.K.; Bataller, R.; Cortez-Pinto, H.; Gao, B.; Gual, A.; Lackner, C.; Tsukamoto, H. Alcoholic liver disease. *Nat. Rev. Dis. Primers* **2018**, *4*, 16. [CrossRef] [PubMed]
15. Shi, Y.; Sun, J.; He, H.; Guo, H.; Zhang, S. Hepatoprotective effects of Ganoderma lucidum peptides against D-galactosamine-induced liver injury in mice. *J. Ethnopharmacol.* **2008**, *117*, 415–419. [CrossRef]
16. Xie, Q.; Guo, F.-F.; Zhou, W. Protective effects of cassia seed ethanol extract against carbon tetrachloride-induced liver injury in mice. *Acta Biochim. Pol.* **2012**, *59*, 265–270. [CrossRef]
17. Liu, W.L.; Chen, X.W.; Li, H.; Zhang, J.; An, J.L.; Liu, X.Q. Anti-Inflammatory Function of Plant-Derived Bioactive Peptides: A Review. *Foods* **2022**, *11*, 2361. [CrossRef]
18. Zhou, C.; Hu, J.; Ma, H.; Yagoub, A.E.; Yu, X.; Owusu, J.; Qin, X. Antioxidant peptides from corn gluten meal: Orthogonal design evaluation. *Food Chem* **2015**, *187*, 270–278. [CrossRef]
19. Wang, Y.; Song, X.; Feng, Y.; Cui, Q. Changes in peptidomes and Fischer ratios of corn-derived oligopeptides depending on enzyme hydrolysis approaches. *Food Chem.* **2019**, *297*, 124931. [CrossRef]
20. Ortiz-Martinez, M.; Winkler, R.; Garcia-Lara, S. Preventive and therapeutic potential of peptides from cereals against cancer. *J. Proteom.* **2014**, *111*, 165–183. [CrossRef]
21. Zhuang, H.; Tang, N.; Dong, S.T.; Sun, B.; Liu, J.B. Optimisation of antioxidant peptide preparation from corn gluten meal. *J. Sci. Food Agric.* **2013**, *93*, 3264–3270. [CrossRef] [PubMed]
22. Kong, X.; Guo, M.; Hua, Y.; Cao, D.; Zhang, C. Enzymatic preparation of immunomodulating hydrolysates from soy proteins. *Bioresour. Technol.* **2008**, *99*, 8873–8879. [CrossRef] [PubMed]
23. Wei, Y.; Zhang, R.; Fang, L.; Qin, X.; Cai, M.; Gu, R.; Wang, Y. Hypoglycemic effects and biochemical mechanisms of Pea oligopeptide on high-fat diet and streptozotocin-induced diabetic mice. *J. Food Biochem.* **2019**, *43*, e13055. [CrossRef] [PubMed]
24. Maemura, K.; Zheng, Q.; Wada, T.; Ozaki, M.; Takao, S.; Aikou, T.; Sun, Z. Reactive oxygen species are essential mediators in antigen presentation by Kupffer cells. *Immunol. Cell Biol.* **2005**, *83*, 336–343. [CrossRef] [PubMed]
25. Zhang, W.N.; Chen, J.; Gong, L.-L.; Su, R.-N.; Yang, R.; Yang, W.-W.; Lu, Y.-M. Structural characterization and in vitro hypoglycemic activity of a glucan from Euryale ferox Salisb. seeds. *Carbohydr. Polym.* **2019**, *209*, 363–371. [CrossRef]
26. Ma, Z.L.; Hou, T.; Shi, W.; Liu, W.W.; Ibrahim, S.A.; He, H. Purification and identification of corn peptides that facilitate alcohol metabolism by semi-preparative high-performance liquid chromatography and nano liquid chromatography with electrospray ionization tandem mass spectrometry. *J. Sep. Sci.* **2016**, *39*, 4234–4242. [CrossRef]
27. Díaz-Gómez, J.L.; Castorena-Torres, F.; Preciado-Ortiz, R.E.; García-Lara, S. Anti-cancer activity of maize bioactive peptides. *Front. Chem.* **2017**, *5*, 44. [CrossRef]
28. Liang, Q.; Chalamaiah, M.; Ren, X.; Ma, H.; Wu, J. Identification of new anti-inflammatory peptides from zein hydrolysate after simulated gastrointestinal digestion and transport in caco-2 cells. *J. Agric. Food Chem.* **2018**, *66*, 1114–1120. [CrossRef]
29. Kany, S.; Vollrath, J.T.; Relja, B. Cytokines in Inflammatory Disease. *Int. J. Mol. Sci.* **2019**, *20*, 6008. [CrossRef]
30. Dunn, W.; Shah, V.H. Pathogenesis of Alcoholic Liver Disease. *Clin. Liver Dis.* **2016**, *20*, 445–456. [CrossRef]
31. Farooq, M.O.; Bataller, R. Pathogenesis and Management of Alcoholic Liver Disease. *Dig. Dis.* **2016**, *34*, 347–355. [CrossRef] [PubMed]
32. Stickel, F.; Datz, C.; Hampe, J.; Bataller, R. Pathophysiology and Management of Alcoholic Liver Disease: Update 2016. *Gut Liver* **2017**, *11*, 173–188. [CrossRef] [PubMed]
33. She, X.; Wang, F.; Ma, J.; Chen, X.; Ren, D.; Lu, J. In vitro antioxidant and protective effects of corn peptides on ethanol-induced damage in HepG2 cells. *Food Agric. Immunol.* **2016**, *27*, 99–110. [CrossRef]
34. Tan, H.K.; Yates, E.; Lilly, K.; Dhanda, A.D. Oxidative stress in alcohol-related liver disease. *World J. Hepatol.* **2020**, *12*, 332–349. [CrossRef] [PubMed]
35. Gluud, L.L.; Dam, G.; Borre, M.; Les, I.; Cordoba, J.; Marchesini, G.; Aagaard, N.K.; Risum, N.; Vilstrup, H. Oral branched-chain amino acids have a beneficial effect on manifestations of hepatic encephalopathy in a systematic review with meta-analyses of randomized controlled trials. *J. Nutr.* **2013**, *143*, 1263–1268. [CrossRef] [PubMed]
36. Tedesco, L.; Corsetti, G.; Ruocco, C.; Ragni, M.; Rossi, F.; Carruba, M.O.; Valerio, A.; Nisoli, E. A specific amino acid formula prevents alcoholic liver disease in rodents. *Am. J. Physiol. Gastrointest. Liver Physiol.* **2018**, *314*, G566–G582. [CrossRef]
37. Luo, W.; Xu, Q.; Wang, Q.; Wu, H.; Hua, J. Effect of modulation of PPAR-gamma activity on Kupffer cells M1/M2 polarization in the development of non-alcoholic fatty liver disease. *Sci. Rep.* **2017**, *7*, 44612. [CrossRef] [PubMed]
38. Li, P.; He, K.; Li, J.; Liu, Z.; Gong, J. The role of Kupffer cells in hepatic diseases. *Mol. Immunol.* **2017**, *85*, 222–229. [CrossRef]
39. Zimmermann, H.W.; Trautwein, C.; Tacke, F. Functional role of monocytes and macrophages for the inflammatory response in acute liver injury. *Front. Physiol.* **2012**, *3*, 56. [CrossRef]
40. Tacke, F. Targeting hepatic macrophages to treat liver diseases. *J. Hepatol.* **2017**, *66*, 1300–1312. [CrossRef]
41. Wu, T.; Liu, T.; Zhang, L.; Xing, L.J.; Zheng, P.Y.; Ji, G. Chinese medicinal formula, Qinggan Huoxue Recipe protects rats from alcoholic liver disease via the lipopolysaccharide-Kupffer cell signal conduction pathway. *Exp. Med.* **2014**, *8*, 363–370. [CrossRef] [PubMed]
42. Xiao, J.; Zhang, R.; Wu, Y.; Wu, C.; Jia, X.; Dong, L.; Liu, L.; Chen, Y.; Bai, Y.; Zhang, M. Rice Bran Phenolic Extract Protects against Alcoholic Liver Injury in Mice by Alleviating Intestinal Microbiota Dysbiosis, Barrier Dysfunction, and Liver Inflammation Mediated by the Endotoxin–TLR4–NF-κB Pathway. *J. Agric. Food Chem.* **2020**, *68*, 1237–1247. [CrossRef] [PubMed]

Article

Immunomodulatory Effects of Chicken Broth and Histidine Dipeptides on the Cyclophosphamide-Induced Immunosuppression Mouse Model

Jian Zhang [1,2], Xixi Wang [1,3], He Li [1,*], Cunshe Chen [1] and Xinqi Liu [1,*]

[1] Beijing Advanced Innovation Center for Food Nutrition and Human Health, School of Food and Health, Beijing Technology and Business University, Beijing 100048, China
[2] Beijing Advanced Innovation Center for Food Nutrition and Human Health, Department of Nutrition and Health, China Agricultural University, Beijing 100193, China
[3] China Animal Disease Control Center, Beijing 102618, China
[*] Correspondence: lihe@btbu.edu.cn (H.L.); liuxinqi@btbu.edu.cn (X.L.)

Abstract: The carnosine and anserine, which represent histidine dipeptides (HD), are abundant in chicken broth (CB). HD are endogenous dipeptide that has excellent antioxidant and immunomodulatory effects. The immunomodulatory effect of CB hydrolysate (CBH) and HD in cyclophosphamide (CTX)-induced immunosuppressed mice was examined in this study. CBH and HD were given to mice via oral gavage for 15 days, accompanied by intraperitoneal CTX administration to induce immunosuppression. CBH and HD treatment were observed to reduce immune organ atrophy ($p < 0.05$) and stimulate the proliferation of splenic lymphocytes ($p < 0.05$) while improving white blood cell, immunoglobulin M (IgM), IgG, and IgA levels ($p < 0.05$). Moreover, CBH and HD strongly stimulated interleukin-2 (IL-2) and interferon-gamma (IFN-γ) production by up-regulating IL-2 and IFN-γ mRNA expression ($p < 0.05$) while inhibiting interleukin-10 (IL-10) overproduction and IL-10 mRNA expression ($p < 0.05$). In addition, CBH and HD prevented the inhibition of the nitric oxide (NP)/cyclic guanosine monophosphate-cyclic adenosine monophosphate (cGMP-cAMP)/protein kinase A (PKA) signaling pathway ($p < 0.05$). These results indicate that CBH and HD have the potential to prevent immunosuppression induced by CTX. Our data demonstrate that CBH can effectively improve the immune capacity of immunosuppressed mice similar to the same amount of purified HD, which indicates that CBH plays its role through its own HD.

Keywords: chicken broth hydrolysate; histidine dipeptides; carnosine; anserine; immunomodulatory; cyclophosphamide; immunosuppressed mice

Citation: Zhang, J.; Wang, X.; Li, H.; Chen, C.; Liu, X. Immunomodulatory Effects of Chicken Broth and Histidine Dipeptides on the Cyclophosphamide-Induced Immunosuppression Mouse Model. *Nutrients* **2022**, *14*, 4491. https://doi.org/10.3390/nu14214491

Academic Editor: Kay Rutherfurd-Markwick

Received: 30 August 2022
Accepted: 22 October 2022
Published: 25 October 2022

1. Introduction

Chicken broth (CB) is an ideal food for those recovering from illness since it is nutritious, easily digestible, and contains high levels of proteins, free amino acids, and polysaccharides [1,2]. Since it is easy to prepare, healthy, and delicious, CB has become a typical dish on family and restaurant tables. In Asia, CB has long been used as a nutritional supplement because it boosts immunity, combats fatigue, and prevents colds [3].

Histidine dipeptides (HD) represent a class of water-soluble dipeptides typically found in the skeletal muscles and brain tissue of many vertebrates and primarily include carnosine (CAR), anserine (ANS), and balenine [4]. The CAR and ANS structures are shown in Figure 1A. CAR and ANS can be synthesized in the body or ingested through diet. CAR and its derivatives are present in substantial levels in red and white meats (beef, chicken, and pork), as well as fish, in the human diet [5]. The order of the HD content in various common animals are as follows: Turkey > chicken > horse > pig > rabbit > cattle [6]. The HD content of meat products can be altered via Specific food processing technology. Papain and flavourzyme protease are often used in food processing to improve CB flavor and protein concentration, and the subsequent CB exhibited better taste and flavor than traditional chicken soup [7]. In addition, chicken meat extract contains high CAR and ANS concentrations at a 1:2 to 1:3 ratio [8].

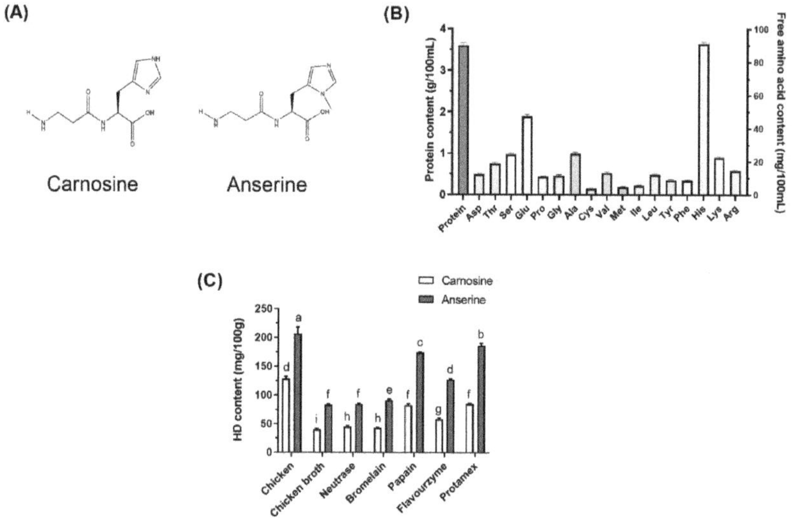

Figure 1. The structure of CAR and ANS (**A**). The protein and free amino acid content in CB (**B**). The effect of different enzymatic hydrolytic treatments on the HD content in the CB (**C**). Different letters indicated significant differences among the groups ($p < 0.05$).

HD can fulfill many physiological functions in the human body. The antioxidant and anti-inflammatory characteristics of HD are responsible for the majority of its beneficial effects, including immunological modulation, anti-aging, anti-neurodegenerative illness, anti-diabetes, and others. The ability of HD to act as an antioxidant is its their primary function [9]. CAR supplementation can decrease levels of advanced glycation end-products, malondialdehyde, protein carbonyl, and advanced oxidized protein product, as well as the generation of reactive oxygen species in the serum and liver of elderly rats [10]. Antonini et al. [11] proved that a diet rich in meat and CAR increased the antioxidant activity of human serum. Additionally, CAR has been shown to play an anti-aging role by suppressing telomere shortening, antioxidant activity, carbonyl scavenging, glycolysis suppression, the upregulation of mitochondrial activity, and the rejuvenation of senescent cells [12]. It also improves collagen content in the skin and may prevent skin aging [13]. Daily ANS/CAR supplementation is beneficial to the memory, cognitive function, and physical activity of elderly people [14]. In addition, CAR was indicated to possess immune regulation effects [15]. During the last 5 years, the ability of CAR to modulate different activities of immune cells such as macrophages has been demonstrated [16,17]. It has been demonstrated that the interaction of CAR with particular receptors on the cell membrane could modify macrophage function by enhancing their phagocytotic activity [18,19]. It has also been demonstrated that HD can influence NO production and macrophage polarization [20,21]. CAR maintained spleen lymphocyte number by inhibiting lymphocyte apoptosis and stimulating lymphocyte proliferation, thus preventing immunocompromise in mice [22]. Deng et al. [23] suggested that CAR can protect murine bone marrow cells from cyclophosphamide (CTX)-induced DNA damage via its antioxidant activity. Thus, HD has a significant impact on immune response regulation.

HD is a beneficial functional component of CB. The resistance immunosuppression function of CB may be played through HD. Therefore, we increased the HD content of CB by enzymatic hydrolysis pretreatment with suitable enzymes in this study. Additionally, in order to understand the immunity-enhancing effect of CB hydrolysate (CBH) and identify its major functional constituents, we studied the immunomodulatory effects of CBH and HD on CTX-treated mice. If CBH aids in the recovery of immunosuppression, it could be a safe and effective booster for postoperative patients.

2. Materials and Methods

2.1. Materials and Chemicals

Hetian chicken was purchased from the Hetian Chicken Development Co., Ltd. (Fujian, China). Neutrase, Bromelain, Papain, Flavourzyme, and Protamex were obtained from the Shanghai Yuanye Bio-Technology Co., Ltd. (Shanghai, China). CTX was purchased from the Jiangsu Hengrui Medicine Co., Ltd. (Jiangsu, China). The CAR and ANS standards were acquired from the Shanghai Yuanye Bio-Technology Co., Ltd. (Shanghai, China). Levamisole was purchased from the Novozymes (China) Investment Co., Ltd. (Beijing, China). Roswell Park Memorial Institute (RPMI) 1640 medium and fetal bovine serum (FBS) were procured from Gibco (Grand Island, NY, USA). Concanavalin A (con A) and 3[4,5-dimethylththiazoyl-2-yl]2,5-diphenyltetrazolium bromide (MTT) were purchased from Sigma Aldrich Co. (St Louis, MO, USA). The enzyme-linked immunosorbent assay (ELISA) kits for the immunoglobulin M (IgM), immunoglobulin G (IgG), immunoglobulin A (IgA), interleukin-2 (IL-2), interleukin-10 (IL-10), interferon-gamma (IFN-γ), nitric oxide (NO), cyclic guanosine monophosphate (cGMP), cyclic adenosine monophosphate (cAMP), and protein kinase A (PKA) were purchased from the Meimian Biotechnology Co. Ltd. (Jiangsu, China). All chemical reagents were of analytical grade.

2.2. Preparation of the CB

2.2.1. Extracting the HD from the Chicken Breast

A meat grinder was used to mince the chopped chicken breast meat. Following that, 100 g of the minced chicken breast flesh was combined with 100 mL of deionized water and maintained in a water bath at 100 °C for 15 min to inactivate the CAR enzymes, followed by 30 min of ultrasonic extraction. Next, the mixture was centrifuged at 12,000 r/min for 8 min at 4 °C. The supernatant was then filtered to a constant volume of 1000 mL with a 0.22 μm nylon filter.

2.2.2. Traditional CB

The chicken breast was cut into small pieces of 1 cm^3. Next, 200 mL deionized water was added to 100 g minced chicken breast flesh. The mixture was added to a steamer and stewed for 4 h at a pressure of 0.07 MPa at 100 °C. The prepared CB was centrifuged at 4 °C for 8 min at 12,000 r/min. The supernatant was then filtered to a constant volume of 1000 mL with a 0.22 μm nylon filter.

2.2.3. Enzymolysis of the CB

The chicken breast was cut into small pieces of 1 cm^3. Subsequently, 200 mL deionized water was added to 100 g minced chicken breast flesh. The mixture was hydrolyzed using Neutrase (pH 7, 45 °C), Papain (pH 6.5, 55 °C), Bromelain (pH 6.5, 45 °C), Flavourzyme (pH 7.5, 50 °C), and Protamex (Papain: Flavourzyme 1:1, pH 7, 45 °C) according to the amount of enzyme added at 1000 U/g, respectively, adjusted to the optimum temperature and pH of the enzyme. In the process of enzymatic hydrolysis, 1 mol/L NaOH was added to adjust the pH value of the system to remain constant, and the system was enzymatic for 2 h. After enzymatic hydrolysis, the mixture was heated to 100 °C for 10 min to inactivate the enzymes. Next, the mixture was placed in a steamer and stewed for 4 h at a pressure of 0.07 MPa at 100 °C. The CBH was cooled and centrifuged at 4 °C for 30 min at 12,000 r/min. The supernatant was then filtered to a constant volume of 1000 mL with a 0.22 μm nylon filter.

2.3. Protein and Free Amino Acid Analysis of the CB

The Kjeldahl method (Kjeltec 8000, FOSS Analytical A/S, Denmark) was used to assess the protein concentrations in the CB. The CB was centrifuged for 10 min at 9600 r/min under 4 °C. The fat from the supernatant was removed and lyophilized to produce CB Powder. Separately, CB Powder was redissolved by Sodium Loading Buffer. The solutions were analyzed with an automatic amino acid analyzer (Biochrom 30+, Biochrom Ltd., Cambridge, England) after filtration through a 0.45 μm nylon filter membrane (Cleman, Beijing, China). Absorbance was recorded at 570 nm and 440 nm.

2.4. UPLC Analysis of the HD

UPLC was used to analyze the CAR and ANS in chicken breast, CB, and CBH. A 10 μL sample was injected into the UPLC system (Agilent, 1290 Infinity II), and the separation was performed using the Poroshell 120 HILIC-Z column (2.1 mm × 100 mm, 2.7 μm). The mobile phase was composed of 90% acetonitrile and 0.1% trifluoroacetic acid at a liquid flow rate of 0.2 mL/min. The detection wavelength was 210 nm.

2.5. Animals

In this case, 50 eight-week-old ICR male mice (SPF Beijing Biotechnology Co., Ltd., Beijing, China) weighing 35 ± 2 g were used. All mice were housed under controlled temperature (22 ± 2 °C) and humidity (50–60%), with a 12 h light: 12 h dark cycle. The experimental protocol was approved by the Institutional Animal Care and Use Committee at the Pony Testing International Group Co., Ltd., Beijing, China (No. PONY-2021-FL-03).

2.6. Experimental Procedures

After a one-week acclimatization period, all mice were randomly divided into five groups (10 per group): (1) Normal group, (2) CTX group, (3) Levamisole group, (4) CBH group, and (5) HD group. Mice in the Levamisole group received Levamisole daily by oral gavage for 15 days at a dose of 10 mg/kg body weight. Mice in the CBH group and HD group received CBH and HD daily by oral gavage for 15 days. After negative pressure concentration, the HD concentration in the CBH was 30 mg/mL, including 9.41 mg/mL CAR and 20.59 mg/mL ANS. The HD doses were equivalent to the HD concentration and proportion in the CBH. Whereas mice in the Normal group and CTX group were treated with the same volume of distilled water. Each group of mice had a gavage volume of 0.1 mL $(10 \text{ g·bw})^{-1}$. The immunosuppressed mouse model was established by intraperitoneal injection of CTX. On days 13 to 15, the mice in groups (2) to (5) were intraperitoneally administered with CTX (80 mg/kg/d). The identical volume of physiological saline was given intraperitoneally to group (1). The mice were fasted for 12 h after the last injection and then executed by cervical dislocation.

2.7. Determination of the Body Weight and Immune Organ Index

Take measurements of the mice's body weight before and after the experiment. After the mice were sacrificed, the thymus and spleen tissues were immediately dissected, washed in pre-cooled normal saline at 4 °C, dried using filter paper, and weighed. The organ index was calculated as follows:

$$\text{Organ index (mg/g)} = \text{Organ weight (mg)} / \text{Bodyweight (g)} \tag{1}$$

2.8. Determination of Splenic Lymphocytes Proliferation

The spleens of the sacrificed mice were aseptically dissected and placed in cold Hank's solution. The spleens were then pulverized using a sterilized glass rod and gently pressed through a 200 mesh, sterile metal strainer, after which the single-cell suspension was obtained. The suspension was centrifuged at 1000 r/min for 10 min at 4 °C after being rinsed twice with Hank's solution. The cells were then moved to an RPMI-1640 medium with 10% FBS after the supernatant fluid was discarded. The cell concentration was adjusted to 3×10^6 cells/mL. The cells were seeded at a density of 3×10^6 cells per well in a 24-well flat-bottomed plate with or without Con A (7.5 µg/mL) and incubated with 5% CO_2 for 72 h at 37 °C. The plates were then centrifuged at $200\times g$ for 10 min, after which the splenocyte culture supernatants were collected and stored at -80 °C until analysis for cytokines. Next, the splenic lymphocyte proliferation was detected by MTT assay.

2.9. Hematological Analyses

Blood was collected from the eyes of the mice 12 h after the last drug administration. The white blood cells (WBC), neutrophils (NEU), lymphocytes (LYM), red blood cells (RBC), hemoglobin (HGB) concentration, and platelet (PLT) number were determined using a hematology analyzer (HEMAVET 950, Drew Scientific Group, Dallas, TX, USA).

2.10. Assay of Immunoglobulins in the Serum

Serum was extracted from the blood taken from the eyes of mice after centrifugation. The serum IgM, IgG, and IgA concentrations were determined using ELISA kits.

2.11. Measurement of the Cytokines

The IL-2, IL-10, IFN-γ, and NO levels in the splenocyte culture supernatants were detected using ELISA Kits.

2.12. Quantitative Real-Time Polymerase Chain Reaction (qRT-PCR) Analysis

In this case, qRT-PCR was used to evaluate the mRNA expression levels of IL-2, IL-10, and IFN-γ. The mice's spleens were extracted in a sterile setting and washed twice in PBS. The spleens' total RNA was isolated using Trizol reagent. The RNA concentration and purity were determined using an Ultra-micro spectrophotometer (Nanodrop 2000c, Thermo Fisher Scientific, Chicago, IL, USA). The total RNA (2 µg) was converted into cDNA using PrimeScript RT Master Mix (Takara-bio, Shiga, Japan), while qRT-PCR amplification was performed using Probe qPCR Mix (Takara-bio, Japan). The 18s rRNA was used to normalize the IL-2, IL-10, and IFN-γ mRNA expression. The relative expression levels of the target genes were calculated based on $2^{-\Delta\Delta Ct}$.

2.13. Measurement of cAMP/cGMP Levels and PKA Activity

After an incubation period of 72 h, the splenocytes were rinsed three times with PBS and centrifuged at 1500 r/min for 5 min at 4 °C. RIPA buffer was used to lyse the cells for 30 min. The splenocyte supernatants were collected via centrifugation at 12,000 rpm for 5 min at 4 °C. The cAMP/cGMP levels and PKA activity were detected using ELISA kits.

2.14. Statistical Analysis

Data were expressed as the mean $\pm$ standard deviation (SD). The results were analyzed with one-way analysis of variance (ANOVA) followed by Tukey's method, using SPSS 23 software, while the graphs were created using Graph Pad Prism 7.

3. Results and Discussion

3.1. Characterization of the CBH

As shown in Figure 1B, CB had a protein content of 3.59 ± 0.07 g/100 mL. CB contained 334.43 mg/100 mL of total free amino acids, with His being the most abundant, followed by Glu, Ala, and Ser. In this study, chicken breasts containing a high level of HD were selected to prepare the CB. Enzymatic hydrolysis is a standard method used during chicken processing to improve the flavor and taste of the products [7]. It also increases the hydrolytic degree of chicken protein and produces more peptides that are easily absorbed. Different proteases were used in the experiment to prepare the CBH. The results showed that different enzymatic hydrolysis treatments significantly increased the HD content in the CBH. As shown in Figure 1B, ANS content in the CBH ranked as follows: Protamex > Papain > Flavourzyme > Bromelain > Neutrase. The ANS content in the CB hydrolyzed with

Protamex was the highest and significantly exceeded that in the other four enzymolytic CB groups ($p < 0.05$). Furthermore, CAR content was considerably greater in the CB hydrolyzed with Papain and Protamex than in the other three enzymolytic CB groups ($p < 0.05$). Therefore, after enzymatic hydrolysis using Protamex, the HD content of the CBH can be effectively increased during the cooking process.

3.2. The Effect of CBH on the Body Weight and Immune Organ Index

Bodyweight was a good predictor of the mice's growth status. The mice were randomly assigned to five groups based on body weight after a week of acclimatization using the randomized design method. Table 1 showed that although no significant differences were evident between the final weight of the mice from each group, it was higher than the initial weight. During the experiment, the group treated with CTX gained more weight than the other four groups, although this difference was not statistically significant ($p > 0.05$). Within three days of CTX exposure, there was no significant difference in body weight between the CTX group and the sample-treated group; however, following 5 days of CTX exposure, the difference in weight gain became apparent [24]. Similarly, three days following the CTX injection, there was no significant difference between groups in terms of weight gain in this study. The results suggested that intragastric administration of CBH and HD did not affect the growth in immunosuppressed mice.

Table 1. The effect of CHB and HD on the body weight and hematological parameters of CTX-treated mice.

	Normal	CTX	Levamisole	CBH	HD
Initial weight (g)	36.30 ± 0.78 [a]	34.57 ± 1.35 [b]	36.04 ± 1.21 [a]	35.86 ± 1.43 [ab]	35.85 ± 1.84 [ab]
Final weight (g)	38.49 ± 1.31 [a]	36.91 ± 1.85 [a]	37.69 ± 1.51 [a]	36.87 ± 2.31 [a]	37.16 ± 2.45 [a]
Weight gain (g)	2.19 ± 0.99 [a]	2.34 ± 0.90 [a]	1.65 ± 1.02 [a]	1.01 ± 1.75 [a]	1.31 ± 1.04 [a]
WBC (K/μL)	7.72 ± 2.80 [a]	1.46 ± 0.67 [b]	2.50 ± 0.46 [c]	3.04 ± 0.56 [c]	2.70 ± 0.43 [c]
NEU (K/μL)	2.01 ± 1.13 [a]	0.50 ± 0.34 [b]	0.80 ± 0.17 [bc]	1.08 ± 0.34 [ac]	0.76 ± 0.17 [bc]
LYM (K/μL)	5.21 ± 1.92 [a]	0.79 ± 0.34 [b]	1.32 ± 0.27 [c]	1.80 ± 0.46 [c]	1.65 ± 0.36 [c]
RBC (M/μL)	10.18 ± 1.76 [ac]	11.69 ± 1.80 [b]	11.22 ± 1.14 [ab]	10.28 ± 0.54 [ac]	9.68 ± 1.07 [c]
HGB (g/dL)	16.20 ± 2.97 [ac]	18.66 ± 2.52 [b]	17.58 ± 1.74 [ab]	16.29 ± 0.80 [ac]	15.19 ± 1.42 [c]
PLT (K/μL)	1104.90 ± 166.68 [a]	712.10 ± 223.60 [b]	855.50 ± 235.86 [bc]	939.60 ± 215.84 [ac]	1029.22 ± 154.29 [ac]

Values are expressed as means ± SD ($n = 10$). WBC, white blood cell count; NEU, neutrophils; LYM, lymphocytes; RBC, red blood cell count; HGB, hemoglobin concentration; PLT, platelet count. Means in the same row with different letters differ significantly ($p < 0.05$).

The immune organ indexes can reflect the immunity moderation level to a certain extent. CTX is a widely used drug in chemotherapy, displaying a significant anti-cancer ability. As an inducer commonly used in immunosuppression models, CTX can damage the immune defense of the host and suppress immune organs, immune cells, and immune molecules [25]. As shown in Figure 2A,B, the spleen and thymus indexes in the CTX group were significantly lower ($p < 0.05$) than in the Normal group, indicating that intraperitoneal injection of CTX resulted in immune organ atrophy, successfully establishing the immuno-suppression model. As previously reported, CTX can significantly increase the apoptotic rate of spleen and thymus cells [26]. Furthermore, levamisole, CBH, and HD significantly increased the spleen and thymus indexes ($p < 0.05$) compared to the Model group, while no significant differences were apparent among the three groups. Levamisole was initially developed as an anthelmintic. As an immunomodulator, the drug garnered considerable attention later on. According to the reports, levamisole has numerous immunomodula-tory effects, including the enhancement of antibody production to various antigens, the enhancement of several cellular immune responses, synergistic activity with T-lymphocyte mitogens, the enhancement of chemotaxis and the enhancement of the phagocytic activity of polymorphonuclear and mononuclear phagocytes [27,28]. Similar to levamisole, the results demonstrated that CBH and HD supplements could prevent and treat the immune organ atrophy caused by CTX. Yu et al. [29] found that peptides from *Nibea japonica* skin could significantly improve the spleen and thymus indexes of CTX-induced immunosuppressed mice. Banerjee et al. [30] revealed that CAR treatment prevented spleen atrophy caused by immunosuppression. Liu et al. [31] intragastrically administered phenylalanine dipeptide to mice, significantly improving the spleen and thymus indexes of the mice. Based on the results mentioned above, CBH effectively improved the atrophy of immune organs caused by CTX. Considering that this has the same effect as HD supplementation, it can be inferred that HD is the primary substance in CBH, preventing the immune organ atrophy caused by CTX. Therefore, low molecular peptides, such as dipeptides, display immunomodulatory activity to prevent immune organ atrophy.

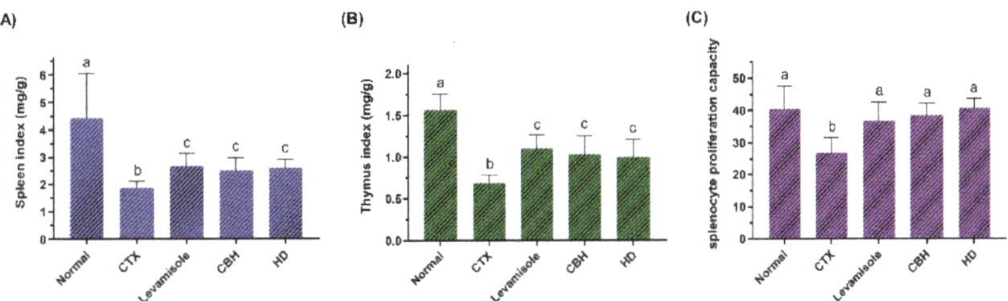

Figure 2. The effect of CHB and HD on the spleen index (**A**), thymus index (**B**), and the proliferation of splenocytes stimulated with Con A (7.5 µg/mL) (**C**). Data are presented as means ± SD (n = 10). Different letters for the same index among the groups represent significant differences at $p < 0.05$.

3.3. The Effect of CBH on the Proliferation of Splenic LYM

The spleen is a unique organ that combines innate and adaptive immune systems. The spleen is important for immunoregulation as well as playing a part in the immunological response [32]. The spleen is the place where mature LYM settle, with B cells comprising around 60% of the total amount of spleen LYM and T cells comprising around 40%. The spleen receives a higher volume of LYM on a daily basis than all other secondary lymphoid organs combined [33]. Consequently, these splenocytes could represent in vivo systemic immune responses. LYM proliferation is an important event during the activation process of the adaptive immune system. T LYM is primarily responsible for regulating the cellular immunity of the body. The most striking feature of T LYM activation is the development of mitotic proliferation. T LYM immunity is frequently detected using Con A-induced LYM proliferation [34].

In this study, splenocytes isolated from all the groups were cultured with Con A, and their proliferation is presented in Figure 2C. With Con A treatment, the LYM proliferative ratio was lower in the CTX group ($p < 0.05$), whereas the splenocyte proliferation in CBH and HD groups, was higher than in the CTX group (142.22 and 150.98%, respectively). Moreover, there were no discernible changes among the Normal, Levamisole, CBH, and HD groups. These results indicated that through enhancing cellular immune systems, CBH and HD could promote Con A-induced LYM proliferation. Since the same concentrations and proportions of HD were present in the CBH and HD groups, HD may represent the primary functional components in CBH, preventing the decrease in splenic LYM proliferation caused by CTX. Furthermore, some investigations have found that the food supply influences splenocyte growth. Many low molecular peptides can enhance the proliferative ability of splenic LYM and display immunomodulatory activity. Yang et al. [35] showed that the peptides in *Pseudostellaria heterophylla* protein hydrolysate could restore splenic LYM proliferation in mice treated with CTX. Hou et al. [36] isolated three immunomodulatory peptides (including two pentapeptides and one dipeptide) from the Alaska pollock frame enzymatic hydrolysates, which significantly increased the proliferation rate of splenic LYM in mice. The results of this study indicated that CBH protects cellular immunity during CTX-induced immune damage.

3.4. The Effect of CBH on the Hematological Parameters

WBCs, such as NEU and LYM, represent crucial immune cells, the number of which directly reflects the humoral immunity of the body [37]. NEUs are the most common leukocytes in peripheral blood, accounting for 50–70% of all the circulating leukocytes in humans and 20% of all circulating leukocytes in mice. NEU performs a vital phagocytic function and represents the initial line of defense against microbial pathogen invasion. LYM, such as T and B LYM and natural killer cells, make up roughly 20–30% of all the circulating leukocytes in humans and are mainly involved in the specific immune response of the body [38]. Previous studies have shown that CTX caused protein functional groups to be alkylated, inhibiting medulla hemopoietic functionality and reducing the number of WBCs in the blood [39].

Table 1 demonstrates that CTX reduced the WBC, NEU, LYM, and PLT counts in the blood ($p < 0.05$), while increasing the RBC and HGB content ($p < 0.05$), showing that the immunity of the mice was suppressed. One of the most noticeable side effects of chemotherapy drugs is bone marrow suppression. WBC are believed to be involved in immunological and defensive mechanisms, have a short lifetime, and require bone marrow stem cells to develop continuously [40]. The second day of CTX treatment deeply affected bone marrow architecture, and recovery began on day 5. Thus, there was a dramatic drop in white blood cell count in the blood of mice 3 days after the CTX injection. Juaristi et al. [41] suggest that the proliferation and differentiation of erythroid progenitor cells after the acute early injury inflicted by CTX, is associated with changes in EPO-R expression during spontaneous recovery. Therefore, the increased RBC and HGB levels in the CTX group may be related to EPO-R expression promotion. This study is consistent with previous studies that CTX significantly reduced the amount of WBC and PLT in the blood while increasing the content of RBC and hemoglobin HGB [42]. Compared to the CTX group, CBH prevented a decrease in WBC, NEU, LYM, and PLT ($p < 0.05$), alleviating the RBC and HGB ($p < 0.05$) increase in the blood. Consequently, the immunosuppression caused by CTX was improved. The HD group had considerably greater WBC, LYM, and PLT counts than the CTX group ($p < 0.05$). RBC and HGB levels, on the other hand, were considerably lower in the HD group than in the CTX group ($p < 0.05$). These results suggest that CBH and HD prevented and treated CTX-induced WBCs decline, with an effect comparable to levamisole. Therefore, HD may represent the main functional components in CBH that prevent the decrease in WBCs caused by CTX. Previous studies have shown that immunoregulatory selenium-enriched peptides from soybean can restore the CTX-induced decrease of WBCs [43]. Lis et al. [44] revealed that a low molecular weight dipeptide bestatin significantly improved the CTX-induced decrease in the number of peripheral blood LYM. These results indicated that the administration of CBH and HD restored normal blood indices, implying that CBH and HD protect against immunosuppression induced by CTX.

3.5. The Effect of CBH on the Immunoglobulin Levels

Serum immunoglobulins are critical markers of humoral immunity and are involved in immune response and regulation [45]. Secreted IgM, IgG, and IgA represent the primary antibody components in serum and are vital effector molecules during the humoral immune response. During the early stages of the first humoral immune response, IgM is the primary antibody produced. It is the largest molecular weight immunoglobulin with a strong bactericidal, bacteriolytic, hemolytic, phagocytotic, and anti-infection ability [45]. In the phagocytotic process of monocytes, IgG, the most abundant and prominent antibody in serum, plays a crucial role [46]. IgG is more likely to diffuse through the capillary wall to the interstitial space, playing an essential anti-infection, toxin neutralization, and conditioning role. IgA, being the primary antibody class in the secretions that bathe these mucosal surfaces, serves as an important first-line of defense. IgA acts against various microbial antigens and can neutralize toxins and viruses [47].

To examine the effect of CBH and HD on the humoral immunity of CTX-treated mice, the serum IgM, IgG, and IgA levels of each group were determined. The results are shown in Figure 3A–C, respectively. CTX significantly reduced serum IgM, IgG, and IgA levels as compared to the Normal group ($p < 0.05$). Compared with the CTX group, the IgM and IgA levels were substantially higher in the serum of the Levamisole, CBH, and HD groups ($p < 0.05$), while no significant differences were evident between these three groups. Furthermore, no significant differences were apparent between the serum IgA levels of the CBH and Normal groups ($p > 0.05$). This indicated that CBH effectively prevented a decrease in the serum IgA content caused by CTX, maintaining it at a normal level. Although the IgG levels in the Levamisole, CBH, and HD groups exceeded that in the CTX group, the differences were not significant ($p > 0.05$). The results showed that the supplementation of CBH or HD prevented a CTX-induced decrease in the immunoglobulin content and that the effect was similar to that of levamisole. Both pretreatment and treatment mediated the effect of CBH and HD on increasing immunoglobulin content in this study. The findings also suggest that HD may represent the main functional components in CBH responsible for preventing a CTX-induced decrease in serum immunoglobulin.

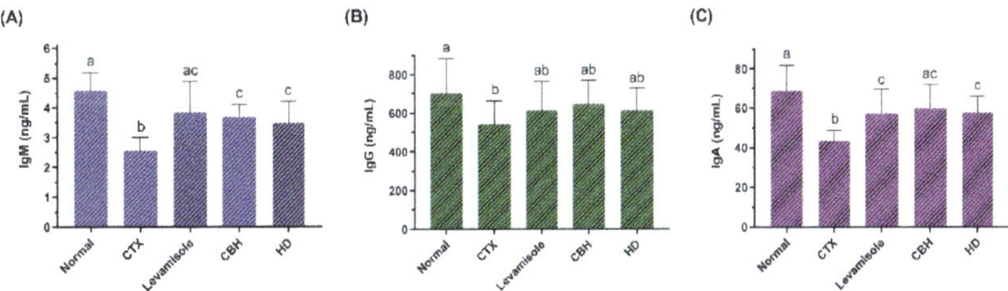

Figure 3. The effect of CHB and HD on the IgM (**A**), IgG (**B**), and IgA (**C**) content in the serum of CTX-treated mice. Data are presented as means ± SD ($n = 10$). Different letters for the same index among the groups represent significant differences at $p < 0.05$.

B LYM primarily participates in the humoral immune response and is transformed into plasma cells via antigen stimulation. Major humoral immune components include immunoglobulins, which are secreted by plasma cells (differentiated B cells). Immunoglobulins interact with immune response mediators and specific cell receptors to mediate a variety of protective activities [26]. Peptides stimulate the receptors on the surface of the B cells, causing them to proliferate and differentiate, transforming them into plasma cells to produce immunoglobulin [48]. Yu et al. [49] investigated the in vivo immunomodulatory activity of a novel pentadecapeptide RVAPEEHPVEGRYLV (SCSP) from a cellular and humoral immunity perspective in a CTX-induced immunosuppression mouse model. The results indicated that SCSP significantly enhanced the serum IgA, IgM, and IgG levels of the mice. In this study, HD may also increase the number and activity of B cells by acting on the receptors on the B cell surfaces, increasing the level of serum immunoglobulin.

3.6. The Effect of CBH on the Cytokine Levels and Gene Expression

Cytokines are synthesized and secreted by immune cells (B cells, T cells, and NK cells) and non-immune cells (endothelial cells, epidermal cells, osteoblasts, neurons, and fibroblasts), which can regulate immune functions [50,51]. Furthermore, cytokines play a crucial function in the intercellular communication of the immune system. They regulate the maturation, proliferation, and responsiveness of certain cell populations, as well as the balance between humoral and cell-based immune responses. The secretion of cytokines is a crucial indicator of immune function in the body. IL-2 has been shown to improve T cell killing activity, cause T cells to secrete IFN-γ, increase NK cell differentiation and activation, and promote B LYM proliferation and differentiation as well as immunoglobulin synthesis [52]. IFN-γ is a key element for the immune system to effectively act against infections and is produced mostly by NK cells, Th1, and CTL. IFN-γ can activate macrophages, enhance their phagocytotic ability, improve NK cell activity, increase antibacterial and anti-tumor ability, and stimulate antigen presentation [53]. Negative regulators are also important for immune response maintenance. IL-10 is a key Th2 negative regulatory cytokine. IL-10 plays a crucial immunosuppressive and anti-apoptotic role by suppressing the release of inflammatory mononuclear macrophage factors. In addition, IL-10 may inhibit the activation of inflammatory cytokines (such as IL-1, IL-12, and TNF-α) and chemokines (MCP family) secreted by antigen-presenting cells, indirectly inhibiting the function of T cells [54]. Cytokines are often used as indicators for evaluating immune response regulation in experiments. IL-2 can promote the activation and proliferation of T cells and NK cells, resulting in the secretion of IFN-γ and the enhancement of immune function [55]. IL-10 can reflect immunosuppression levels. TNF-α, which is secreted primarily by mononuclear macrophages, is an essential component of the host defense system and can induce the expression of other immunoregulatory and inflammatory mediators to eliminate the tumor cell. TNF-α is not secreted primarily by splenic lymphocytes; therefore, TNF-α levels were not measured. In this study, IL-2, IL-10, and IFN-γ were measured in the supernatants of splenic lymphocyte cultures to investigate the regulatory effect of CBH and HD on immune suppression.

It is generally acknowledged that splenic LYM represents the principal immune response effector cells, synthesizing various immunomodulation factors, including cytokines and cell adhesion molecules. To further illustrate the immunomodulatory action of CBH and HD, the levels of IL-2, IFN-γ, and IL-10 in the splenic LYM were measured. As shown in Figure 4A,B, the levels of IL-2 and IFN-γ in the splenic LYM of the CTX group were markedly reduced compared with the Normal group ($p < 0.05$), while these reductions were observed to a lesser extent in the Levamisole, CBH, and HD groups ($p < 0.05$). Figure 5C demonstrates that the levels of IL-10 in the splenic LYM of the CTX group were considerably higher than the Normal group ($p < 0.05$). However, the concentration of IL-10 in the Levamisole, CBH, and HD groups was markedly lower than in the CTX group ($p < 0.05$). In addition, the effect of CBH on the downregulation of IL-10 level exceeded that of HD, and no differences were evident between the CBH and Normal groups. Although this may be attributed to the contribution of other functional peptides in the CBH to the downregulation of IL-10, HD provided 72% of its down-regulation ability. These findings indicated that CBH improved the immune responses by increasing IL-2 and IFN-γ secretion while decreasing IL-10 secretion in the splenic LYM of the CTX-induced immunosuppressed mice. Moreover, HD represents the main functional CBH component for regulating cytokine secretion. Jia et al. [56] found that peptides extracted from calf spleens can improve the immune function of CTX-induced immunosuppressed mice by regulating the levels of cytokines, such as IL-2, IL-10, and IFN-γ. Xu et al. [57] revealed that Gly-Gln dipeptide could significantly increase the IL-2 content secreted by the blood and splenic LYM of mice.

To further confirm the effect of CBH and HD on the regulation of the three cytokines secreted by the splenic LYM, the induction of the IL-2, IFN-γ, and IL-10 transcriptional regulation was investigated using qRT-PCR. As shown in Figure 4D,E, the mRNA expression of IL-2 and IFN-γ were substantially lower in the splenic LYM of the CTX group than in the Normal group ($p < 0.05$), while these reductions were observed to a lesser extent in the Levamisole, CBH, and HD groups ($p < 0.05$). However, the mRNA expression of IFN-γ was substantially higher in the CBH group than in the HD group. Figure 4E illustrates that the IL-10 mRNA expression in the splenic LYM of the CTX group was considerably higher than in the Normal group ($p < 0.05$). However, the mRNA expression of IL-10 in the Levamisole, CBH, and HD groups was markedly lower than in the CTX group ($p < 0.05$). IL-10 mRNA expression was substantially lower in the CBH group than in the HD group. The results showed that CBH and HD enhanced the IL-2 and IFN-γ mRNA expression while inhibiting IL-10 mRNA expression in the splenic LYM. Therefore, the IL-2 and IFN-γ secretion was increased, while that of IL-10 was reduced. Yoo et al. [58] found that *Phellinus baumii* extract contained HD that increased the IFN-γ mRNA expression and other cytokines in the spleen of immunosuppressed mice induced by CTX. In this study, CBH and HD played an immunomodulatory role by regulating the mRNA expression of IL-2, IFN-γ, and IL-10 in the splenic LYM, regulating the content of the three cytokines secreted by the splenic LYM. It is speculated that HD represents the functional component in CBH with significant immunomodulatory ability. In this study, CBH and HD mediated their regular effects on cytokine secretion from splenic LYM through a combination of pretreatment before CTX injection and therapeutic effects after CTX injection.

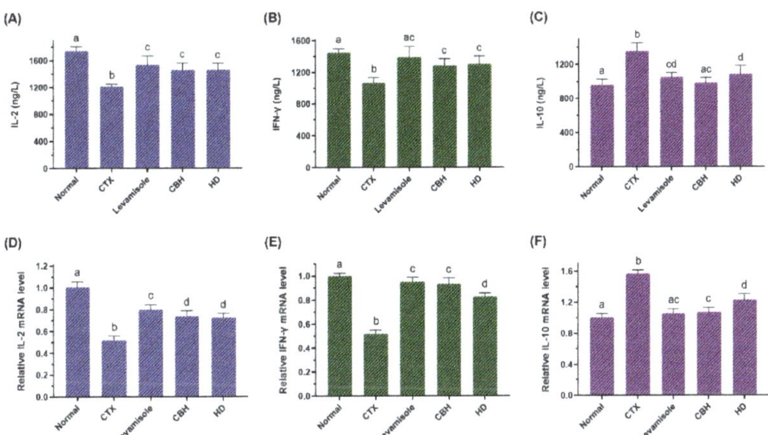

Figure 4. The effect of CHB and HD on the IL-2 (**A**), IFN-γ (**B**), and IL-10 (**C**) levels in the splenocytes, and the mRNA expression levels of IL-2 (**D**), IFN-γ (**E**), and IL-10 (**F**) in the spleens of CTX-treated mice. Data are presented as means $\pm$ SD (n = 10). Different letters for the same index among the groups represent significant differences at $p < 0.05$.

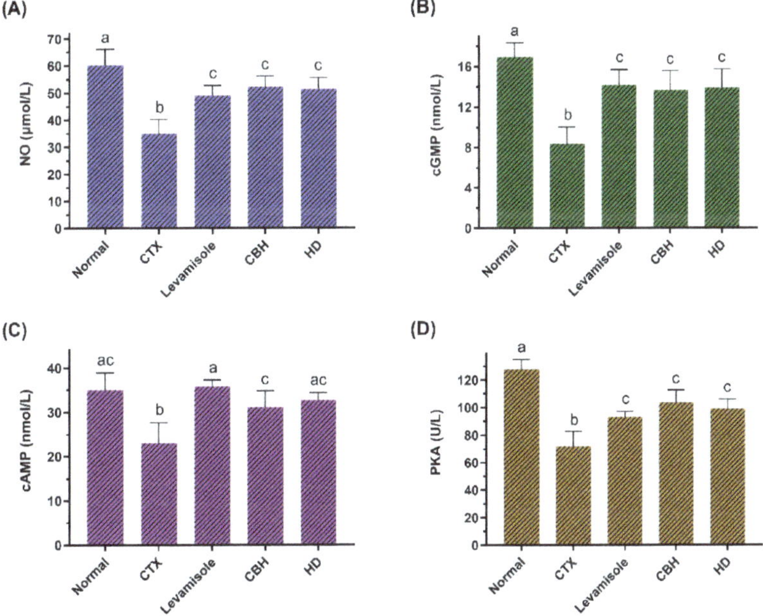

Figure 5. The effect of CHB and HD on the NO (**A**), cGMP (**B**), and cAMP (**C**) content, as well as the PKA (**D**) activity in the splenocytes of CTX-treated mice. Data are presented as means $\pm$ SD (n = 10). Different letters for the same index among the groups represent significant differences at $p < 0.05$.

3.7. The Effect of CBH on the NO/cGMP and cAMP/PKA Signaling Pathways

When external stimuli act on the receptors on LYM surfaces, second messengers are activated via transmembrane transmission. The second messenger transmits the signal to the downstream protein kinase, where it becomes phosphorylated to regulate gene expression and cell function. cAMP and cGMP are the primary second messengers in cells catalyzed by adenylate cyclase (AC) and guanylate cyclase (GC), respectively [59].

During the immune response, cGMP is involved in cell differentiation, chemotaxis, cell proliferation, and the release of soluble mediators. NO plays a crucial role in inflammation and immunity, with numerous physiologic and pathophysiologic effects [60]. NO is generated from the L-arginine and catalyzed by one of three NO synthase (NOS) isoforms: neuronal NOS (nNOS), endothelial NOS (eNOS), and inducible NOS (iNOS) [61]. The only isoform of NOS really implicated in the immune response is the iNOS, while nNOS and eNOS are constitutively activated [21,62]. The primary physiological stimulant of cGMP synthesis is NO, which activates soluble GC (sGC). In addition, activation of the sGC pathway results in the production of cGMP. Once produced, cGMP exerts its function through protein kinase G (PKG).

This study determined the NO and cGMP content in the splenic LYM of the mice. Figure 5A,B show that the NO and cGMP levels in the CTX group were significantly lower ($p < 0.05$) than in the Normal group. The NO and cGMP levels in the Levamisole, CBH, and HD groups were considerably higher than in the CTX group ($p < 0.05$), and no significant differences ($p > 0.05$) were apparent between the three groups. The results showed that the CBH and HD could significantly increase the NO and cGMP content in the splenic LYM of immunosuppressed mice. The activation of the NO/cGMP pathway can promote cell proliferation [63]. Carvalho et al. [64] showed that decreased cGMP and NO levels in T LYM decreased the cytokines mRNA levels, such as IFN-γ and IL-2. In this study, CBH and HD regulated the proliferation of splenic LYM via the NO/cGMP signaling pathway and regulated cytokine gene transcription, affecting cytokine secretion and fulfilling an immunomodulatory function.

Signal transduction via the cAMP/PKA pathway is essential for many processes in a variety of cells. cAMP is a crucial component in the G protein-mediated signaling pathway. When exposed to external stimuli, the intracellular cAMP content can rapidly increase over a short time, forming signal molecules. PKA is made up of tetramers consisting of two regulatory and two catalytic subunits that are found in every cell and govern a variety of functions. When two cAMP molecules are bound by the regulatory subunits, the catalytic subunits are released, phosphorylating their target proteins.

This study determined the cAMP and PKA activity in the splenic LYM of the mice. Figure 5C,D show that the cGMP levels and PKA activity in the CTX group were significantly lower ($p < 0.05$) than in the Normal group. The cGMP levels and PKA activity in the Levamisole, CBH, and HD groups were substantially higher than in the CTX group ($p < 0.05$). The results showed that CBH and HD could significantly increase the cAMP content and PKA activity in the splenic LYM of immunosuppressed mice. Liopeta et al. [65] demonstrated the suppressive effect of cAMP/PKA on the IL-10 levels. Zuo et al. [66] found that an increase in the cAMP and PKA levels in rat spleen cells reduced the serum IL-10 levels. In this study, the CBH and HD may regulate the transcription and secretion of cytokines via the cAMP/PKA signaling pathway. The activation of the NO/cGMP and cAMP/PKA intracellular signaling cascades in immune cells suggests that pretreatment and treatment with CBH and HD display immunomodulatory activity.

4. Conclusions

In summary, this study provides in vivo evidence that CBH and HD have an immunomodulatory effect on CTX-induced immunosuppression in mice. CBH and HD significantly increase the spleen and thymus indexes. Moreover, peripheral WBC, RBC, HGB, and PLT show that CBH and HD treatment inhibits CTX-induced immunosuppression in mice, restoring splenocyte proliferation. Furthermore, IgM, IgG, and IgA secretion are modified following CBH and HD treatment, significantly increasing IL-2 and IFN-γ production and mRNA expression in splenic LYM. CBH and HD also significantly decrease IL-10 production and mRNA expression in splenic LYM. Moreover, CBH and HD play an immunomodulatory role in splenic LYM via the cAMP/PKA and NO/cGMP signaling pathways. The immunotherapy effects of CBH and HD in this study included both a preventive effect from 0 to 12 days and a therapeutic effect after CTX injection. Although, in theory, many molecules can potentially be responsible for the immunomodulatory effect, this study suggests that HD plays a pivotal role in CBH. These observations indicate that CBH supplements exhibit considerable potential for preventing immune system suppression.

Author Contributions: Data curation, J.Z. and H.L.; funding acquisition, H.L.; investigation, J.Z., X.W. and H.L.; software, X.W.; supervision, H.L. and C.C.; validation, C.C. and X.L.; visualization, J.Z.; writing—original draft, J.Z. and X.W.; writing—review and editing, H.L., C.C. and X.L. All authors have read and agreed to the published version of the manuscript.

Funding: This research was funded by the National Key Research and Development Program of China (No. 2021YFC2101400).

Institutional Review Board Statement: The study was conducted according to the guidelines of the Declaration of Helsinki, and approved by the Institutional Animal Care and Use Committee at the Pony Testing International Group Co., Ltd. (No. PONY-2021-FL-03).

Informed Consent Statement: Not applicable.

Data Availability Statement: The data that support the findings of this study are available from the corresponding author upon reasonable request.

Conflicts of Interest: The authors declare no conflict of interest.

References

1. Jayasena, D.D.; Jung, S.; Alahakoon, A.U.; Nam, K.C.; Lee, J.H.; Jo, C. Bioactive and Taste-related Compounds in Defatted Freeze-dried Chicken Soup Made from Two Different Chicken Breeds Obtained at Retail. *J. Poultry Sci.* **2015**, *52*, 156–165. [CrossRef]
2. Yang, Y.; Feng, Y.; Li, Z.; Xie, X.; Miao, F.; Fang, G.; Tu, H.; Wen, J. Effect of sex and diet nutrition on the contents of flavor precursors in fujian hetian chicken. *Acta Vet. Zootech. Sin.* **2006**, *37*, 242–249. [CrossRef]
3. Zhang, L.Z.; Rong, J.H.; Jian, H.U.; Zhao, S.M. Effect of Pre-treatment on Nutritional Characteristics of Chicken Soup. *Food Sci.* **2009**, *30*, 83–87. [CrossRef]
4. Caruso, G.; Fresta, C.G.; Grasso, M.; Santangelo, R.; Lazzarino, G.; Lunte, S.M.; Caraci, F. Inflammation as the Common Biological Link Between Depression and Cardiovascular Diseases: Can Carnosine Exert a Protective Role? *Curr. Med. Chem.* **2020**, *27*, 1782–1800. [CrossRef] [PubMed]
5. Gil-Agusti, M.; Esteve-Romero, J.; Carda-Broch, S. Anserine and carnosine determination in meat samples by pure micellar liquid chromatography. *J. Chromatogr. A* **2008**, *1189*, 444–450. [CrossRef] [PubMed]
6. Peiretti, P.G.; Medana, C.; Visentin, S.; Giancotti, V.; Zunino, V.; Meineri, G. Determination of carnosine, anserine, homocarnosine, pentosidine and thiobarbituric acid reactive substances contents in meat from different animal species. *Food Chem.* **2011**, *126*, 1939–1947. [CrossRef]
7. Kong, Y.; Yang, X.; Ding, Q.; Zhang, Y.; Sun, B.; Chen, H.; Sun, Y. Comparison of non-volatile umami components in chicken soup and chicken enzymatic hydrolysate. *Food Res. Int.* **2017**, *102*, 559–566. [CrossRef]
8. Yeum, K.-J.; Orioli, M.; Regazzoni, L.; Carini, M.; Rasmussen, H.; Russell, R.M.; Aldini, G. Profiling histidine dipeptides in plasma and urine after ingesting beef, chicken or chicken broth in humans. *Amino Acids* **2010**, *38*, 847–858. [CrossRef]
9. Min, J.; Senut, M.-C.; Rajanikant, K.; Greenberg, E.; Bandagi, R.; Zemke, D.; Mousa, A.; Kassab, M.; Farooq, M.U.; Gupta, R.; et al. Differential neuroprotective effects of carnosine, anserine, and N-acetyl carnosine against permanent focal ischemia. *J. Neurosci. Res.* **2008**, *86*, 2984–2991. [CrossRef]
10. Binguel, I.; Yilmaz, Z.; Aydin, A.F.; Coban, J.; Dogru-Abbasoglu, S.; Uysal, M. Antiglycation and anti-oxidant efficiency of carnosine in the plasma and liver of aged rats. *Geriatr. Gerontol. Int.* **2017**, *17*, 2610–2614. [CrossRef]
11. Antonini, F.M.; Petruzzi, E.; Pinzani, P.; Orlando, C.; Poggesi, M.; Serio, M.; Pazzagli, M.; Masotti, G. The meat in the diet of aged subjects and the antioxidant effects of carnosine. *Arch. Gerontol. Geriatr.* **2002**, *35*, 7–14. [CrossRef]
12. Hipkiss, A.R.; Baye, E.; de Courten, B. Carnosine and the processes of ageing. *Maturitas* **2016**, *93*, 28–33. [CrossRef] [PubMed]
13. Kim, Y.; Kim, E.; Kim, Y. L-histidine and L-carnosine accelerate wound healing via regulation of corticosterone and PI3K/Akt phosphorylation in D-galactose-induced aging models in vitro and in vivo. *J. Funct. Foods* **2019**, *58*, 227–237. [CrossRef]

14. Caruso, G.; Caraci, F.; Jolivet, R.B. Pivotal role of carnosine in the modulation of brain cells activity: Multimodal mechanism of action and therapeutic potential in neurodegenerative disorders. *Prog. Neurobiol.* **2019**, *175*, 35–53. [CrossRef] [PubMed]

15. Naghshvar, F.; Abianeh, S.M.; Ahmadashrafi, S.; Hosseinimehr, S.J. Chemoprotective effects of carnosine against genotoxicity induced by cyclophosphamide in mice bone marrow cells. *Cell Biochem. Funct.* **2012**, *30*, 569–573. [CrossRef]

16. Caruso, G.; Fresta, C.G.; Fidilio, A.; O'Donnell, F.; Musso, N.; Lazzarino, G.; Grasso, M.; Amorini, A.M.; Tascedda, F.; Bucolo, C.; et al. Carnosine Decreases PMA-Induced Oxidative Stress and Inflammation in Murine Macrophages. *Antioxidants* **2019**, *8*, 281. [CrossRef]

17. Fresta, C.G.; Fidilio, A.; Lazzarino, G.; Musso, N.; Grasso, M.; Merlo, S.; Amorini, A.M.; Bucolo, C.; Tavazzi, B.; Lazzarino, G.; et al. Modulation of Pro-Oxidant and Pro-Inflammatory Activities of M1 Macrophages by the Natural Dipeptide Carnosine. *Int. J. Mol. Sci.* **2020**, *21*, 776. [CrossRef]

18. Li, X.; Yang, K.; Gao, S.; Zhao, J.; Liu, G.; Chen, Y.; Lin, H.; Zhao, W.; Hu, Z.; Xu, N. Carnosine Stimulates Macrophage-Mediated Clearance of Senescent Skin Cells Through Activation of the AKT2 Signaling Pathway by CD36 and RAGE. *Front. Pharmacol.* **2020**, *11*, 593832. [CrossRef]

19. Guiotto, A.; Calderan, A.; Ruzza, P.; Borin, G. Carnosine and carnosine-related antioxidants: A review. *Curr. Med. Chem.* **2005**, *12*, 2293–2315. [CrossRef]

20. Fresta, C.G.; Chakraborty, A.; Wijesinghe, M.B.; Amorini, A.M.; Lazzarino, G.; Lazzarino, G.; Tavazzi, B.; Lunte, S.M.; Caraci, F.; Dhar, P.; et al. Non-toxic engineered carbon nanodiamond concentrations induce oxidative/nitrosative stress, imbalance of energy metabolism, and mitochondrial dysfunction in microglial and alveolar basal epithelial cells. *Cell Death Dis.* **2018**, *9*, 245. [CrossRef]

21. Caruso, G.; Fresta, C.G.; Martinez-Becerra, F.; Antonio, L.; Johnson, R.T.; de Campos, R.P.S.; Siegel, J.M.; Wijesinghe, M.B.; Lazzarino, G.; Lunte, S.M. Carnosine modulates nitric oxide in stimulated murine RAW 264.7 macrophages. *Mol. Cell. Biochem.* **2017**, *431*, 197–210. [CrossRef] [PubMed]

22. Li, Y.; He, R.; Tsoi, B.; Li, X.; Li, W.; Abe, K.; Kurihara, H. Anti-Stress Effects of Carnosine on Restraint-Evoked Immunocompromise in Mice through Spleen Lymphocyte Number Maintenance. *PLoS ONE* **2012**, *7*, e33190. [CrossRef] [PubMed]

23. Deng, J.; Zhong, Y.; Wu, Y.; Luo, Z.; Sun, Y.; Wang, G.; Kurihara, H.; Li, Y.; He, R. Carnosine attenuates cyclophosphamide-induced bone marrow suppression by reducing oxidative DNA damage. *Redox Biol.* **2018**, *14*, 1–6. [CrossRef] [PubMed]

24. Jiang, X.; Yang, F.; Zhao, Q.; Tian, D.; Tang, Y. Protective effects of pentadecapeptide derived from Cyclaina sinensis against cyclophosphamide-induced hepatotoxicity. *Biochem. Biophys. Res. Commun.* **2019**, *520*, 392–398. [CrossRef] [PubMed]

25. Huyan, X.; Lin, Y.; Gao, T.; Chen, R.; Fan, Y. Immunosuppressive effect of cyclophosphamide on white blood cells and lymphocyte subpopulations from peripheral blood of Balb/c mice. *Int. Immunopharmacol.* **2011**, *11*, 1293–1297. [CrossRef]

26. Gao, S.; Hong, H.; Zhang, C.; Wang, K.; Zhang, B.; Han, Q.; Liu, H.; Luo, Y. Immunomodulatory effects of collagen hydrolysates from yak (Bos grunniens) bone on cyclophosphamide-induced immunosuppression in BALB/c mice. *J. Funct. Foods* **2019**, *60*, 103420. [CrossRef]

27. Stevenson, H.C.; Green, I.; Hamilton, J.M.; Calabro, B.A.; Parkinson, D.R. Levamisole-known effects on the immune-system, clinical-results, and future applications to the treatment of cancer. *J. Clin. Oncol.* **1991**, *9*, 2052–2066. [CrossRef]

28. Sajid, M.S.; Iqbal, Z.; Muhammad, G.; Iqbal, M.U. Immunomodulatory effect of various anti-parasitics: A review. *Parasitology* **2006**, *132*, 301–313. [CrossRef]

29. Yu, F.; He, K.; Dong, X.; Zhang, Z.; Wang, F.; Tang, Y.; Chen, Y.; Ding, G. Immunomodulatory activity of low molecular-weight peptides from Nibea japonica skin in cyclophosphamide-induced immunosuppressed mice. *J. Funct. Foods* **2020**, *68*, 103888. [CrossRef]

30. Banerjee, S.; Sinha, K.; Chowdhury, S.; Sil, P.C. Unfolding the mechanism of cisplatin induced pathophysiology in spleen and its amelioration by carnosine. *Chem. Biol. Interact.* **2018**, *279*, 159–170. [CrossRef]

31. Liu, Q.; Ge, M.; Xu, X.; Cheng, L.; Su, C.; Yu, S.; Huang, Z. Immunifaction Accommodation of Phenylalanine Dipeptide Compound Y101 in Normal Mice. *Pharm. J. Chin. People's Lib. Army* **2011**, *27*, 215–217. [CrossRef]

32. Zhao, L.; Liu, L.; Guo, B.; Zhu, B. Regulation of adaptive immune responses by guiding cell movements in the spleen. *Front. Microbiol.* **2015**, *6*, 645. [CrossRef] [PubMed]

33. Bajenoff, M.; Glaichenhaus, N.; Germain, R.N. Fibroblastic reticular cells guide T lymphocyte entry into and migration within the splenic T cell zone. *J. Immunol.* **2008**, *181*, 3947–3954. [CrossRef] [PubMed]

34. Lozano-Ojalvo, D.; Molina, E.; Lopez-Fandino, R. Hydrolysates of egg white proteins modulate T- and B-cell responses in mitogen-stimulated murine cells. *Food Funct.* **2016**, *7*, 1048–1056. [CrossRef] [PubMed]

35. Yang, Q.; Huang, M.; Cai, X.; Jia, L.; Wang, S. Investigation on activation in RAW264.7 macrophage cells and protection in cyclophosphamide-treated mice of Pseudostellaria heterophylla protein hydrolysate. *Food Chem. Toxicol.* **2019**, *134*, 110816. [CrossRef] [PubMed]

36. Hou, H.; Fan, Y.; Li, B.; Xue, C.; Yu, G.; Zhang, Z.; Zhao, X. Purification and identification of immunomodulating peptides from enzymatic hydrolysates of Alaska pollock frame. *Food Chem.* **2012**, *134*, 821–828. [CrossRef]

37. Pantoja, P.; Perez-Guzman, E.X.; Rodriguez, I.V.; White, L.J.; Gonzalez, O.; Serrano, C.; Giavedoni, L.; Hodara, V.; Cruz, L.; Arana, T.; et al. Zika virus pathogenesis in rhesus macaques is unaffected by pre-existing immunity to dengue virus. *Nat. Commun.* **2017**, *8*, 15674. [CrossRef]

38. Zhou, H.; Sun, F.; Li, H.; Zhang, S.; Liu, Z.; Pei, J.; Liang, C. Effect of recombinant Ganoderma lucidum immunoregulatory protein on cyclophosphamide-induced leukopenia in mice. *Immunopharmacol. Immunotoxicol.* **2013**, *35*, 426–433. [CrossRef]

39. Chen, Q.; Zhang, S.; Ying, H.; Dai, X.; Li, X.; Yu, C.; Ye, H. Chemical characterization and immunostimulatory effects of a polysaccharide from Polygoni Multiflori Radix Praeparata in cyclophosphamide-induced anemic mice. *Carbohydr. Polym.* **2012**, *88*, 1476–1482. [CrossRef]

40. Arslan, S.; Ozyurek, E.; Gunduz-Demir, C. A color and shape based algorithm for segmentation of white blood cells in peripheral blood and bone marrow images. *Cytom. Part A* **2014**, *85A*, 480–490. [CrossRef]

41. Juaristi, J.A.; Aguirre, M.V.; Todaro, J.S.; Alvarez, M.A.; Brandan, N.C. EPO receptor, Bax and Bcl-x(L) expressions in murine erythropoiesis after cyclophosphamide treatment. *Toxicology* **2007**, *231*, 188–199. [CrossRef] [PubMed]

42. Chu, Q.; Zhang, Y.; Chen, W.; Jia, R.; Yu, X.; Wang, Y.; Li, Y.; Liu, Y.; Ye, X.; Yu, L.; et al. Apios americana Medik flowers polysaccharide (AFP) alleviate Cyclophosphamide-induced immunosuppression in ICR mice. *Int. J. Biol. Macromol.* **2020**, *144*, 829–836. [CrossRef] [PubMed]

43. Zhang, J.; Gao, S.; Li, H.; Cao, M.; Li, W.; Liu, X. Immunomodulatory effects of selenium-enriched peptides from soybean in cyclophosphamide-induced immunosuppressed mice. *Food Sci. Nutr.* **2021**, *9*, 6322–6334. [CrossRef] [PubMed]

44. Lis, M.; Obminska-Mrukowicz, B. Modulatory effects of bestatin on T and B lymphocyte subsets and the concentration of cytokines released by Th1/Th2 lymphocytes in cyclophosphamide-treated mice. *Cent. Eur. J. Immunol.* **2013**, *38*, 42–53. [CrossRef]

45. Ehrenstein, M.R.; Notley, C.A. The importance of natural IgM: Scavenger, protector and regulator. *Nat. Rev. Immunol.* **2010**, *10*, 778–786. [CrossRef]

46. Nimmerjahn, F.; Ravetch, J.V. Antibody-mediated modulation of immune responses. *Immunol. Rev.* **2010**, *236*, 265–275. [CrossRef]

47. Woof, J.M.; Kerr, M.A. The function of immunoglobulin A in immunity. *J. Pathol.* **2006**, *208*, 270–282. [CrossRef] [PubMed]

48. Maestri, E.; Marmiroli, M.; Marmiroli, N. Bioactive peptides in plant-derived foodstuffs. *J. Proteom.* **2016**, *147*, 140–155. [CrossRef]

49. Yu, F.; Zhang, Z.; Ye, S.; Hong, X.; Jin, H.; Huang, F.; Yang, Z.; Tang, Y.; Chen, Y.; Ding, G. Immunoenhancement effects of pentadecapeptide derived from Cyclina sinensis on immune-deficient mice induced by Cyclophosphamide. *J. Funct. Foods* **2019**, *60*, 103408. [CrossRef]

50. Musso, N.; Caruso, G.; Bongiorno, D.; Grasso, M.; Bivona, D.A.; Campanile, F.; Caraci, F.; Stefani, S. Different Modulatory Effects of Four Methicillin-Resistant Staphylococcus aureus Clones on MG-63 Osteoblast-Like Cells. *Biomolecules* **2021**, *11*, 72. [CrossRef]

51. Fujiwara, M.; Ozono, K. Cytokines and osteogenesis. *Clin. Calcium* **2014**, *24*, 845–851. [PubMed]

52. Chan, C.C.; Caspi, R.; Mochizuki, M.; Diamantstein, T.; Gery, I.; Nussenblatt, R.B. Cyclosporine and dexamethasone inhibit T-lymphocyte MHC class II antigens and IL-2 receptor expression in experimental autoimmune uveitis. *Immunol. Investig.* **1987**, *16*, 319–331. [CrossRef] [PubMed]

53. Ivashkiv, L.B. IFN gamma: Signalling, epigenetics and roles in immunity, metabolism, disease and cancer immunotherapy. *Nat. Rev. Immunol.* **2018**, *18*, 545–558. [CrossRef] [PubMed]

54. Castillo, P.; Kolls, J.K. IL-10: A Paradigm for Counterregulatory Cytokines. *J. Immunol.* **2016**, *197*, 1529–1530. [CrossRef] [PubMed]

55. Zhang, W.; Gong, L.; Liu, Y.; Zhou, Z.; Wan, C.; Xu, J.; Wu, Q.; Chen, L.; Lu, Y.; Chen, Y. Immunoenhancement effect of crude polysaccharides of Helvella leucopus on cyclophosphamide-induced immunosuppressive mice. *J. Funct. Foods* **2020**, *69*, 103942. [CrossRef]

56. Jia, D.; Lu, W.; Wang, C.; Sun, S.; Cai, G.; Li, Y.; Wang, G.; Liu, Y.; Zhang, M.; Wang, D. Investigation on Immunomodulatory Activity of Calf Spleen Extractive Injection in Cyclophosphamide-induced Immunosuppressed Mice and Underlying Mechanisms. *Scand. J. Immunol.* **2016**, *84*, 20–27. [CrossRef]

57. Xu, P.; Fang, X.; Shu, G.; Zhu, X.; Luo, Z.; Jiang, Q.; Gao, P.; Zhang, Y. Prokaryotic Expression of Gly-Gln Dipeptide and Its Bioactive Analysis: A Novel Method for Short Peptide Production. *Agric. Sci. China* **2010**, *9*, 736–744. [CrossRef]

58. Yoo, J.-H.; Lee, Y.-S.; Ku, S.; Lee, H.-J. Phellinus baumii enhances the immune response in cyclophosphamide-induced immuno-suppressed mice. *Nutr. Res.* **2020**, *75*, 15–31. [CrossRef]

59. Kurelic, R.; Krieg, P.F.; Sonner, J.K.; Bhaiyan, G.; Ramos, G.C.; Frantz, S.; Friese, M.A.; Nikolaev, V.O. Upregulation of Phosphodiesterase 2A Augments T Cell Activation by Changing cGMP/cAMP Cross-Talk. *Front. Pharmacol.* **2021**, *12*, 748798. [CrossRef]

60. Tripathi, P.; Tripathi, P.; Kashyap, L.; Singh, V.; Bogdan, C. The role of nitric oxide in inflammatory reactions. *FEMS Immunol. Med. Microbiol.* **2012**, *66*, 449. [CrossRef]

61. Lechner, M.; Lirk, P.; Rieder, J. Inducible nitric oxide synthase (iNOS) in tumor biology: The two sides of the same coin. *Semin. Cancer Biol.* **2005**, *15*, 277–289. [CrossRef] [PubMed]

62. Xue, Q.; Yan, Y.; Zhang, R.; Xiong, H. Regulation of iNOS on Immune Cells and Its Role in Diseases. *Int. J. Mol. Sci.* **2018**, *19*, 3805. [CrossRef] [PubMed]

63. Gao, J.; Xu, Y.; Lei, M.; Shi, J.; Gong, Q. Icariside II, a PDE5 inhibitor from Epimedium brevicornum, promotes neuron-like pheochromocytoma PC12 cell proliferation via activating NO/cGMP/PKG pathway. *Neurochem. Int.* **2018**, *112*, 18–26. [CrossRef] [PubMed]

64. Carvalho, M.U.W.B.; Vendramini, P.; Kubo, C.A.; Soreiro-Pereira, P.V.; de Albuquerque, R.S.; Antunes, E.; Condino-Neto, A. BAY 41-2272 inhibits human T lymphocyte functions. *Int. Immunopharmacol.* **2019**, *77*, 105976. [CrossRef]

65. Liopeta, K.; Boubali, S.; Virgilio, L.; Thyphronitis, G.; Mavrothalassitis, G.; Dimitracopoulos, G.; Paliogianni, F. cAMP regulates IL-10 production by normal human T lymphocytes at multiple levels: A potential role for MEF2. *Mol. Immunol.* **2009**, *46*, 345–354. [CrossRef]
66. Zuo, L.; Shi, L.; Yan, F. The reciprocal interaction of sympathetic nervous system and cAMP-PKA-NF-kB pathway in immune suppression after experimental stroke. *Neurosci. Lett.* **2016**, *627*, 205–210. [CrossRef]

Article

Oral Administration of Branched-Chain Amino Acids Attenuates Atherosclerosis by Inhibiting the Inflammatory Response and Regulating the Gut Microbiota in ApoE-Deficient Mice

Ziyun Li [1,†], Ranran Zhang [1,†], Hongna Mu [1], Wenduo Zhang [2], Jie Zeng [3], Hongxia Li [1], Siming Wang [1], Xianghui Zhao [1], Wenxiang Chen [1,3], Jun Dong [1] and Ruiyue Yang [1,*]

1   The Key Laboratory of Geriatrics, Beijing Institute of Geriatrics, Institute of Geriatric Medicine, Chinese Academy of Medical Sciences, Beijing Hospital/National Center of Gerontology of National Health Commission, Beijing 100730, China
2   Department of Cardiology, Beijing Hospital, National Center of Gerontology, Institute of Geriatric Medicine, Chinese Academy of Medical Sciences, Beijing 100730, China
3   National Center for Clinical Laboratories, Institute of Geriatric Medicine, Chinese Academy of Medical Sciences, Beijing Hospital, National Center of Gerontology, Beijing 100730, China
*   Correspondence: ruiyue_yang@163.com
†   These authors contributed equally to this work.

Abstract: Atherosclerosis (AS) is a chronic inflammatory disease that serves as a common pathogenic underpinning for various cardiovascular diseases. Although high circulating branched-chain amino acid (BCAA) levels may represent a risk factor for AS, it is unclear whether dietary BCAA supplementation causes elevated levels of circulating BCAAs and hence influences AS, and the related mechanisms are not well understood. Here, ApoE-deficient mice (ApoE$^{-/-}$) were fed a diet supplemented with or without BCAAs to investigate the effects of BCAAs on AS and determine potential related mechanisms. In this study, compared with the high-fat diet (HFD), high-fat diet supplemented with BCAAs (HFB) reduced the atherosclerotic lesion area and caused a significant decrease in serum cholesterol (TC) and low-density lipoprotein cholesterol (LDL-C) levels. BCAA supplementation suppressed the systemic inflammatory response by reducing macrophage infiltration; lowering serum levels of inflammatory factors, including monocyte chemoattractant protein-1 (MCP-1), tumor necrosis factor-α (TNF-α), interleukin-1β (IL-1β) and interleukin-6 (IL-6); and suppressing inflammatory related signaling pathways. Furthermore, BCAA supplementation altered the gut bacterial beta diversity and composition, especially reducing harmful bacteria and increasing probiotic bacteria, along with increasing bile acid (BA) excretion. In addition, the levels of total BAs, primary BAs, 12α-hydroxylated bile acids (12α-OH BAs) and non-12α-hydroxylated bile acids (non-12α-OH BAs) in cecal and colonic contents were increased in the HFB group of mice compared with the HFD group. Overall, these data indicate that dietary BCAA supplementation can attenuate atherosclerosis induced by HFD in ApoE$^{-/-}$ mice through improved dyslipidemia and inflammation, mechanisms involving the intestinal microbiota, and promotion of BA excretion.

Keywords: atherosclerosis; branched-chain amino acids; inflammation; gut microbiota; bile acids

Citation: Li, Z.; Zhang, R.; Mu, H.; Zhang, W.; Zeng, J.; Li, H.; Wang, S.; Zhao, X.; Chen, W.; Dong, J.; et al. Oral Administration of Branched-Chain Amino Acids Attenuates Atherosclerosis by Inhibiting the Inflammatory Response and Regulating the Gut Microbiota in ApoE-Deficient Mice. *Nutrients* **2022**, *14*, 5065. https://doi.org/10.3390/nu14235065

Academic Editor: Philip J. Atherton

Received: 4 November 2022
Accepted: 25 November 2022
Published: 28 November 2022

1. Introduction

Atherosclerosis (AS) is a chronic inflammatory disease that narrows the arterial lumen through complex atherosclerotic plaque formation. AS is the common pathological basis of many cardiovascular and peripheral vascular diseases [1,2]. Cardiovascular disease (most commonly coronary atherosclerotic disease) is the leading cause of death in both developed and developing regions [3–5]. Early detection and early intervention represent effective strategies to reduce cardiovascular morbidity and mortality.

Atherosclerosis is often accompanied by dyslipidemia, including elevated serum cholesterol (TC), triglycerides (TGs), low-density lipoprotein cholesterol (LDL-C) levels and decreased high-density lipoprotein cholesterol (HDL-C) levels. Excess serum LDL-C deposition accumulates in the arterial wall endothelium and is oxidized to oxidized low-density lipoprotein (ox-LDL). Macrophages can recognize and engulf ox-LDL and convert it into foam cells, which accumulate to form streaks or lipid patches [1,6,7]. Atherosclerosis encompasses more than just LDL deposition in the artery wall; it is also a chronic inflammatory disease, and the inflammatory response is critical to AS progression. Endothelial damage, which frequently occurs during the early stages of AS, enhances the inflammatory response [8–10]. In addition, activated endothelial cells release intercellular cell adhesion molecule-1 (ICAM-1), vascular cell adhesion molecule-1 (VCAM-1) and other inflammatory factors, which help recruit monocytes. Numerous inflammatory molecules including interleukin (IL), chemokines, and tumor necrosis factor (TNF) are released throughout these processes, exacerbating inflammation [11]. Massive macrophages and inflammatory factors infiltrate the vessel wall in the later stages of AS. Similarly, macrophages release extracellular metalloproteases that hydrolyze extracellular matrix collagen fibers and increase plaque instability, increasing the risk of cardiovascular disease [12].

The diversity and homeostasis of the gut microbiota are closely related to the nutrition, metabolism, diseases, and other physiological processes of the host. Dysbiosis is closely associated with various metabolic diseases, such as obesity, metabolic-associated fatty liver disease and AS [13–15]. Several studies suggest that the gut microbiota plays an important role in the development of AS [16,17].

Bile acids (BAs) are catabolic metabolites of cholesterol and lipids, and their dysregulated synthesis and metabolism are associated with diseases, including insulin resistance, dyslipidemia, and AS [18–20]. Promoting BA synthesis can improve cholesterol metabolism, which may help to prevent and alleviate AS. Furthermore, BAs may affect host physiological activity and intestinal flora composition via the immune response and Farnesoid X receptor [20–22].

Many epidemiological studies have shown that elevated serum BCAA levels are independently associated with a high risk of coronary and cerebrovascular atherosclerotic disease [23–26]. However, animal studies have found that dietary BCAA supplementation significantly lowers TG levels in macrophages by decreasing very low-density lipoprotein (VLDL) uptake and subsequently inhibiting lipid accumulation and macrophage foam cell formation [27]. In addition, studies have found that gut bacteria can regulate BCAA biosynthesis, transport, and metabolism. Our previous study demonstrated that oral administration of BCAAs ameliorates high-fat diet-induced metabolic-associated fatty liver disease via gut microbiota-associated mechanisms [28]. Although high circulating BCAA levels may represent a risk factor for AS, it is unclear whether dietary BCAA supplementation causes elevated levels of circulating BCAAs and hence influences AS. It is also unknown whether dietary BCAA supplementation influences the development of AS through gut microbiota-related mechanisms. Therefore, the purpose of this study was to determine the effects of dietary BCAA supplementation on AS in ApoE$^{-/-}$ mice and the underlying mechanisms, with a focus on gut microbiota remodeling and the inflammatory response.

2. Materials and Methods

2.1. Animal Studies

ApoE$^{-/-}$ mice (9 weeks old, male) were purchased from Vital River Laboratory Animal Technology Co., Ltd. (Beijing, China). The mice were raised (2–3 mice per cage) in a special room with 22 °C, 40–70% humidity, and a 12-h light/dark cycle. After one week of adaptation to a normal diet, mice were randomly divided into the following four groups: normal chow diet group (ND, $n = 10$), normal chow diet supplemented with BCAAs group (NB, $n = 10$), high-fat diet group (HFD, $n = 10$), and high-fat diet supplemented with BCAAs group (HFB, $n = 10$). The normal chow diet (12% of kcal fat, 1025), normal chow diet supplemented with BCAAs, high-fat diet (41% of kcal fat with an extra supplement of 0.15% (*w/w*) cholesterol, H10141), and high-fat diet supplemented with BCAAs were purchased from Beijing HFK Bioscience Co., Ltd. (Beijing, China). Valine, L-leucine, and L-isoleucine were purchased from Nanjing Jingzhu Biotechnology Co., Ltd. (Nanjing, China). The extra amounts of L-leucine, L-isoleucine, and valine per 100 g diet supplemented with BCAAs were 0.56 g, 0.40 g, and 0.40 g, respectively. All mice had free access to food and water. Throughout the experiment, the padding and water were changed once a week, whereas the food of the HFD and HFB groups was changed twice a week to prevent sudor production via fat oxidation, which would subsequently affect model establishment. The amount of food consumed and the body weight were recorded every two weeks.

2.2. Sample Collection

After 12 weeks of feeding, the mice were anesthetized with pentobarbital (30 mg/kg) and sacrificed after blood samples were collected by cardiac puncture. The whole aorta and aortic root were separated for protein extraction and staining. The livers were collected and frozen immediately in liquid nitrogen and then stored at $-80\,^{\circ}\text{C}$. The contents of the cecum and colon were collected in a sterile manner, immediately frozen in liquid nitrogen, and then stored at $-80\,^{\circ}\text{C}$ until use.

The animal experiments were approved by Peking University Biomedical Ethics Commitment Experimental Animal Ethics Branch (LA2022010) and complied with the Guide for the Care and Use of Laboratory Animals established by the National Institutes of Health (NIH).

2.3. Histological Analysis

The aortas were carefully obtained from the mice and stained with Oil Red O. The lesion area was calculated as a percentage of the total aorta area. The fresh aortic roots were collected and immediately fixed in 4% paraformaldehyde solution, dehydrated, embedded, and then sectioned. Paraffin-embedded sections (5 μm) were used for hematoxylin and eosin (HE) staining, Masson staining, and immunohistochemical staining; frozen sections (8 μm) were used for Oil Red O staining. Immunohistochemistry of aortic roots was performed with an anti-F4/80 antibody (Cell Signaling Technology, Inc., Beverly, MA, USA) and following the instructions provided.

2.4. Serum Biochemical Assays

Serum concentrations of total TC, TG, HDL-C, LDL-C, and glucose (Glu) were measured using a 7180 automatic biochemical analyzer (Hitachi Ltd., Tokyo, Japan), according to the manufacturer's instructions. Wako Pure Chemical Industries, Ltd. reagents (Osaka, Japan) and their matching calibration and quality control were used to measure HDL-C by direct method, LDL-C by direct method, TG by deglycerin method, and TC by enzyme method, respectively. DiaSys Diagnostic Systems GmbH (Holzheim, Germany) reagents and their matching calibration and quality control were used to measure glucose by hexokinase method.

2.5. Assay of BCAAs in Serum

The previously published LC−MS/MS method was used to determine the concentrations of BCAAs (Val, Leu, Ile) in serum [29]. In brief, 50 μL of serum or standard solution was added to an equal volume of isotopically labeled internal standard solution; acetonitrile (containing 0.1% formic acid) was added for precipitation and extraction, and the supernatant was transferred to vials for LC−MS/MS detection. The separation was performed on an Agilent 1260 HPLC system equipped with a Kinetex HILIC column (2.6 μm, 2.1 mm × 150 mm), and mass spectrometry data were collected using an API 5500 system (AB Sciex, Framingham, MA, USA) in ESI positive ion mode and MRM mode.

2.6. Enzyme-Linked Immunosorbent Assay (ELISA)

ELISA analysis was used to determine serum interleukin-1β (IL-1β), tumor necrosis factor-α (TNF-α), monocyte chemoattractant protein-1 (MCP-1), and interleukin-6 (IL-6) levels according to the instructions provided by the kit. The mouse IL-1β ELISA kit, mouse TNF-α ELISA kit, and mouse MCP-1 ELISA kit were obtained from R&D Systems (Minneapolis, MN, USA), and the mouse IL-6 ELISA kit was purchased from Novus (St Charles, MO, USA).

2.7. Western Blotting

Protein expression was determined by Western blot assay. Protein lysate (Sigma-Aldrich, Taufkirchen, Germany) was used to extract total protein from aorta tissues and liver tissues, and total protein concentrations were determined using a BCA protein assay kit (Applygen Technologies, Beijing, China). A total of 30 micrograms of protein was separated in SDS−PAGE gels and transferred onto PVDF membranes (Roche, Indianapolis, IN, USA). After washing with TBST, the membranes were blocked with TBST (containing 5% skimmed milk) for 2 h at room temperature. In brief, membranes were incubated overnight with primary antibodies at 4 °C, washed with TBST, and incubated with secondary antibodies for 2 h at ambient temperature. The primary antibodies against phospho-AKT, AKT, phospho-NF-κB p65, NF-κB p65, phospho-BCKDH-E1α, BCKDH-E1α, and β-actin were purchased from Cell Signaling Technology Inc. (Beverly, MA, USA). Bands were visualized using an ECL detection kit (Applygen Technologies, Beijing, China). β-Actin served as an internal reference protein. Band intensity was assessed by densitometry and expressed as the mean density area using ImageJ software.

2.8. 16S rRNA Gene Sequencing

To determine cecal and colonic content, 16S ribosomal RNA (rRNA) gene sequencing was performed. Amplification of the high variant region 3 (V3) and the high variant region 4 (V4) of bacterial 16S rRNA was performed using the 338F primer (5′-ACTCCTACGGAGG-GCAGCA-3′) and the 806R primer (5′-GGACTACHVGGGTWTCTAAT-3′) for Illumina deep sequencing. High-fidelity DNA polymerase was used for PCR amplification. The 16S gene sequencing procedure was performed using Beijing Biomarker Technologies Co., Ltd. (Beijing, China), and all results were based on sequencing reads and operational taxonomic units (OTUs).

2.9. Assessment of Bile Acids in Intestines

A UPLC−MS/MS method was used to assess sixteen bile acids, including cholic acid (CA), chenodeoxycholic acid (CDCA), deoxycholic acid (DCA), α-muricholic acid (α-MCA), β-muricholic acid (β-MCA), ursodeoxycholic acid (UDCA), hyodeoxycholic acid (HDCA), lithocholic acid (LCA), taurocholic acid (TCA), glycocholic acid (GCA), taurodeoxycholic acid (TDCA), glycodeoxycholic acid (GDCA), tauro-α-muricholic acid/tauro-β-muricholic acid (Tα-MCA/Tβ-MCA), tauroursodeoxycholic acid (TUDCA), glycoursodeoxycholic acid (GUDCA), taurohyodeoxycholic acid (THDCA) and taurolithocholic acid (TLCA), in cecal and colonic contents. After thawing at 4 °C for 30 min, the sample (~50 mg) was placed in a 1.5-mL centrifuge tube and weighed accurately. A total of 200 μL of water was added to a test tube, and the mixture was ultrasonicated for 5 min. Then, 1 mL of methanol was added. The sample was vortexed for 5 min followed by ultrasonic treatment for 1 h and centrifugation (4 °C, 12,000 rpm) for 5 min. The supernatant (100 μL) and isotope standard solutions (10 μL, 2 μg/mL CA-d4 and LCA-d4) were loaded into a 1.5-mL centrifuge tube and vortexed for 2 min. The mixture was then transferred to vials for LC−MS/MS testing. A Waters ACQUITY UPLC I-CLASS chromatograph (Waters, Milford, USA) equipped with a UPLC BEH Amide column (2.1 mm × 100 mm, 1.7 μm, Waters, USA) was used for separation. The column temperature was 50 °C. The mobile phase consisted of water/acetonitrile (10:1, *v*/*v*, containing 1 mmol/L ammonium acetate, phase A) and acetonitrile/isopropanol (1:1, *v*/*v*, phase B) with a flow rate of 0.26 mL/min. Analyte isolation was performed by elution gradient, and the injection volume was 5.0 µL. The MS data were collected using a QTRAP 6500+ LOW MASS system (AB Sciex, Framingham, MA, USA) in negative ion mode. The analysis was performed based on the following parameters: ion source voltage 2.0 kV, ion source temperature 150 °C, desolvation temperature 600 °C, desolvation gas flow 1000 L/h, cone voltage 21 V, cone gas flow 10 L/h.

Total bile acid concentrations were calculated as the sum of the concentrations of all sixteen bile acids. The primary BAs included CA, TCA, GCA, CDCA, α-MCA, β-MCA, and Tα-MCA/Tβ-MCA. The secondary BAs included DCA, GDCA, LCA, HDCA, UDCA, TLCA, THDCA, GUDCA, and TUDCA. 12α-Hydroxylated bile acids (12α-OH BAs) included CA, DCA, TCA, GCA, GDCA, and TDCA. Non-12α-hydroxylated bile acids (non-12α-OH BAs) included CDCA, α-MCA, β-MCA, LCA, HDCA, UDCA, Tα-MCA/Tβ-MCA, TLCA, THDCA, GUDCA, and TUDCA. The concentrations of primary BAs, secondary BAs, 12α-OH BAs, and non-12α-OH BAs were calculated based on the sum of the concentrations of each type of bile acid mentioned above.

2.10. Statistical Analysis

The experimental results are expressed as the mean ± standard error (SEM). The statistical software SPSS 26.0 (Armonk, NY, USA), GraphPad Prism 8.0 (San Diego, CA, USA), and R 3.6.1 (Vienna, Austria) were applied for data analysis, processing, and graph generation. Two-group comparisons were performed using a *t* test. Multiple group comparisons were made using ANOVA and Tukey's post hoc test, and repeated measures data were statistically analyzed using multivariate analysis of variance (MANOVA). The alpha diversity indices, including ACE, Chao1, Simpson, and Shannon indices, and beta diversity indices, including principal coordinate analysis (PCoA), were evaluated by QI-IME. Data were analyzed and plotted using q-values ≤ 0.05 (obtained after *p* value correction) as a criterion for differential colony screening. Correlation analysis was performed using Spearman correlation analysis. A *p* value < 0.05 was considered to be statistically significant.

3. Results

3.1. BCAA Supplementation Reduces Atherosclerotic Plaques Induced by a High-Fat Diet in ApoE$^{-/-}$ Mice

AS is characterized by aortic plaque formation, and the lesion degree can be evaluated based on the size of atherosclerotic plaques. Therefore, the aorta was separated and stained with Oil Red O after feeding ApoE$^{-/-}$ mice for 12 weeks to assess the atherosclerotic plaque, and the results are displayed in Figure 1A. The plaque area of whole aorta in the NB group was slightly increased compared with that in the ND group, but the difference was not statistically significant. The plaque area of whole aorta in the HFD group was significantly larger than that in the ND group, whereas the plaque area in the HFB group was significantly smaller than that in the HFD group. The aortic root atherosclerotic plaque area is also commonly utilized to assess atherosclerotic lesions. As shown in Figure 1B,C, the plaque area of the aortic root in the HFD group was greatly enlarged compared with that in the ND group, and the plaque area in the HFB group was markedly less than that in the HFD group. The aorta root intima in the ND group was smooth with neatly arranged endothelial cells and no atherosclerotic plaque formation, whereas very few atherosclerotic plaques were noted in the NB group. In the HFD group, apparent plaque lesions were noted with thickening aortic intima, an increased number of foam-like cells, and willow-shaped crystals inside the plaque. Compared with the HFD group, the plaque area and number of foam-like cells were remarkably decreased in the HFB group. These results suggest that dietary BCAA supplementation reduced atherosclerotic plaques induced by a high-fat diet in ApoE$^{-/-}$ mice. Plaque stability was examined in this research by Masson staining of aortic roots. In Figure 1D, the collagen fiber content was apparently lower in the HFB group than in the HFD group.

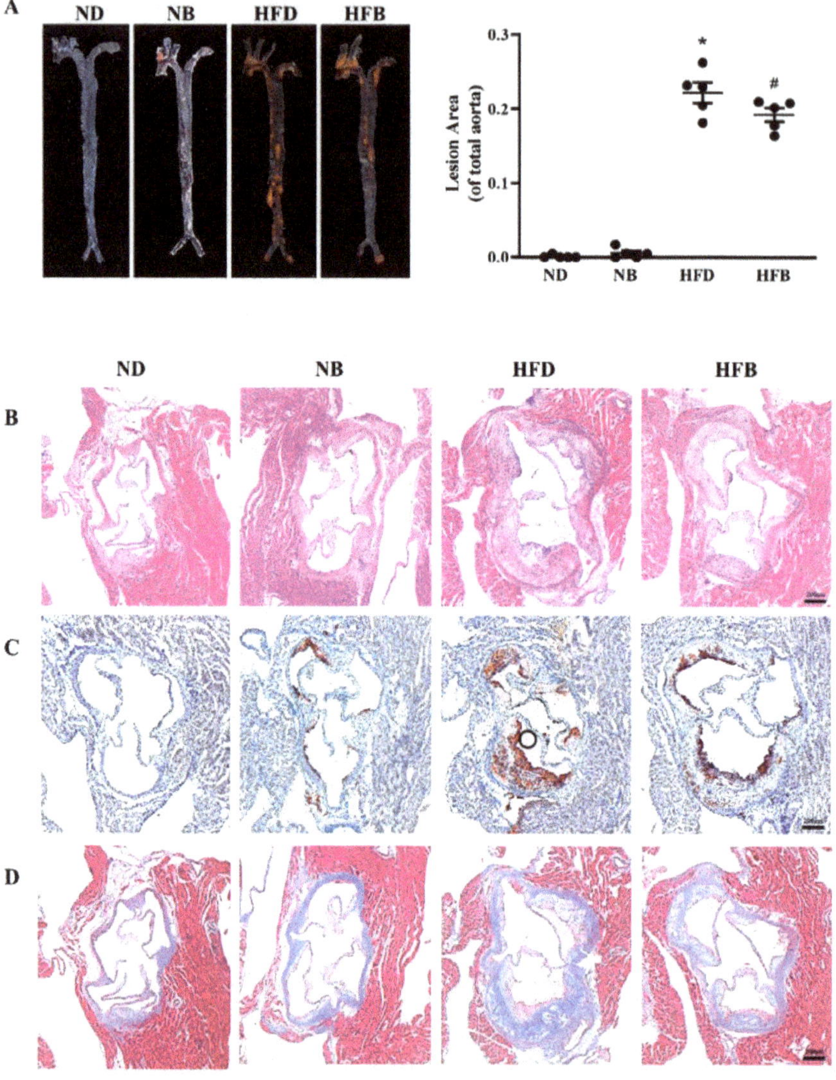

Figure 1. BCAA supplementation reduces atherosclerotic plaques induced by a high-fat diet in ApoE$^{-/-}$ mice. (**A**) Representative images of Oil Red O staining of the aorta (left) and lesion area (right). The lesion area was expressed as a percentage of the total aorta area ($n = 5$ for each group). (**B–D**) Representative images of hematoxylin and eosin staining (**B**), Oil Red O staining (**C**), and Masson staining (**D**) of the aortic root (bar = 200 μm). The data are expressed as the mean ± standard error of the mean (SEM). * $p < 0.05$ vs. ND group; # $p < 0.05$ vs. HFD group.

3.2. BCAA Supplementation Ameliorates Dyslipidemia Induced by a High-Fat Diet in ApoE$^{-/-}$ Mice

The difference in food intake between the ND and NB groups was not significant. The HFD group exhibited greater food intake compared with that of the ND group, and HFB group mice had significantly higher food intake compared with HFD group mice over the first 5 weeks (Figure 2A). However, neither a high-fat diet nor BCAA supplementation resulted in significant changes in body weight (Figure 2B).

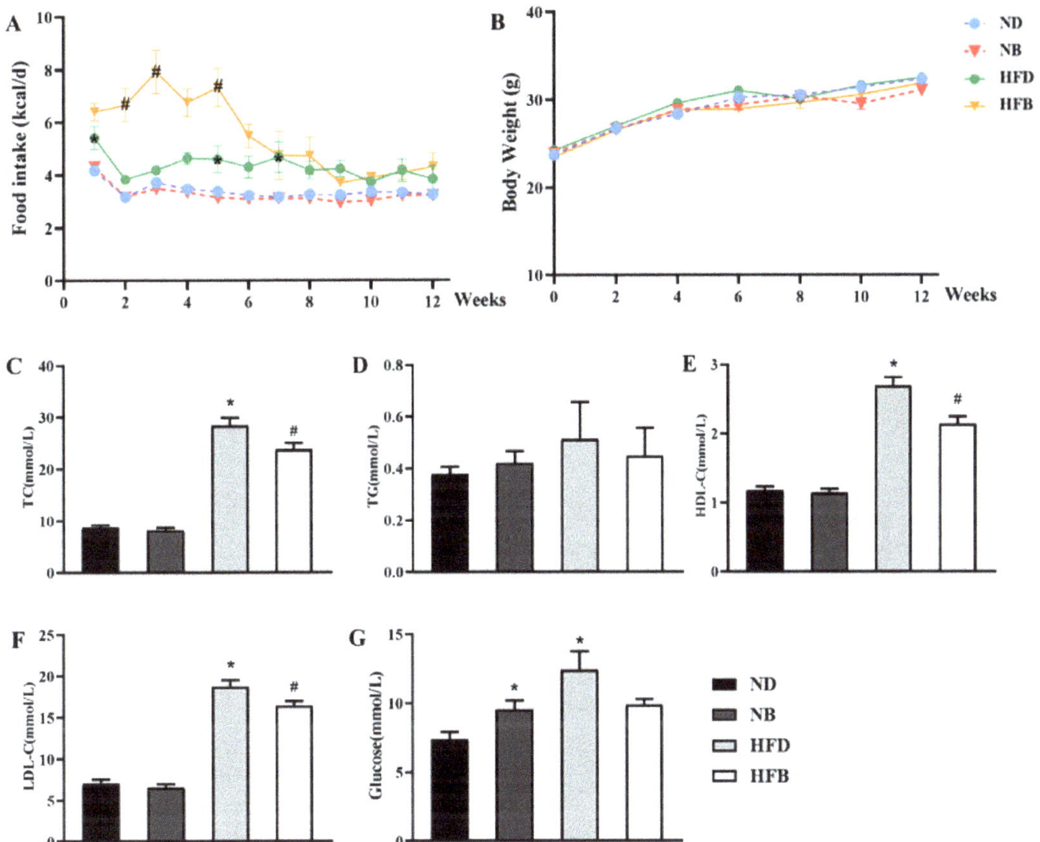

Figure 2. BCAA supplementation ameliorates dyslipidemia induced by a high-fat diet in ApoE$^{-/-}$ mice. (**A,B**) The food intake (**A**) and body weight growth curve (**B**) of ApoE$^{-/-}$ mice (0–12 weeks). The data were analyzed by two-way ANOVA. (**C–G**) Serum concentrations of (**C**) TC, (**D**) TG, (**E**) HDL-C, (**F**) LDL-C, and (**G**) glucose. The data are expressed as the mean $\pm$ SEM ($n = 8$ for each group). * $p < 0.05$ vs. ND group; # $p < 0.05$ vs. HFD group.

As shown in Figure 2C–F, serum TC, LDL-C, and HDL-C levels were significantly increased in the HFD group compared with the ND group. The levels in the HFB group were significantly reduced compared with those in the HFD group. TC, TG, HDL-C, and LDL-C levels did not significantly differ between the NB group and the ND group. This finding suggests that dietary supplementation with BCAAs improves dyslipidemia induced by a high-fat diet in ApoE$^{-/-}$ mice. Additionally, the blood glucose levels of the NB and HFD groups were significantly higher than those in the ND group. Compared with the HFD group, the blood glucose levels in the HFB group decreased, but the difference was not significant (Figure 2G).

3.3. Effects of BCAA Supplementation on Serum Levels of BCAAs and BCAA Metabolic Enzymes in ApoE$^{-/-}$ Mice

As shown in Figure 3A, compared with the ND group, the total levels of the three BCAAs were elevated in the NB group, but the concentrations of Val, Ile, and Leu did not significantly differ between these two groups. Similarly, only the total levels of three BCAAs increased in the HFB group compared with the HFD group. BCAAs cannot be synthesized in mice and must be obtained through dietary supplementation. However,

BCAAs can be catabolized in vivo, so BCAA circulating levels are also affected by BCAA catabolism. BCKDH is the rate-limiting enzyme in BCAA catabolism, and its activity is dependent on the level of phosphorylation of its E1α subunit. The expression levels of hepatic p-BCKDH-E1α and BCKDH-E1α were examined using Western blotting. The p-BCKDH-E1α/BCKDH-E1α ratio did not differ significantly among the four groups (Figure 3B,C).

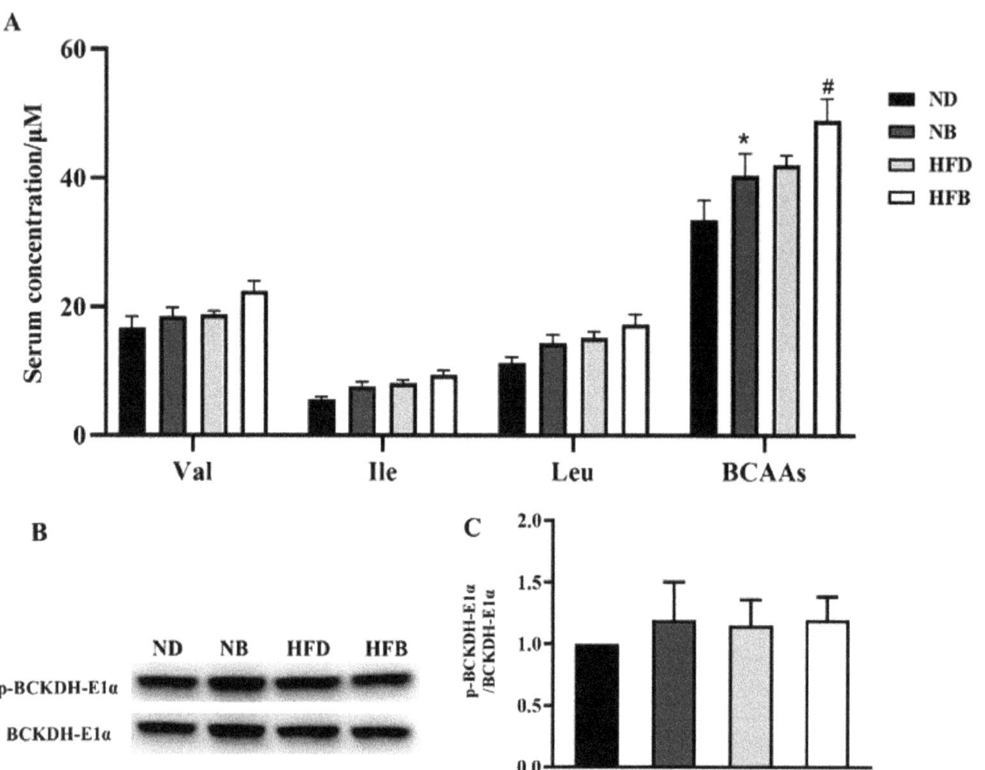

Figure 3. Effects of BCAA supplementation on serum levels of BCAAs and BCAA metabolic enzymes in ApoE$^{-/-}$ mice. (**A**) Serum Val, Ile, Leu and total BCAA concentrations were determined by an LC−MS/MS method. (**B**) Western blotting was used to assess p-BCKDH-E1α and BCKDH-E1α protein expression in the liver. (**C**) The relative quantitative values for p-BCKDH-E1α/BCKDH-E1α are indicated. Densitometry values were normalized to β-actin. The data are expressed as the mean $\pm$ SEM (n = 8 for each group). * $p < 0.05$ vs. ND group; # $p < 0.05$ vs. HFD group.

3.4. BCAA Supplementation Relieves Inflammation Induced by a High-Fat Diet in ApoE$^{-/-}$ Mice

F4/80 immunohistochemistry staining was applied to assess macrophage infiltration in aortic roots (Figure 4A). Macrophage infiltration was increased in the NB and HFD groups compared with the ND group, and was decreased in the HFB group compared with the HFD group. Immunofluorescence staining was performed on aortic root sections to examine the expression of intercellular cell adhesion molecule-1 (ICAM-1) and vascular cell adhesion molecule-1 (VCAM-1), which are important indicators of the inflammatory response (Figure 4B,C). Compared with the ND group, ICAM-1 and VCAM-1 expression was increased at the aortic root in the NB group and markedly elevated in the HFD group. These levels were dramatically downregulated in the HFB group compared with the HFD group.

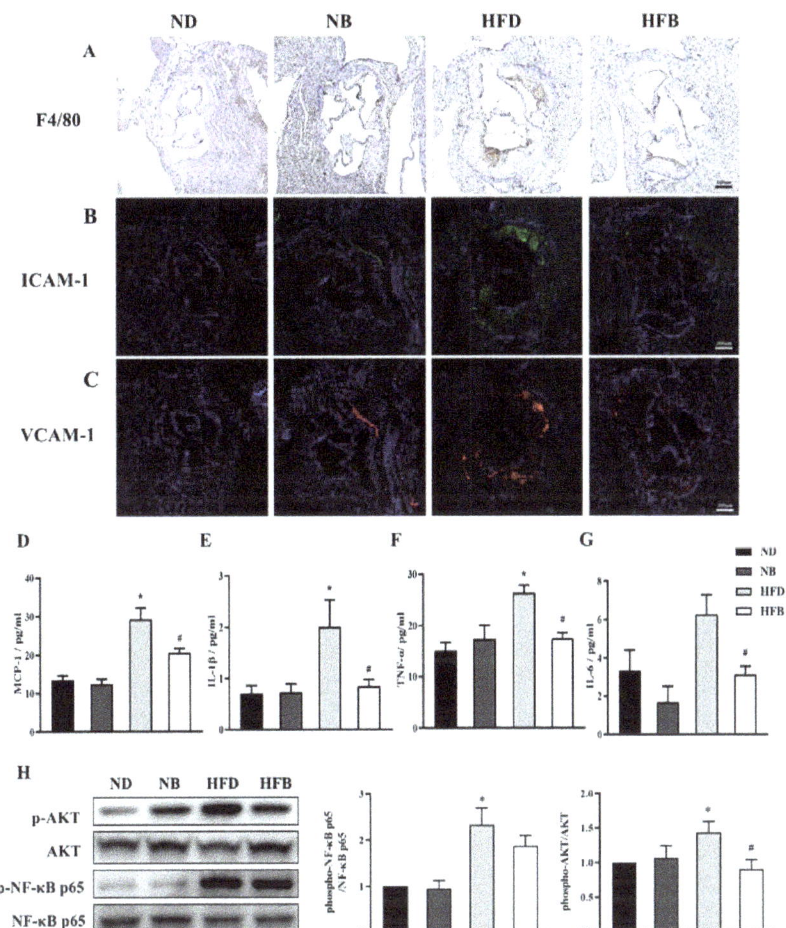

Figure 4. BCAA supplementation relieves inflammation induced by a high-fat diet in ApoE$^{-/-}$ mice. (**A**) Representative F4/80 immunohistochemical sections of aortic root (bar = 200 μm). (**B,C**) ICAM-1 (**B**) and VCAM-1 (**C**) expression at the aortic root was determined by immunofluorescence staining. (**D–G**) The expression of inflammatory cytokines, including MCP-1 (**D**), IL-1β (**E**), TNF-α (**F**), and IL-6 (**G**), was measured using ELISA kits. (**H**) Representative western blot bands of AKT, *p*-AKT, NF-κB, and *p*-NF-κB in the aortic root (left). The relative quantitative data for AKT/*p*-AKT (middle) and NF-κB/*p*-NF-κB (right) are indicated. The data are expressed as the mean ± SEM (n = 8 for each group). * $p < 0.05$ vs. ND group; # $p < 0.05$ vs. HFD group.

In addition, as shown in Figure 4D–G, the levels of MCP-1, IL-1β, and TNF-α were significantly increased in the HFD group compared with the ND group. In addition, the levels of IL-6 were increased, but the result was not statistically significant. MCP-1, IL-1β, TNF-α, and IL-6 levels were significantly decreased in the HFB group compared with the HFD group.

The expression levels of protein kinase B (AKT) and nuclear factor κB (NF-κB), the upstream regulatory molecules of inflammatory factors, were further analyzed, and the data are provided in Figure 4H. The *p*-AKT/AKT ratio was significantly higher in the HFD group than in the ND group. The *p*-AKT/AKT ratio was significantly lower in the HFB group than in the HFD group. The *p*-NF-κB/NF-κB ratio in the NB group did not differ significantly from that in the ND group, whereas it increased significantly in the HFD group. The p-NF-κB/NF-κB ratio in the HFB group tended to be lower than that in the HFD group, but the difference was not statistically significant. These results suggest that dietary BCAA supplementation could alleviate the inflammatory response induced by a high-fat diet by potentially affecting the NF-κB/AKT pathway.

3.5. BCAA Supplementation Alters Intestinal Flora Diversity in ApoE$^{-/-}$ Mice

Here, 16S rRNA gene sequencing was performed to examine and analyze the intestinal bacteria in the cecum and colon, respectively, to investigate the effect of dietary supplementation with BCAAs on the diversity of intestinal flora in ApoE$^{-/-}$ mice (Figure 5). Microbial diversity can be measured by two parameters: alpha diversity and beta diversity. Alpha diversity is generally applied to reveal the abundance and diversity of gut microbiota, with indicators including OTU, ACE, Chao1, Shannon indexes, and Simpson indexes. The alpha-diversity analysis (Figure 5A–E) shows that a high-fat diet increased the richness and diversity of intestinal flora, whereas BCAA supplementation had no significant effect on the alpha diversity of intestinal flora. Beta diversity analysis was employed to analyze the overall structural changes in the gut microbiota. PCoA based on UniFrac distance revealed distinct and separate clustering of microbiota compositions between HFD mice and ND mice, and between HFD mice and HFD supplemented with BCAA, and between ND mice and ND supplemented with BCAA. (Figure 5F,G). Overall, the beta diversity of intestinal flora in the cecum and colon of ApoE$^{-/-}$ mice was changed by BCAA supplementation in both the normal and high-fat diets, and this alteration was more obvious when mice were fed a high-fat diet.

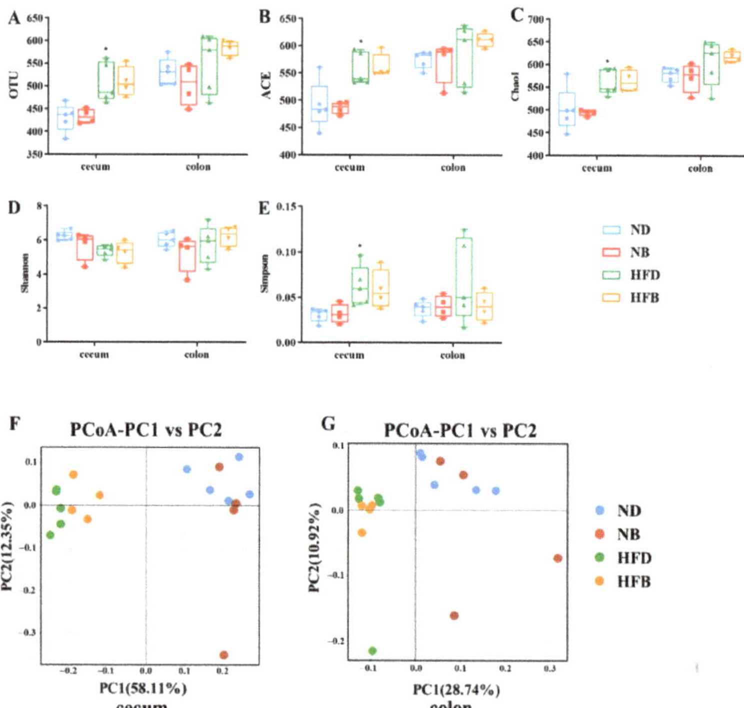

Figure 5. BCAA supplementation alters gut microbiota diversity in ApoE$^{-/-}$ mice. (**A–E**) Alpha diversity of gut microbiota in the cecum and colon was assessed by OTU (**A**), ACE index (**B**), Chao1 index (**C**), Shannon index (**D**), Simpson index (**E**). The data are presented as the mean ± SEM (n = 4–5 for each group). (**F**) PCoA based on UniFrac distance of gut microbiota in the cecum (**F**) and colon (**G**), n = 4−5 for each group. * $p < 0.05$ vs. ND group.

3.6. BCAA Supplementation Alters Gut Microbiota Composition in ApoE$^{-/-}$ Mice

After ApoE$^{-/-}$ mice were fed for 12 weeks, the effects of dietary supplementation with BCAAs on gut microbiota composition were investigated. The results of the heatmap and clustering analysis based on the relative abundance of intestinal flora are shown in Figure 6. The graphs were based on the relative abundance of bacteria in the cecum, as shown in Figure 6A. Compared with the ND group, the relative abundances of *Papillibacter*, *Prevotellaceae_NK3B31_group*, *Family_XIII_UCG-001* and *Ileibacterium* were upregulated, whereas the relative abundances of *Alistipes* and *Marvinbryantia* were downregulated in the NB group. Compared with the ND group, the relative abundances of *GCA-900066225*, *Ruminiclostridium_5* and *uncultured_bacterium_f_Peptococcaceae* at the genus level were significantly upregulated in the HFD group, and the relative abundances of *Papillibacter*, *Prevotellaceae_NK3B31_group* and *Family_XIII_UCG-001* were significantly downregulated at the genus level. The relative abundances of *Ruminiclostridium_5*, *Faecalibaculum* and *uncultured_bacterium_f_ Lachnospiraceae* were upregulated in the HFB group compared to the HFD group.

Heatmaps based on the relative abundance of colonic bacteria in mice are presented in Figure 6B. The relative abundances of the 24 different bacterial groups at the genus level revealed differences in the abundance and composition of the intestinal flora in each group. The relative abundance of *Prevotellaceae_NK3B31_group* was significantly higher in the NB group than in the ND group. Compared to the ND group, the relative abundances of *Prevotellaceae_UCG-001*, *Muribaculum*, *[Eubacterium]_xylanophilum_group* and *uncultured_bacterium_f_Muribaculaceae* were significantly lower, whereas *Lactococcus*, *Photobacterium*, *uncultured_bacterium_f_Mycoplasmataceae*, *Bifidobacterium* and *Leuconostoc* had significantly higher relative abundances in the HFD group. In addition, 10 genus levels of *Succinivibrio*, *Ochrobactrum*, and *Acidipila* also exhibited elevated intestinal flora abundances in the HFD group. Compared with the HFD group, the relative abundance of *Lactococcus*, *Photobacterium*, *uncultured_bacterium_f_Mycoplasmataceae* and *Bifidobacterium* decreased in the HFB group, and that of *uncultured_bacterium_f_ Gemmatimonadaceae*, *uncultured_bacterium_f_Ilumatobacteraceae* and *Succinivibrio* showed significantly higher levels of 14 genera in the intestinal flora.

To further determine the changes in flora among different groups, the flora with significantly altered relative abundances were analyzed, and the statistical results are shown in Figure 7. In the cecum, the relative abundances of *Desulfovibrio*, *Faecalibaculum*, *Prevotella_9*, *Ruminiclostridium_5*, *uncultured_bacterium_f_Lachnospiraceae* and *Weissella* were significantly higher in the HFD group compared with the ND group, whereas the relative abundances of *Ileibacterium*, *Lachnospiraceae_NK4A136_group* and *Muribaculum* were significantly lower than that in the ND group. The relative abundance of *Desulfovibrio* in the HFB group was significantly higher than that in the HFD group, whereas the relative abundance of *Faecalibaculum* was significantly lower than that in the HFD group. In addition, the relative abundances of *Lactococcus* and *Photobacterium* in the HFD group in the colon were significantly higher than that in the ND group, and the relative abundances noted in the HFB group were significantly lower than that in the HFD group. The relative abundance of *Prevotellaceae_UCG-003* in the HFB group was significantly higher than that in the HFD group.

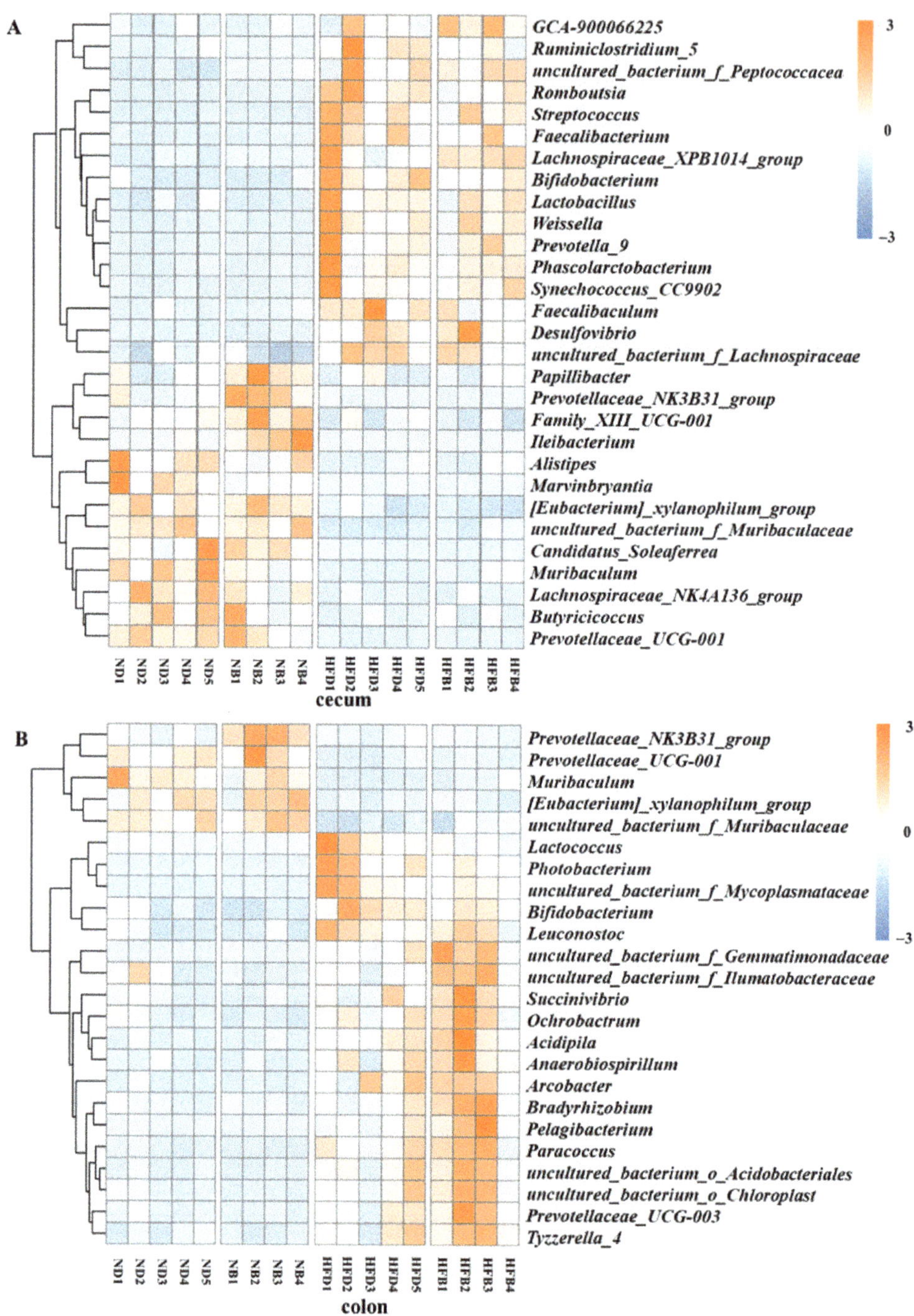

Figure 6. BCAA supplementation alters gut microbiota composition in ApoE$^{-/-}$ mice. Heatmap of gut microbiota composition at the genus level for each group in the cecum (**A**) and colon (**B**), *n* = 4–5 per group. The scale bar indicates the standardized Z value of the microbial relative abundance. Red and blue colors indicate higher and lower average expression, respectively.

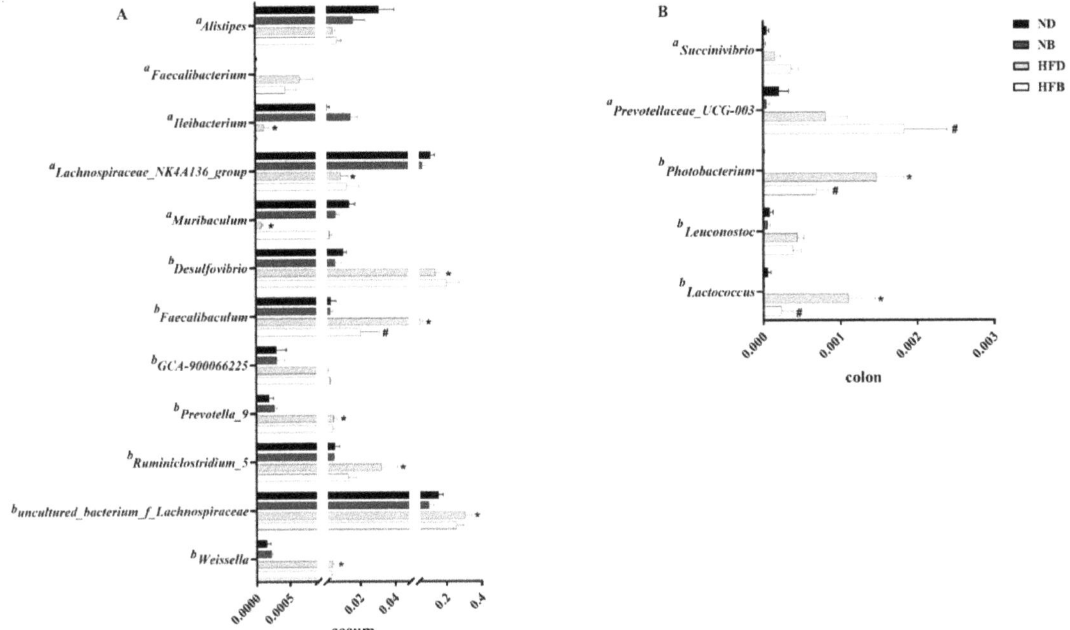

Figure 7. BCAA supplementation alters gut microbiota relative abundance in ApoE$^{-/-}$ mice. The relative abundance of gut microbiota at the genus level in the cecum (**A**) and colon (**B**) The data are presented as the mean ± SEM (*n* = 4–5 for each group). Letters "a" and "b" indicate possible probiotics and pathogenic bacteria previously reported, respectively. * $p < 0.05$ vs. ND group; # $p < 0.05$ vs. HFD group.

3.7. BCAA Supplementation Promotes Bile Acid Excretion in ApoE$^{-/-}$ Mice

Dyslipidemia is an important risk factor for AS, and cholesterol can be metabolized to BAs in the liver, representing the main pathway of cholesterol metabolism. The concentrations of 16 types of BAs in the intestinal contents were assayed using LC−MS/MS. As shown in Figure 8, the levels of total BAs, 12α-OH BAs, non-12α-OH BAs, primary BAs, and secondary BAs and the ratio of 12α-OH/non-12α-OH BAs were significantly increased in the HFD group compared with the ND group. We also found that BCAA supplementation in the high-fat diet markedly increased the levels of total BAs, 12α-OH BAs, non-12α-OH BAs, primary BAs, and the ratio of primary/secondary BAs in the intestinal contents, but no significant differences in secondary BAs and the ratio of 12α-OH/non-12α-OH BAs were observed between the HFB and HFD groups. These results suggest that dietary supplementation with BCAAs may facilitate cholesterol catabolism by promoting BA excretion.

3.8. Correlation of Intestinal Flora with Blood Glucose, Lipid Indices and Bile Acids

The above results showed that dietary supplementation with BCAAs altered lipid profiles, BA levels, and intestinal flora diversity in ApoE$^{-/-}$ mice; thus, the correlation of intestinal flora with blood glucose, lipid indices, and BA levels in mice was further analyzed. Spearman correlation analysis (Figure 9) revealed that the relative abundances of *Eubacterium xylanophilum group*, *Prevotellaceae*, *Lachnospiraceae NK4A136 group*, and *Muribaculum* were significantly negatively correlated with HDL-C, LDL-C, and TC levels. The relative abundances of *Desulfovibrio* and *Arcobacter* were significantly and positively correlated with GLU, HDL, LDL, and TC levels. The relative abundance of the intestinal flora of 5 genera, including *uncultured_bacterium_f_Gemmatimonadaceae*, was positively correlated with HDL-C, LDL-C, and TC levels. The relative abundance of intestinal flora of 10 genera, including *Streptococcus,* was significantly and positively correlated with LDL and TC levels. The relative abundance of *Prevotellaceae_UCG-003* was significantly and positively correlated with HDL-C and LDL-C levels. In addition, the heatmap correlation analysis (Figure 9B) showed that the relative abundance of intestinal flora of 8 genera, including *Muribaculum,* was significantly and negatively correlated with the levels of multiple BAs, while the relative abundance of intestinal flora of 17 genera, including *Romboutsia,* was significantly positively correlated with the levels of several BAs.

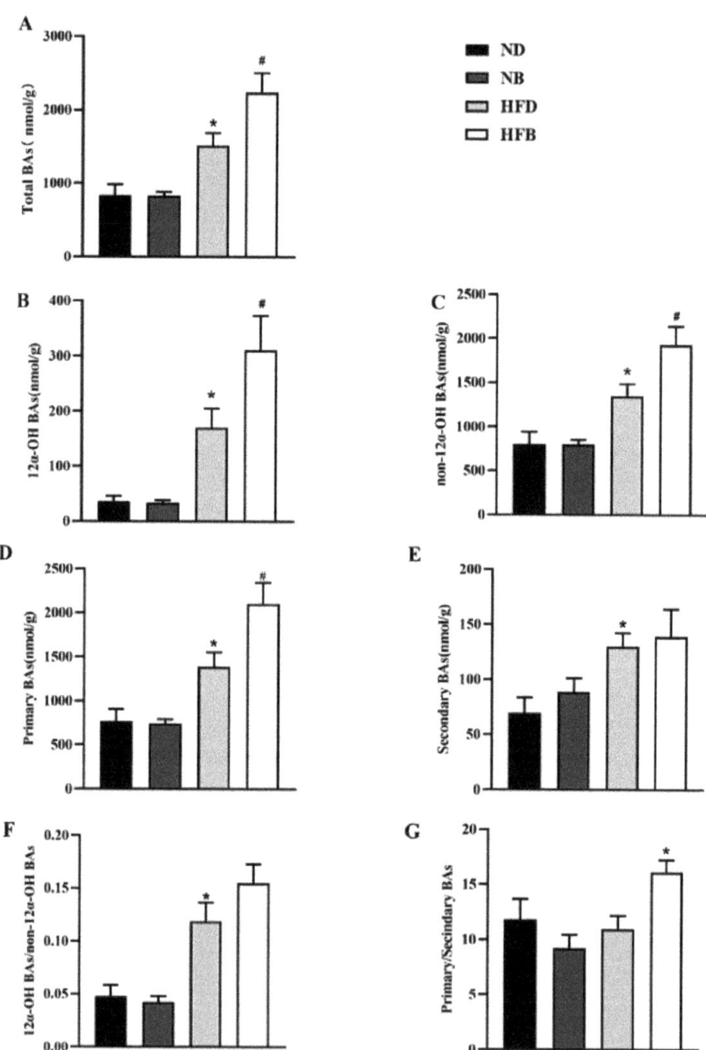

Figure 8. BCAA supplementation alters the bile acid composition in intestinal contents in ApoE$^{-/-}$ mice. Sixteen molecular species of BAs were determined in cecal and colonic contents by an LC−MS/MS method. (**A**) Total bile acid concentrations. (**B**) 12α-OH BA concentrations. (**C**) Non-12α-OH BA concentrations. (**D**) Primary bile acid concentrations. (**E**) Secondary bile acid concentrations. (**F,G**) The ratios of 12α-OH BAs to non-12α-OH BAs (**F**) and ratios of primary BAs to secondary BAs (**G**). The data are presented as the mean ± SEM (n = 7–8 for each group). * $p < 0.05$ vs. ND group; # $p < 0.05$ vs. HFD group.

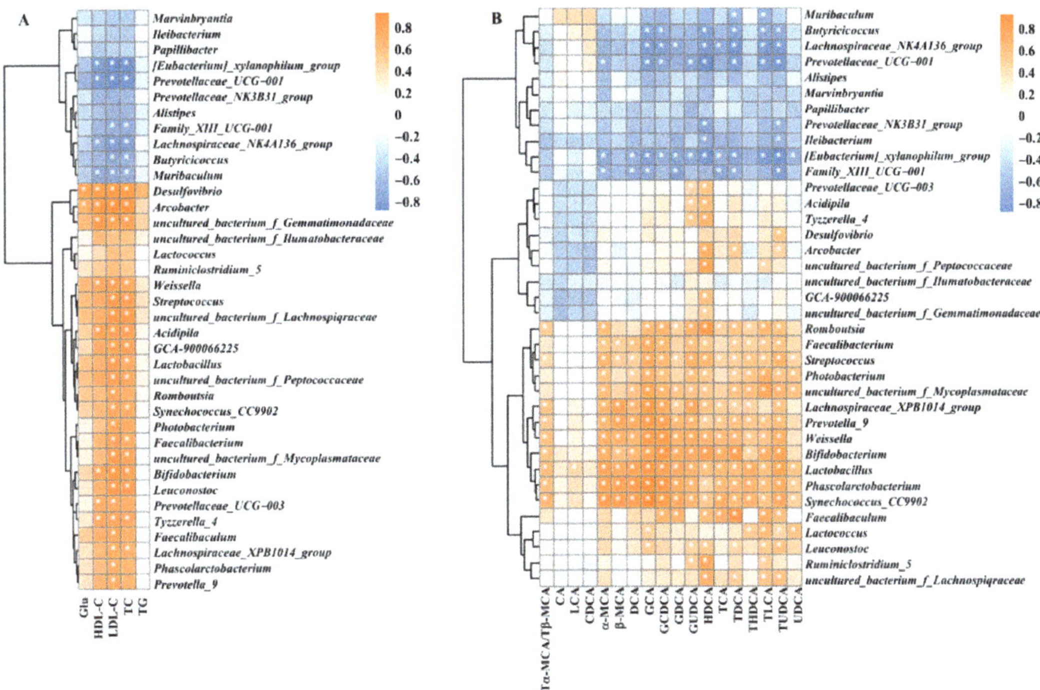

Figure 9. Spearman rank correlation analysis. (**A**) Spearman rank correlation analyses between intestinal microbiota and serum biochemical parameters, *n* = 4–5 per group. (**B**) Spearman rank correlation analyses between intestinal microbiota and bile acids in the cecum and colon, *n* = 4–5 per group. Red indicates positive correlation coefficients, and blue indicates negative correlation coefficients. Significant correlations are indicated by * $p < 0.05$.

4. Discussion

AS is a common pathophysiological foundation for several types of cardiovascular and cerebrovascular diseases that involves complicated interactions among various factors. The pathogenesis of AS is sophisticated but not completely elucidated. In the present study, we investigated the effect of BCAA supplementation on AS and its mechanism in ApoE$^{-/-}$ mice fed with a high-fat diet. BCAA supplementation substantially reduced the formation of aortic plaques, ameliorated dyslipidemia by enhancing the excretion of BAs, and alleviated the inflammatory response by potentially affecting the inflammatory related pathway. In addition, BCAA supplementation altered the gut bacterial beta diversity and gut microbiota composition and abundance, especially reducing harmful bacterial abundance and increasing probiotic abundance.

Atherosclerotic plaque formation is the most significant lesion characteristic of AS, and the plaque area and size directly indicate the extent of atherosclerotic lesions. In this study, high-fat feeding ApoE$^{-/-}$ mice had massive plaques on the interior of the aorta, suggesting that the AS model had been successfully established. The lesion area dramatically decreased in the HFB group, indicating that BCAA dietary supplementation could attenuate AS induced by a high-fat diet. Furthermore, HE and Oil Red O-stained aortic root slices revealed that BCAA supplementation reduced the atherosclerotic plaque area. Zhao et al. also found that adding leucine to drinking water also significantly reduced the size of aortic atherosclerotic lesions in ApoE$^{-/-}$ mice [30]. Plaque stability is another important indicator when evaluating the risk of plaque formation. It has been demonstrated that an elevated collagen fiber content plays an important role in enhancing plaque stability [31]. Masson staining results from this study revealed a decrease in collagen content in the HFB group compared to the HFD group. Considering that except for collagen fibers, plaque stability is also affected by factors such as lipids, macrophages, and smooth muscle cells [32–34], further experiments are required to verify the effect of BCAAs on plaque stability.

Dyslipidemia is widely recognized as an important risk factor for AS. In vivo, LDL-C is the vehicle for cholesterol metabolic transport, and excessive LDL-C deposition can cause subintimal migration and aggregation, damaging the vascular endothelium, increasing the number of foam cells, and ultimately resulting in AS [35,36]. A cross-sectional study showed that plasma BCAA levels are positively correlated with LDL-C levels, and increased BCAA levels are also positively associated with metabolic dyslipidemia [37]. Oren Rom et al. demonstrated that leucine reducedTG content in macrophages by inhibiting VLDL uptake in vitro, thereby inhibiting lipid accumulation and macrophage foam cell formation [38]. Zhao et al. demonstrated that adding BCAAs (leucine) to drinking water improved the lipid profile by promoting ATP-binding cassette transporter G5 (ABCG5)/ABCG 8-mediated hepatic cholesterol efflux in ApoE$^{-/-}$ mice [30]. Similarly, in this study we revealed that dietary BCAA supplementation alleviates high-fat diet-induced dyslipidemia by reducing serum TC, LDL-C, and HDL-C levels in AS model mice. These studies indicate that dietary supplementation with BCAAs has a beneficial effect on the lipid profile, which can help prevent AS.

AS is a chronic inflammatory disease, and the inflammatory response is involved in the development of AS [8,9]. The AKT/NF-κB signaling pathway is a crucial regulatory pathway of the inflammatory response. AKT performs a crucial function in cellular metabolism and the inflammatory response by regulating the activity of many target proteins [39]. The NF-κB pathway is regulated by phosphorylated AKT. NF-κB is a key transcriptional regulator of most inflammatory genes and is involved in the inflammatory response by coordinating the expression of multiple genes that control inflammation, tissue injury, and immune response. Several studies have shown that the NF-κB signaling pathway plays an important role in AS [40–42]. This study showed that dietary supplementation with BCAAs inhibited the activation of AKT and NF-κB.

Inflammatory factors, including IL-6, TNF-α, and IL-1β, are released from the organism when NF-κB is activated [43]. IL-6 is an important upstream inflammatory biomarker that can promote the development of AS by driving vascular smooth muscle cell (VSMC) migration and inducing intracellular cholesterol accumulation [44,45]. TNF-α is an essential cytokine that is primarily released by T cells and activated monocyte macrophages. TNF-α promotes atherosclerosis by facilitating platelet aggregation and enhancing LDL transport in endothelial cells [46]. IL-1β is an inflammatory cytokine that enhances the expression of numerous proinflammatory cytokines, and monoclonal antibodies that target IL-1β may protect against cardiovascular disease [47–49]. In addition, the increased IL-1β level was positively correlated with the extent of coronary AS [50]. Additionally, MCP-1, a member of the CC chemokine family, is a strong chemokine for monocytes, controlling the migration and infiltration of monocytes and macrophages. Thus, MCP-1 plays a significant role in the development of AS. It was also demonstrated that MCP-1 could increase macrophage numbers and promote oxidized lipids to aggravate AS in ApoE$^{-/-}$ mice [51]. In the current study, serum MCP-1, IL-6, TNF-α, and IL-1β levels were significantly elevated in ApoE$^{-/-}$ mice fed with a high-fat diet, whereas dietary supplementation with BCAAs dramatically decreased the serum levels of these inflammatory factors. Several previous animal studies have similarly demonstrated that supplementation with Leu dramatically reduced serum levels of inflammatory factors, including IL-6, TNF-α, MCP-1, and leptin in mice [30,52], decreased macrophage infiltration at the aorta, and suppressed systemic inflammatory responses to ameliorate AS [38]. Numerous studies have reported that ICAM-1 and VCAM-1 enhance monocyte and endothelial cell adhesion, which promotes AS [53,54]. Aortic root section staining confirmed that dietary supplementation with BCAAs reduced the inflammatory response at aortic lesions. However, it has also been shown that ingestion of BCAAs enhances human platelet activity and promotes arterial thrombosis in mice [55].

Activated platelets initiate inflammatory responses and participate in AS through multiple pathways. Results that are inconsistent with the above studies may be observed because the effects of dietary supplementation with BCAAs on AS are influenced by the ratio and dose of BCAAs and the approach of administration. Further studies are needed to investigate these differences in the future.

Branched chain amino acid transaminases (BCATs) and particularly Branched chain alpha-ketoacid dehydrogenase (BCKDH) (the rate-limiting enzyme in BCAA catabolism), which are essential for maintaining BCAA homeostasis, regulate BCAA catabolism in vivo. BCKDH-E1α is primarily responsible for BCKDH activity, which is regulated via phosphorylation and dephosphorylation [56]. BCKD kinase-mediated phosphorylation of BCKDH-E1α can inactivate BCKDH, increasing BCAA levels in serum. Monitoring BCAA levels can serve as an indirect indicator of the catabolism of BCAAs, which is mostly based on dietary supplementation and catabolism of BCAAs in vivo. Therefore, serum BCAA levels and liver p-BCKDH-E1α and BCKDH-E1α expression levels were examined in this study to analyze BCAA metabolism. Our previous study showed that the enzymatic activity of BCKDH was impaired by HFD feeding and that dietary supplementation with BCAAs boosted the enzymatic activity and counteracted the accumulation of BCAAs in circulation [28]. However, in this study, total serum BCAA levels were significantly increased in the dietary BCAA supplementation group, but no significant differences in the levels of each BCAA or BCKDH-E1α enzyme activity was noted among the four groups, suggesting that the increased serum BCAA levels in ApoE$^{-/-}$ mice may primarily be attributable to dietary BCAAs supplementation. Considering that BCAA homeostasis in vivo is related to the balance between BCAA intake and BCAA metabolism, our studies indicated that BCAA supplementation does not have a dramatic effect on this balance, which is consistent with a previous study in which dietary intake of BCAA was only moderately positively correlated with circulating BCAA levels in humans [57]. Thus, unlike the elevated circulation of BCAA levels being risk factors for AS, BCAA supplementation may have beneficial effects on AS, since BCAA supplementation does not necessarily lead to the impairment of BCAA catabolism and the accumulation of BCAA in circulation and then cause diseases. In the future, whether BCAA can be used as a supplement for disease prevention and treatment in humans needs to be confirmed by randomized clinical trials.

Recently, a growing number of studies have shown that gut microbiota dysbiosis plays an important role in many diseases, including AS. Dysbiosis can promote AS development by enhancing the systemic inflammatory response. Furthermore, during AS progression, some bacteria can promote AS development, whereas others can inhibit AS formation. *GCA-900066225* is associated with abnormal lipid metabolism, and *Prevotella_9*, *Weissella*, *Leuconostoc* and *Photobacterium* are opportunistic bacteria [58,59]. In the present study, a high-fat diet increased the relative abundance of these harmful bacteria. *Alistipes* plays a protective role in cardiovascular disease, and *Ileibacterium* and *Muribaculum* are associated with attenuating the inflammatory response [60–62]. *Desulfovibrio*, also known as sulfate-reducing bacteria, metabolizes and produces hydrogen sulfide, which is harmful to the intestinal epithelium and impairs the mucosal barrier [63]. Similar to previous studies, a high-fat diet dramatically increased the relative abundance of *Desulfovibrio*. We found that a high-fat diet reduced the relative abundance of these beneficial bacteria. In summary, a high-fat diet may exacerbate AS by increasing the abundance of harmful bacteria while reducing the abundance of beneficial bacteria.

In addition, dietary supplementation with BCAAs may have protective benefits for AS by reducing the abundance of harmful bacteria and increasing the abundance of probiotics. These effects may be beneficial for intestinal health. Dietary supplementation with BCAAs decreased the relative abundance of *Faecalibaculum* and *Fusobacterium*, two genera of opportunistic pathogenic bacteria that boost inflammatory responses and impair the intestinal barrier. Duan et al. [64] reported that overexpression of *Prevotellaceae_UCG-003* suppresses the inflammatory response, and dietary supplementation with BCAAs significantly increased its relative abundance in the current study. The abundance of *Succinivibrio* in patients with metabolic disorders was much lower than that in healthy patients [65], suggesting that *Succinivibrio* is a potential metabolic probiotic and that dietary supplementation with BCAAs increased its relative abundance.

Cholesterol is converted to BAs to maintain plasma cholesterol levels. Therefore, the upregulation of BA biosynthesis and the excretion of BAs are closely related to low serum cholesterol levels [66]. Cholesterol is the primary factor used in the biosynthesis of BAs, and two main pathways are involved. The classical pathway produces mainly 12α-OH BAs, whereas the alternative pathway produces mainly non-12α-OH BAs. A low proportion of non-12α-OH BAs has been shown to have beneficial metabolic effects [67–69]. In addition, primary BAs play critical roles in cholesterol metabolism, lipid digestion, and host-microbe interactions, and can be converted to secondary BAs by the intestinal flora. High-fat diets increase some secondary BAs, which are risk factors for colonic inflammation and cancer [70]. Our results similarly showed that a high-fat diet caused an increase in secondary BAs and a decrease in the proportion of non-12α-OH BAs, indicating that a high-fat diet has a negative effect on ApoE$^{-/-}$ mouse metabolism. In addition, the levels of total BAs, primary BAs, 12α-OH and non-12α-OH BAs in cecal and colonic contents were all increased in the HFB group compared with the HFD group. This finding indicates that dietary supplementation with BCAAs may enhance the excretion of total BAs, mainly increasing primary BAs, which can affect lipid metabolism and exert a metabolic benefit.

Correlation analysis revealed that lipid-related parameters were significantly negatively associated with the relative abundance of *Family_XIII_UCG-001* and slightly negatively associated with the relative abundance of *Ileibacterium* and *Papillibacter*. In contrast, dietary supplementation with BCAAs significantly increased the relative abundance of the three bacteria. In addition, serum lipids showed a significant positive correlation with the relative abundance of *Photobacterium* and a mild positive correlation with the relative abundance of *Lactococcus*. In contrast, dietary supplementation with BCAAs significantly reduced the relative abundance of both. It has been reported that *Bacteroides* is associated with intestinal catabolism of BCAAs, whereas *Streptococcus* is associated with intestinal biosynthesis of BCAAs [71,72]. However, a correlation between *Bacteroides* and *Streptococcus* and serum BCAA levels was not observed in this study, which may be due to the use of different animal models and individual differences in animals. We also found that BA levels were correlated with probiotic bacteria, such as *Bifidobacterium*, *Lactobacillus*, *Phascolarctobacterium*, and *Prevotellaceae_UCG-003*, consistent with previous reports [73–76]. Overall, dietary supplementation with BCAAs may promote BA metabolism by upregulating the relative abundance of probiotics, such as *Prevotellaceae_UCG-003*.

5. Conclusions

In conclusion, this study demonstrates that dietary supplementation with BCAAs may not only alleviate AS by suppressing the inflammatory response, but may also regulate bile acid excretion by altering intestinal flora in ApoE-deficient mice. This research supported the beneficial effects of BCAAs on inflammation and elucidated their potential mechanisms, providing different perspectives on the molecular mechanisms of BCAA supplementation to alleviate AS.

Author Contributions: Conceptualization, Z.L., R.Z., H.M., J.D. and R.Y.; Data curation, Z.L. and R.Z.; Formal analysis, Z.L. and R.Z.; Funding acquisition, H.M., J.D. and R.Y.; Investigation, Z.L., R.Z., H.M., W.Z., J.Z., H.L. and X.Z.; Methodology, H.M., W.Z., J.Z. and S.W.; Project administration, Z.L, R.Z. and R.Y.; Resources, W.Z., J.Z., H.L. and S.W.; Supervision, W.C., J.D. and R.Y.; Validation, Z.L. and R.Z.; Visualization, Z.L., R.Z. and X.Z.; Writing—original draft, Z.L. and R.Z.; Writing—review & editing, H.M. and R.Y. All authors have read and agreed to the published version of the manuscript.

Funding: This research was funded by the National Natural Science Foundation of China (grant number 81672075), the Beijing Natural Science Foundation (grant number 7222156 and 7214250), the CAMS Innovation Fund for Medical Sciences (grant number 2021-I2M-1-050) and the National Key R&D Program of China (grant number 2021YFE0114200).

Institutional Review Board Statement: The study was conducted according to the guidelines of the Declaration of Helsinki, and approved by the Peking University Biomedical Ethics Committee Experimental Animal Ethics Branch in 11 January 2022 (Approval No. LA2022010).

Informed Consent Statement: Not applicable.

Data Availability Statement: The data that support the findings of this study are available from the corresponding author upon reasonable request.

Conflicts of Interest: The authors declare no conflict of interest.

References

1. Libby, P.; Buring, J.E.; Badimon, L.; Hansson, G.K.; Deanfield, J.; Bittencourt, M.S.; Tokgözoğlu, L.; Lewis, E.F. Atherosclerosis. *Nat. Rev. Dis. Prim.* **2019**, *5*, 56. [CrossRef] [PubMed]
2. Kobiyama, K.; Ley, K. Atherosclerosis: A chronic inflammatory disease with an autoimmune component. *Circ. Res.* **2018**, *123*, 1118. [CrossRef] [PubMed]
3. Powell-Wiley, T.M.; Poirier, P.; Burke, L.E.; Després, J.-P.; Gordon-Larsen, P.; Lavie, C.J.; Lear, S.A.; Ndumele, C.E.; Neeland, I.J.; Sanders, P.; et al. Obesity and Cardiovascular Disease: A Scientific Statement From the American Heart Association. *Circulation* **2021**, *143*, e984–e1010. [CrossRef]
4. Andersson, C.; Vasan, R.S. Epidemiology of cardiovascular disease in young individuals. *Nat. Rev. Cardiol.* **2018**, *15*, 230–240. [CrossRef]
5. Benjamin, E.J.; Blaha, M.J.; Chiuve, S.E.; Cushman, M.; Das, S.R.; Deo, R.; De Ferranti, S.D.; Floyd, J.; Fornage, M.; Gillespie, C. Heart disease and stroke statistics-2016 update a report from the American Heart Association. *Circulation* **2017**, *133*, e38–e48.
6. Miller, Y.I.; Choi, S.-H.; Wiesner, P.; Fang, L.; Harkewicz, R.; Hartvigsen, K.; Boullier, A.; Gonen, A.; Diehl, C.J.; Que, X.; et al. Oxidation-Specific Epitopes Are Danger-Associated Molecular Patterns Recognized by Pattern Recognition Receptors of Innate Immunity. *Circ. Res.* **2011**, *108*, 235–248. [CrossRef] [PubMed]
7. Navab, M.; Ananthramaiah, G.M.; Reddy, S.T.; Van Lenten, B.J.; Ansell, B.J.; Fonarow, G.C.; Vahabzadeh, K.; Hama, S.; Hough, G.; Kamranpour, N.; et al. Thematic review series: The Pathogenesis of Atherosclerosis The oxidation hypothesis of atherogenesis: The role of oxidized phospholipids and HDL. *J. Lipid Res.* **2004**, *45*, 993–1007. [CrossRef] [PubMed]
8. Wolf, D.; Ley, K. Immunity and Inflammation in Atherosclerosis. *Circ. Res.* **2019**, *124*, 315–327. [CrossRef]
9. Zhu, Y.; Xian, X.; Wang, Z.; Bi, Y.; Chen, Q.; Han, X.; Tang, D.; Chen, R. Research Progress on the Relationship between Atherosclerosis and Inflammation. *Biomolecules* **2018**, *8*, 80. [CrossRef]
10. Geovanini, G.R.; Libby, P. Atherosclerosis and inflammation: Overview and updates. *Clin. Sci.* **2018**, *132*, 1243–1252. [CrossRef]
11. Ranjit, N.; Diez-Roux, A.V.; Shea, S.; Cushman, M.; Seeman, T.; Jackson, S.A.; Ni, H. Psychosocial Factors and Inflammation in the Multi-Ethnic Study of Atherosclerosis. *Arch. Intern. Med.* **2007**, *167*, 174–181. [CrossRef] [PubMed]
12. Liu, Y.; Yu, H.; Zhang, Y.; Zhao, Y. TLRs are important inflammatory factors in atherosclerosis and may be a therapeutic target. *Med. Hypotheses* **2008**, *70*, 314–316. [CrossRef]
13. Fan, Y.; Pedersen, O. Gut microbiota in human metabolic health and disease. *Nat. Rev. Microbiol.* **2021**, *19*, 55. [CrossRef] [PubMed]
14. Witkowski, M.; Weeks, T.L.; Hazen, S.L. Gut Microbiota and Cardiovascular Disease. *Circ. Res.* **2020**, *127*, 553–570. [CrossRef] [PubMed]
15. Sanchez-Rodriguez, E.; Egea-Zorrilla, A.; Plaza-Díaz, J.; Aragón-Vela, J.; Muñoz-Quezada, S.; Tercedor-Sánchez, L.; Abadia-Molina, F. The Gut Microbiota and Its Implication in the Development of Atherosclerosis and Related Cardiovascular Diseases. *Nutrients* **2020**, *12*, 605. [CrossRef]
16. Zhu, L.; Zhang, D.; Zhu, H.; Zhu, J.; Weng, S.; Dong, L.; Liu, T.; Hu, Y.; Shen, S. Berberine treatment increases Akkermansia in the gut and improves high-fat diet-induced atherosclerosis in Apoe$^{-/-}$ mice. *Atherosclerosis* **2018**, *268*, 117. [CrossRef]
17. Gregory, J.C.; Buffa, J.A.; Org, E.; Wang, Z.; Levison, B.S.; Zhu, W.; Wagner, M.A.; Bennett, B.J.; Li, L.; DiDonato, J.A.; et al. Transmission of Atherosclerosis Susceptibility with Gut Microbial Transplantation. *J. Biol. Chem.* **2015**, *290*, 5647–5660. [CrossRef]
18. Wang, P.-X.; Deng, X.-R.; Zhang, C.-H.; Yuan, H.-J. Gut microbiota and metabolic syndrome. *Chin. Med. J.* **2020**, *133*, 808–816. [CrossRef]
19. Molinaro, A.; Wahlström, A.; Marschall, H.U. Role of Bile Acids in Metabolic Control. *Trends Endocrinol. Metab. TEM* **2018**, *29*, 31. [CrossRef]
20. Wahlström, A.; Sayin, S.I.; Marschall, H.-U.; Bäckhed, F. Intestinal Crosstalk between Bile Acids and Microbiota and Its Impact on Host Metabolism. *Cell Metab.* **2016**, *24*, 41–50. [CrossRef]
21. Fuchs, C.D.; Trauner, M. Role of bile acids and their receptors in gastrointestinal and hepatic pathophysiology. *Nat. Rev. Gastroenterol. Hepatol.* **2022**, *19*, 432–450. [CrossRef] [PubMed]
22. Fiorucci, S.; Distrutti, E. Bile Acid-Activated Receptors, Intestinal Microbiota, and the Treatment of Metabolic Disorders. *Trends Mol. Med.* **2015**, *21*, 702. [CrossRef] [PubMed]
23. Hu, W.; Liu, Z.; Yu, W.; Wen, S.; Wang, X.; Qi, X.; Hao, H.; Lu, Y.; Li, J.; Li, S.; et al. Effects of *PPM1K* rs1440581 and rs7678928 on serum branched-chain amino acid levels and risk of cardiovascular disease. *Ann. Med.* **2021**, *53*, 1317–1327. [CrossRef] [PubMed]

24. Tobias, D.K.; Lawler, P.R.; Harada, P.H.; Demler, O.V.; Ridker, P.M.; Manson, J.E.; Cheng, S.; Mora, S. Circulating Branched-Chain Amino Acids and Incident Cardiovascular Disease in a Prospective Cohort of US Women. *Circ. Genom. Precis. Med.* **2018**, *11*, e002157. [CrossRef]

25. Yang, P.; Hu, W.; Fu, Z.; Sun, L.; Zhou, Y.; Gong, Y.; Yang, T.; Zhou, H. The positive association of branched-chain amino acids and metabolic dyslipidemia in Chinese Han population. *Lipids Health Dis.* **2016**, *15*, 120. [CrossRef] [PubMed]

26. Ruiz-Canela, M.; Toledo, E.; Clish, C.B.; Hruby, A.; Liang, L.; Salas-Salvadó, J.; Razquin, C.; Corella, D.; Estruch, R.; Ros, E.; et al. Plasma Branched-Chain Amino Acids and Incident Cardiovascular Disease in the PREDIMED Trial. *Clin. Chem.* **2016**, *62*, 582–592. [CrossRef]

27. Grajeda-Iglesias, C.; Aviram, M. Specific Amino Acids Affect Cardiovascular Diseases and Atherogenesis via Protection against Macrophage Foam Cell Formation: Review Article. *Rambam. Maimonides Med. J.* **2018**, *9*, e0022. [CrossRef]

28. Zhang, R.; Mu, H.; Li, Z.; Zeng, J.; Zhou, Q.; Li, H.; Wang, S.; Li, X.; Zhao, X.; Sun, L.; et al. Oral administration of branched-chain amino acids ameliorates high-fat diet-induced metabolic-associated fatty liver disease via gut microbiota-associated mechanisms. *Front. Microbiol.* **2022**, *13*, 920277. [CrossRef]

29. Wang, M.; Yang, R.; Dong, J.; Zhang, T.; Wang, S.; Zhou, W.; Li, H.; Zhao, H.; Zhang, L.; Wang, S.; et al. Simultaneous quantification of cardiovascular disease related metabolic risk factors using liquid chromatography tandem mass spectrometry in human serum. *J. Chromatogr. B* **2016**, *1009–1010*, 144–151. [CrossRef]

30. Zhao, Y.; Dai, X.-Y.; Zhou, Z.; Zhao, G.; Wang, X.; Xu, M.-J. Leucine supplementation via drinking water reduces atherosclerotic lesions in apoE null mice. *Acta Pharmacol. Sin.* **2016**, *37*, 196–203. [CrossRef]

31. Nadkarni, S.K.; Bouma, B.E.; De Boer, J.; Tearney, G.J. Evaluation of collagen in atherosclerotic plaques: The use of two coherent laser-based imaging methods. *Lasers Med. Sci.* **2009**, *24*, 439–445. [CrossRef] [PubMed]

32. Yurdagul, A. Crosstalk Between Macrophages and Vascular Smooth Muscle Cells in Atherosclerotic Plaque Stability. *Arter. Thromb. Vasc. Biol.* **2022**, *42*, 372–380. [CrossRef] [PubMed]

33. Harman, J.L.; Jørgensen, H.F. The role of smooth muscle cells in plaque stability: Therapeutic targeting potential. *Br. J. Pharmacol.* **2019**, *176*, 3741–3753. [CrossRef] [PubMed]

34. Ohayon, J.; Finet, G.; Le Floc'h, S.; Cloutier, G.; Gharib, A.; Heroux, J.; Pettigrew, R.I. Biomechanics of Atherosclerotic Coronary Plaque: Site, Stability and In Vivo Elasticity Modeling. *Ann. Biomed. Eng.* **2014**, *42*, 269–279. [CrossRef]

35. Zhong, S.; Li, L.; Shen, X.; Li, Q.; Xu, W.; Wang, X.; Tao, Y.; Yin, H. An update on lipid oxidation and inflammation in cardiovascular diseases. *Free. Radic. Biol. Med.* **2019**, *144*, 266–278. [CrossRef] [PubMed]

36. Ference, B.A.; Ginsberg, H.N.; Graham, I.; Ray, K.K.; Packard, C.J.; Bruckert, E.; Hegele, R.A.; Krauss, R.M.; Raal, F.J.; Schunkert, H.; et al. Low-density lipoproteins cause atherosclerotic cardiovascular disease. 1. Evidence from genetic, epidemiologic, and clinical studies. A consensus statement from the European Atherosclerosis Society Consensus Panel. *Eur. Heart J.* **2017**, *38*, 2459–2472. [CrossRef] [PubMed]

37. Fukushima, K.; Harada, S.; Takeuchi, A.; Kurihara, A.; Iida, M.; Fukai, K.; Kuwabara, K.; Kato, S.; Matsumoto, M.; Hirata, A.; et al. Association between dyslipidemia and plasma levels of branched-chain amino acids in the Japanese population without diabetes mellitus. *J. Clin. Lipidol.* **2019**, *13*, 932–939.e2. [CrossRef] [PubMed]

38. Grajeda-Iglesias, C.; Rom, O.; Hamoud, S.; Volkova, N.; Hayek, T.; Abu-Saleh, N.; Aviram, M. Leucine supplementation attenuates macrophage foam-cell formation: Studies in humans, mice, and cultured macrophages. *BioFactors* **2018**, *44*, 245–262. [CrossRef]

39. Manning, B.D.; Toker, A. AKT/PKB Signaling: Navigating the Network. *Cell* **2017**, *169*, 381–405. [CrossRef]

40. Pan, J.X. LncRNA H19 promotes atherosclerosis by regulating MAPK and NF-kB signaling pathway. *Eur. Rev. Med. Pharmacol. Sci.* **2017**, *21*, 322.

41. Simion, V.; Zhou, H.; Pierce, J.B.; Yang, D.; Haemmig, S.; Tesmenitsky, Y.; Sukhova, G.; Stone, P.H.; Libby, P.; Feinberg, M.W. LncRNA VINAS regulates atherosclerosis by modulating NF-κB and MAPK signaling. *JCI Insight* **2020**, *5*, e140627. [CrossRef]

42. Ben, J.; Jiang, B.; Wang, D.; Liu, Q.; Zhang, Y.; Qi, Y.; Tong, X.; Chen, L.; Liu, X.; Zhang, Y. Major vault protein suppresses obesity and atherosclerosis through inhibiting IKK-NF-κB signaling mediated inflammation. *Nat. Commun.* **2019**, *10*, 1801. [CrossRef]

43. Lawrence, T. The nuclear factor NF-kappaB pathway in inflammation. *Cold Spring Harb. Perspect. Biol.* **2009**, *1*, a001651. [CrossRef] [PubMed]

44. Ridker, P.M.; Rane, M. Interleukin-6 Signaling and Anti-Interleukin-6 Therapeutics in Cardiovascular Disease. *Circ. Res.* **2021**, *128*, 1728–1746. [CrossRef] [PubMed]

45. Zhang, N.; Lei, J.; Lei, H.; Ruan, X.; Liu, Q.; Chen, Y.; Huang, W. MicroRNA-101 overexpression by IL-6 and TNF-α inhibits cholesterol efflux by suppressing ATP-binding cassette transporter A1 expression. *Exp. Cell Res.* **2015**, *336*, 33–42. [CrossRef]

46. Zhang, Y.; Yang, X.; Bian, F.; Wu, P.; Xing, S.; Xu, G.; Li, W.; Chi, J.; Ouyang, C.; Zheng, T. TNF-α promotes early atherosclerosis by increasing transcytosis of LDL across endothelial cells: Crosstalk between NF-κB and PPAR-γ. *J. Mol. Cell. Cardiol.* **2014**, *72*, 85. [CrossRef] [PubMed]

47. Buckley, L.F.; Abbate, A. Interleukin-1 blockade in cardiovascular diseases: A clinical update. *Eur. Heart J.* **2018**, *39*, 2063–2069. [CrossRef] [PubMed]

48. Abbate, A.; Van Tassell, B.W.; Biondi-Zoccai, G.G. Blocking interleukin-1 as a novel therapeutic strategy for secondary prevention of cardiovascular events. *BioDrugs Clin. Immunother. Biopharm. Gene Ther.* **2012**, *26*, 217. [CrossRef]

49. Grebe, A.; Hoss, F.; Latz, E. NLRP3 Inflammasome and the IL-1 Pathway in Atherosclerosis. *Circ. Res.* **2018**, *122*, 1722–1740. [CrossRef] [PubMed]

50. Galea, J.; Armstrong, J.; Gadsdon, P.; Holden, H.; Francis, S.E. Holt CM, Interleukin-1 beta in coronary arteries of patients with ischemic heart disease. *Arterioscler. Thromb. Vasc. Biol.* **1996**, *16*, 1000. [CrossRef]

51. Aiello, R.J.; Bourassa, P.-A.K.; Lindsey, S.; Weng, W.; Natoli, E.; Rollins, B.J.; Milos, P.M. Monocyte Chemoattractant Protein-1 Accelerates Atherosclerosis in Apolipoprotein E-Deficient Mice. *Arter. Thromb. Vasc. Biol.* **1999**, *19*, 1518–1525. [CrossRef] [PubMed]

52. Jiao, J.; Han, S.-F.; Zhang, W.; Xu, J.-Y.; Tong, X.; Yin, X.-B.; Yuan, L.-X.; Qin, L.-Q. Chronic leucine supplementation improves lipid metabolism in C57BL/6J mice fed with a high-fat/cholesterol diet. *Food Nutr. Res.* **2016**, *60*, 31304. [CrossRef] [PubMed]

53. Önal, B.; Özen, D.; Demir, B.; Ak, D.G.; Dursun, E.; Demir, C.; Akkan, A.G.; Özyazgan, S. The Anti-Inflammatory Effects of Anacardic Acid on a TNF-α—Induced Human Saphenous Vein Endothelial Cell Culture Model. *Curr. Pharm. Biotechnol.* **2020**, *21*, 710. [CrossRef] [PubMed]

54. Lee, J.; Ha, S.J.; Park, J.; Kim, Y.H.; Lee, N.H.; Hong, Y.-S.; Song, K.-M. Arctium lappa root extract containing L-arginine prevents TNF-α-induced early atherosclerosis in vitro and in vivo. *Nutr. Res.* **2020**, *77*, 85–96. [CrossRef] [PubMed]

55. Xu, Y.; Jiang, H.; Li, L.; Chen, F.; Liu, Y.; Zhou, M.; Wang, J.; Jiang, J.; Li, X.; Fan, X.; et al. Branched-Chain Amino Acid Catabolism Promotes Thrombosis Risk by Enhancing Tropomodulin-3 Propionylation in Platelets. *Circulation* **2020**, *142*, 49. [CrossRef]

56. Holeček, M. Branched-chain amino acids in health and disease: Metabolism, alterations in blood plasma, and as supplements. *Nutr. Metab.* **2018**, *15*, 33. [CrossRef]

57. Tobias, D.K.; Clish, C.; Mora, S.; Li, J.; Liang, L.; Hu, F.B.; Manson, J.E.; Zhang, C. Dietary Intakes and Circulating Concentrations of Branched-Chain Amino Acids in Relation to Incident Type 2 Diabetes Risk Among High-Risk Women with a History of Gestational Diabetes Mellitus. *Clin. Chem.* **2018**, *64*, 1203–1210. [CrossRef]

58. Tung, Y.-C.; Chou, R.-F.; Nagabhushanam, K.; Ho, C.-T.; Pan, M.-H. 3′-Hydroxydaidzein Improves Obesity Through the Induced Browning of Beige Adipose and Modulation of Gut Microbiota in Mice with Obesity Induced by a High-Fat Diet. *J. Agric. Food Chem.* **2020**, *68*, 14513–14522. [CrossRef]

59. Fusco, V.; Quero, G.M.; Cho, G.-S.; Kabisch, J.; Meske, D.; Neve, H.; Bockelmann, W.; Franz, C.M.A.P. The genus Weissella: Taxonomy, ecology and biotechnological potential. *Front. Microbiol.* **2015**, *6*, 155. [CrossRef]

60. Hiraishi, K.; Zhao, F.; Kurahara, L.-H.; Li, X.; Yamashita, T.; Hashimoto, T.; Matsuda, Y.; Sun, Z.; Zhang, H.; Hirano, K. Lactulose Modulates the Structure of Gut Microbiota and Alleviates Colitis-Associated Tumorigenesis. *Nutrients* **2022**, *14*, 649. [CrossRef]

61. Kai, L.; Zong, X.; Jiang, Q.; Lu, Z.; Wang, F.; Wang, Y.; Wang, T.; Jin, M. Protective effects of polysaccharides from Atractylodes macrocephalae Koidz. against dextran sulfate sodium induced intestinal mucosal injury on mice. *Int. J. Biol. Macromol.* **2022**, *195*, 142–151. [CrossRef] [PubMed]

62. Kim, S.; Goel, R.; Kumar, A.; Qi, Y.; Lobaton, G.; Hosaka, K.; Mohammed, M.; Handberg, E.; Richards, E.M.; Pepine, C.J.; et al. Imbalance of gut microbiome and intestinal epithelial barrier dysfunction in patients with high blood pressure. *Clin. Sci.* **2018**, *132*, 701–718. [CrossRef] [PubMed]

63. Verstreken, I.; Laleman, W.; Wauters, G.; Verhaegen, J. Desulfovibrio desulfuricans Bacteremia in an Immunocompromised Host with a Liver Graft and Ulcerative Colitis. *J. Clin. Microbiol.* **2012**, *50*, 199–201. [CrossRef]

64. Duan, J.; Huang, Y.; Tan, X.; Chai, T.; Wu, J.; Zhang, H.; Li, Y.; Hu, X.; Zheng, P.; Ji, P.; et al. Characterization of gut microbiome in mice model of depression with divergent response to escitalopram treatment. *Transl. Psychiatry* **2021**, *11*, 303. [CrossRef] [PubMed]

65. Hull, M.A. Nutritional prevention of colorectal cancer. *Proc. Nutr. Soc.* **2021**, *80*, 59. [CrossRef] [PubMed]

66. Buhman, K.K.; Furumoto, E.J.; Donkin, S.S.; Story, J.A. Dietary Psyllium Increases Fecal Bile Acid Excretion, Total Steroid Excretion and Bile Acid Biosynthesis in Rats. *J. Nutr.* **1998**, *128*, 1199–1203. [CrossRef] [PubMed]

67. Bertaggia, E.; Jensen, K.K.; Castro-Perez, J.; Xu, Y.; Di Paolo, G.; Chan, R.B.; Wang, L.; Haeusler, R.A. *Cyp8b1* ablation prevents Western diet-induced weight gain and hepatic steatosis because of impaired fat absorption. *Am. J. Physiol. Metab.* **2017**, *313*, E121–E133. [CrossRef]

68. Li, P.; Ruan, X.; Yang, L.; Kiesewetter, K.; Zhao, Y.; Luo, H.; Chen, Y.; Gucek, M.; Zhu, J.; Cao, H. A Liver-Enriched Long Non-Coding RNA, lncLSTR, Regulates Systemic Lipid Metabolism in Mice. *Cell Metab.* **2015**, *21*, 455–467. [CrossRef]

69. Haeusler, R.A.; Astiarraga, B.; Camastra, S.; Accili, D.; Ferrannini, E. Human Insulin Resistance Is Associated With Increased Plasma Levels of 12α-Hydroxylated Bile Acids. *Diabetes* **2013**, *62*, 4184–4191. [CrossRef]

70. Zeng, H.; Umar, S.; Rust, B.; Lazarova, D.; Bordonaro, M. Secondary Bile Acids and Short Chain Fatty Acids in the Colon: A Focus on Colonic Microbiome, Cell Proliferation, Inflammation, and Cancer. *Int. J. Mol. Sci.* **2019**, *20*, 1214. [CrossRef]

71. Zeng, S.-L.; Li, S.-Z.; Xiao, P.-T.; Cai, Y.-Y.; Chu, C.; Chen, B.-Z.; Li, P.; Li, J.; Liu, E.-H. Citrus polymethoxyflavones attenuate metabolic syndrome by regulating gut microbiome and amino acid metabolism. *Sci. Adv.* **2020**, *6*, eaax6208. [CrossRef] [PubMed]

72. Kim, G.-L.; Lee, S.; Luong, T.T.; Nguyen, C.T.; Park, S.-S.; Pyo, S.; Rhee, D.-K. Effect of decreased BCAA synthesis through disruption of ilvC gene on the virulence of Streptococcus pneumoniae. *Arch. Pharmacal Res.* **2017**, *40*, 921–932. [CrossRef]

73. Chen, M.-L.; Yi, L.; Zhang, Y.; Zhou, X.; Ran, L.; Yang, J.; Zhu, J.-D.; Zhang, Q.-Y.; Mi, M.-T. Resveratrol Attenuates Trimethylamine-N-Oxide (TMAO)-Induced Atherosclerosis by Regulating TMAO Synthesis and Bile Acid Metabolism via Remodeling of the Gut Microbiota. *mBio* **2016**, *7*, e02210-15. [CrossRef]

74. Liang, J.; Kou, S.; Chen, C.; Raza, S.H.A.; Wang, S.; Ma, X.; Zhang, W.-J.; Nie, C. Effects of Clostridium butyricum on growth performance, metabonomics and intestinal microbial differences of weaned piglets. *BMC Microbiol.* **2021**, *21*, 85. [CrossRef] [PubMed]

75. Hernández-Gómez, J.G.; López-Bonilla, A.; Trejo-Tapia, G.; Ávila-Reyes, S.V.; Jiménez-Aparicio, A.R.; Hernández-Sánchez, H. In Vitro Bile Salt Hydrolase (BSH) Activity Screening of Different Probiotic Microorganisms. *Foods* **2021**, *10*, 674. [CrossRef]

76. Tanaka, H.; Doesburg, K.; Iwasaki, T.; Mierau, I. Screening of Lactic Acid Bacteria for Bile Salt Hydrolase Activity. *J. Dairy Sci.* **1999**, *82*, 2530–2535. [CrossRef] [PubMed]

 nutrients

Article

Soft-Shelled Turtle Peptides Extend Lifespan and Healthspan in *Drosophila*

Qianqian Wang [1], Junhui Zhang [1], Jiachen Zhuang [1], Fei Shen [1], Minjie Zhao [1], Juan Du [2], Peng Yu [3], Hao Zhong [4,*] and Fengqin Feng [1,*]

1. College of Biosystems Engineering and Food Science, Zhejiang University, Hangzhou 310058, China
2. Zhejiang Nuoyan Biotechnology Co., Ltd., Huzhou 313000, China
3. Yuyao Lengjiang Turtle Industry, Ningbo 315400, China
4. College of Food Science and Technology, Zhejiang University of Technology, Hangzhou 310014, China
* Correspondence: zhonghao@zjut.edu.cn (H.Z.); feng_fengqin@hotmail.com (F.F.)

Abstract: In traditional Chinese medicine, soft-shelled turtle protein and peptides serve as a nutraceutical for prolonging the lifespan. However, their effects on anti-aging have not been clarified scientifically in vivo. This study aimed to determine whether soft-shelled turtle peptides (STP) could promote the lifespan and healthspan in *Drosophila melanogaster* and the underlying molecular mechanisms. Herein, STP supplementation prolonged the mean lifespan by 20.23% and 9.04% in males and females, respectively, delaying the aging accompanied by climbing ability decline, enhanced gut barrier integrity, and improved anti-oxidation, starvation, and heat stress abilities, while it did not change the daily food intake. Mechanistically, STP enhanced autophagy and decreased oxidative stress by downregulating the target of rapamycin (TOR) signaling pathway. In addition, 95.18% of peptides from the identified sequences in STP could exert potential inhibitory effects on TOR through hydrogen bonds, van der Walls, hydrophobic interactions, and electrostatic interactions. The current study could provide a theoretical basis for the full exploitation of soft-shelled turtle aging prevention.

Keywords: soft-shelled turtle peptide; lifespan; healthspan; TOR; molecular docking

Citation: Wang, Q.; Zhang, J.; Zhuang, J.; Shen, F.; Zhao, M.; Du, J.; Yu, P.; Zhong, H.; Feng, F. Soft-Shelled Turtle Peptides Extend Lifespan and Healthspan in *Drosophila*. *Nutrients* 2022, 14, 5205. https://doi.org/10.3390/nu14245205

Academic Editor: Maria Luz Fernandez

Received: 25 October 2022
Accepted: 4 December 2022
Published: 7 December 2022

1. Introduction

Aging is defined as biological and physiological processes at the late phase of life, typically characterized by the increased functional decline of organs and subsequent organismal death. The onset of age-related disorders such as atherosclerosis, type 2 diabetes, osteoporosis, neurodegeneration, and cardiovascular disease is frequently accompanied by the decay process [1,2]. Thus, strategies delaying the aging process should increase both lifespan and healthspan because most people are not only concerned about how long they will live but also care more about the length of time they can have good health and the ability to deal with daily activities [3,4].

Bioactive peptides, which are produced by proteolysis from food protein, have been proven to prolong lifespan and suppress the incidences of age-linked illnesses like cancer, cognitive decline, and neurodegeneration [5–9]. Soft-shelled turtles (*Pelodiscus sinensis*) have been widely consumed as a food and medicine with high nutritional value since ancient times. Especially, the traditional remedy of soft-shelled turtles in oriental medicine claims a longevity-promoting effect. A previous study demonstrated that soft-shelled turtle peptide (STP) supplementation could reduce physical exhaustion and oxidative stress as well as enhance exercise endurance and energy metabolism by modulating the oxidative stress-related protein in mice [10]. Due to the relationship between oxidative stress and aging, it is necessary to explore the anti-aging effect of STP. However, few reports explored the longevity-promoting and direct anti-aging effects of STP [11]. Additionally, previous studies primarily concentrated on confirming the function effect of STP, whereas the structural properties of peptides and structure-activity relationship remain unknown [10,12]. Hence, it is essential to identify and classify effective anti-aging peptides from STP.

Previous studies reported that the target of the rapamycin (TOR) signaling pathway was necessary for the lifespan promotion effect of some bioactive peptides [13,14]. TOR, a highly conserved serine/threonine kinase of the phosphatidylinositol kinase-related kinase family, is essential for sensing and responding to nutrient stimuli as well as delivering growth signaling in eukaryotes [15]. TOR complex 1 (TORC1) and TOR complex 2 (TORC2) are the two function complexes for TOR. Multiple TORC1-regulated processes appear to coordinately and overlappingly contribute to the pro-longevity effects of TORC1 inhibition [16]. Studies indicated that TOR signaling is a ROS-sensitive pathway, and the antioxidant effects of substances could affect the expression of TOR [17,18]. Regarding the in vivo antioxidant activity of STP, the idea that the exploration of STP to regulate the expression of TOR to increase autophagy and inhibit oxidative stress for longevity-promoting was practicable.

Here, we utilized *Drosophila* as a model organism to investigate the lifespan and healthspan extension effects of STP, and further identify metabolic pathways that correlate with lifespan. Simultaneously, the possible mechanism underlying the anti-aging effect of STP were elucidated by measuring the gene expression levels of the TOR signaling pathway. Moreover, the peptide profiles of STP were investigated as possible anti-aging peptides by molecular docking.

2. Materials and Methods

2.1. Materials

Wild-type Canton-S strain of *Drosophila melanogaster* was obtained from the Institute of Food Bioscience and Technology of Zhejiang University. The obtained flies were kept under controlled conditions at a temperature of 25 °C, humidity of 65%, and a 12-h light–dark cycle. Soft-shelled turtle peptides were kindly provided by Hangzhou Kangyuan Food Science and Technology Co., Ltd. (Hangzhou, China). The basic chemical components, relative molecular mass distribution, and amino acid composition of STP were shown in our previous study [10].

2.2. Fly Husbandry and Treatment

The parental flies were inoculated at a ratio of 20 males: 40 females and were cultured in the basal medium. Then, the parental flies were released on the fifth day. Afterward, the offspring of flies were reared at a standard larval density of ~300 flies per bottle, and all enclosed adults were collected over a 12-h period to obtain the F1 generation. We repeated this procedure to produce the F2 generation of flies. After the eclosion of the embryos of the F2 generation, male and female flies were collected within 48 h under CO_2 anesthesia for further experiments. We refer to the first day of a dietary treatment as the first day of adulthood for flies. The flies were reared on basal medium or soft-shelled turtle peptide-supplemented medium. The basal medium contained corn meal (10.5%, w/v), yeast (4%, w/v), sucrose (7.5%, w/v), agar (0.75%, w/v), and propionic acid (1%, v/v) (control group, CT). Soft-shelled turtle peptide-supplemented medium was prepared by adding soft-shelled turtle peptide power into a cooled (65 °C) liquid basal medium with a concentration of 0.8% (w/v) STP.

2.3. Lifespan Analysis

Flies (including male and female) were harvested within 48 h after hatching and anesthetized by light CO_2 for sortation, after which the flies were transferred to 30-mL vials containing treatment medium (CT or STP) with a density of 30–35 flies per vial. A total of 10 vials were set up for the CT and STP groups. After every 3–4 days, these flies were again transferred to new vials containing several types of media. All the dead flies were counted until no survivors remained. The mean, median, and maximum lifespan were calculated in accordance with the previously reported method [13].

2.4. Feeding Assay

Flies in the CT and STP groups were kept for starvation in (1%) agar for 24 h on day 30. Afterward, flies were cultured in the darkness for 4 h on agar containing F&D blue No. 1 (0.5%, Shanghai Macklin Biochemical Co., Ltd., Shanghai, China). Then, the flies were frozen on liquid nitrogen immediately and measured [19].

2.5. Climbing Assay

The evaluation of climbing ability was conducted as previously described [20]. On the 40th day, empty tubes with a line 6 cm above the bottom were used to keep both STP and CT groups of flies. After transferring the flies into containers, flies were gently trapped at the bottom; a ten-second timeline was recorded for flies to cross the drawn line.

2.6. Smurf Assay

The intestinal barrier function was evaluated using the smurf assay, as published previously [21]. On the 40th day, flies were transferred into new vials with the medium containing F&D blue No. 1 (2.5%) for 9 h. Flies with gut stains were counted as a smurf.

2.7. Stress Assay

Flies were fed for 30 days before a stress assay. A total of 6 vials (180–210 flies) were set up for CT and STP groups. For the oxidative stress assay, flies were fed H_2O_2 (30%) dissolved in glucose (5%) supplied on filter paper [13]. In the starvation test, flies were placed in tubes with agar (1%) [22]. The survival rate was documented every five hours. For the heat stress assay, flies were placed in an empty tube at 37 °C, and the survival rate was documented every half hour [23]. For the cold stress assay, flies were placed in empty vials bathed on ice at 4 °C for 2 h, and then placed at a temperature of 25 °C for behavior recording for 2 h [24].

2.8. Determination of Biochemical Index

Flies aged 30 days were immersed in liquid nitrogen and then kept at −80 °C for further examination. Commercial kits from the Nanjing Jiancheng Bioengineering Institute, Nanjing, China, with the manuals were used to determine the protein content (Catalog No. A045-3-2), malondialdehyde (MDA, Catalog No. A003-1-2), triacyleglyceride (TAG, Catalog No. A110-1-1), glutathione peroxidase (GSH-PX, Catalog No. A005-1-2), and superoxide dismutase (T-SOD, Catalog No. A001-1-2).

2.9. Untargeted Metabolomics Analysis

The whole flies at 30 days of age were collected, and the sample preparation methods of metabolomics analysis were assessed as reported previously [25]. Briefly, samples were separated using Agilent 1290 UHPLC (Agilent Technologies, Santa Clara, CA, USA) equipped with the ACQUITY UPLC BEH Amide column (1.7 μm, 2.1 mm × 100 mm), and analyzed by Triple 6600 TOF mass spectrometer (AB Sciex, Concord, Toronto, ON, Canada). Metabolites were identified based on the exact mass of their MS and tandem MS spectra, which were then searched and compared using a laboratory database (Shanghai Applied Protein Technology Co., Ltd., Shanghai, China). The initially processed data were enumerated with SIMCA software (V14.1, Umetrics, Ume, Västerbotten, Sweden) for mode identification following normalization to total peak intensity (by weight of the complete flies). Following the collection of valid data, principal component analysis (PCA) and orthogonal partial least-squares discriminant analysis (OPLS-DA) were applied to differentiate STP from the CT group. Variable importance in projection (VIP) > 1 and $p < 0.05$ were employed as criteria for screening potential biomarkers, and the KEGG metabolomics pathway analysis was constructed to reveal the most relevant pathway for STP to exert anti-aging effects.

2.10. Real-Time Quantitative PCR

The detailed protocol of RT-qPCR was performed as reported previously in our laboratory [10]. Primers were provided by Tsingke Biotechnology Co., Ltd. (Beijing, China). RP49 was used as a reference by using the $2^{(-\Delta\Delta Ct)}$ method, and the designed primer sequences are shown in Table S1.

2.11. Peptides Sequence Identification

STP was desalted using a C18 stage tip before lyophilization according to the method of Hu et al. [26] and analyzed by EASY-nLC-orbitrap MS/MS system. The detailed information was shown in our previous study [27].

2.12. Molecular Docking of STP on FKBP12-FRB

The interaction between the FKBP12-FRB (as the receptor, PDB ID: 3FAP) and peptide identified from Section 2.11 (as the ligands) was conducted by molecular docking (Discovery Studio software, Accelrys, San Diego, CA, USA) to predict the potential inhibitory activity of peptides with anti-aging activity [15]. The structure of FKBP12-FRB was optimized via operations cleaning, preparation, dehydration, and hydrogenation operation. The docking program was conducted with special binding sites (coordinates: x = 10.8388, y = 24.8616, z = 36.6092) and a set receptor radius (25.00 Å).

2.13. Statistical Analyses

The statistical analysis was accomplished with GraphPad prism 6.0. All values signify means $\pm$ SEM, and p-value < 0.05 was taken as statistically significant. The Two-tailed unpaired t-test was used to analyze the comparisons between two independent groups.

3. Results

3.1. STP Extended Lifespan and Healthspan in Drosophila

As shown in Figure 1B,C, the survival curves of flies in the STP group were right-skewed compared to the CT group, with the increase of mean, median, and maximum lifespan by 20.23% ($p < 0.001$), 25.48% ($p < 0.01$), and 4.03% in male flies, 9.04% ($p < 0.01$), 6.02% ($p < 0.05$), and 0.31% in females, respectively. Flies in the STP group exhibited slightly increased average food consumption than the CT group, but this was not statistically significant (Figure 1D). Furthermore, STP supplementation increased the climbing ability by 23.19% ($p < 0.001$) for male flies and 10.55% for females ($p = 0.063$) compared to the CT group (Figure 1E). Similarly, STP decreased the number of smurf flies by 37.74% ($p < 0.05$) for male flies and 32.09% ($p < 0.05$) for females compared to the CT group (Figure 1F). Taken together, these results suggested that STP extended the lifespan and healthspan in both male and female flies without limiting food intake. Moreover, the male flies that received STP exhibited better performance than females. Hereafter, our experiments focused on male flies.

3.2. STP Improved Stress Resistance in Drosophila

STP enhanced the survival rate of male flies under oxidative stress, and the mean, median, and maximum lifespan were increased by 9.44%, 27.38% ($p < 0.05$), and −0.89%, respectively, compared to the CT group (Figure 2A). The T-SOD activity increased ($p < 0.01$), whereas the MDA content decreased ($p < 0.05$) (Figure 2B,C). Moreover, STP increased the survival rate of male flies under starvation stress with the extension of mean, median, and maximum lifespan by 35.97% ($p < 0.05$), 60.98% ($p < 0.05$), and 17.79% ($p < 0.05$), respectively (Figure 2D). Accordingly, the TAG level was increased via STP supplementation by 48.26% ($p < 0.05$) (Figure 2E). In addition, STP enhanced the survival rate of male flies under heat stress, which corresponds with the upregulation of *Hsp70* mRNA expression in comparison with the CT group ($p < 0.01$) (Figure 2F). The mean, median, and maximum lifespan in STP under heat stress were prolonged by 29.44%, 32.61%, and 30.72% ($p < 0.05$), respectively (Figure 2G). However, STP supplementation has no significant impact on the resistance capability under cold stress conditions (Figure 2H). These findings implied that STP could improve the stress resistance of flies to oxidation, heat, and starvation.

3.3. STP Impacted the Potential Metabolic Markers in Pathways Associated with Aging in Drosophila

A clear separation of metabolites between the two groups was shown in the PCA plot (Figure 3A), indicating that STP supplementation could induce significant metabolic changes in flies. As shown in Figure 3B, 74 significant differential metabolites were identified mainly including amino acids, peptides, and analogs (8), nucleosides (8), carbohydrates and carbohydrate conjugates (13), lipids and lipid-like molecules (3), organoheterocyclic compounds (12), and organic nitrogen compounds (2). The KEGG pathway enrichment study revealed that STP intervention significantly affected 10 metabolic pathways (Figure 3C). For instance, compared to the CT group, metabolites (hypoxanthine, adenosine, and inosine) closely related to the purine metabolism pathway were downregulated after STP treatment. Nicotinamide and trigonelline, which are involved in the pathway of nicotinate and nicotinamide metabolism, were downregulated, whereas the nicotinate was upregulated. Moreover, compared to the CT group, 2-phospho-D-glycerate, methylglyoxal, and 4-methyl-2-oxopentanoate, crucial to the amino acid metabolism and biosynthesis (including glycine, serine, and threonine metabolism, arginine biosynthesis, valine, leucine, and isoleucine biosynthesis), were found downregulated after STP treatment, while creatine, citrulline, N-acetylglutamate, and L-leucine upregulated. Raffinose, D-galactonate, sucrose, and trehalose were downregulated, which are involved in the carbohydrate metabolism (ascorbate and aldarate, starch and sucrose, and galactose metabolism), and L-ascorbate, L-threonate, and D-galacturonate were upregulated. The main pathways affected by STP were summarized and sketched in Figure 3D.

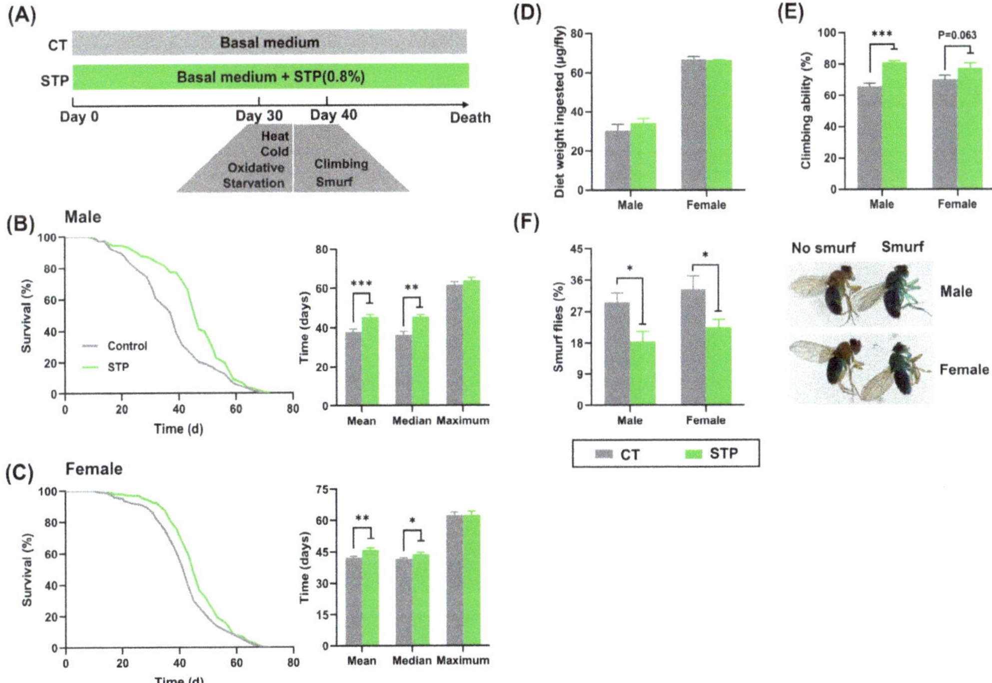

Figure 1. STP extended lifespan and improved healthspan in flies. (**A**) The scheme for animal experiment; Lifespan curves of male flies (**B**) and female flies (**C**); (**D**) Food intake; (**E**) Climbing ability; (**F**) Smurf flies. Data are shown as mean ± SEM. Statistical test: two-tailed unpaired *t*-test (* $p < 0.05$, ** $p < 0.01$, and *** $p < 0.001$ vs. CT group).

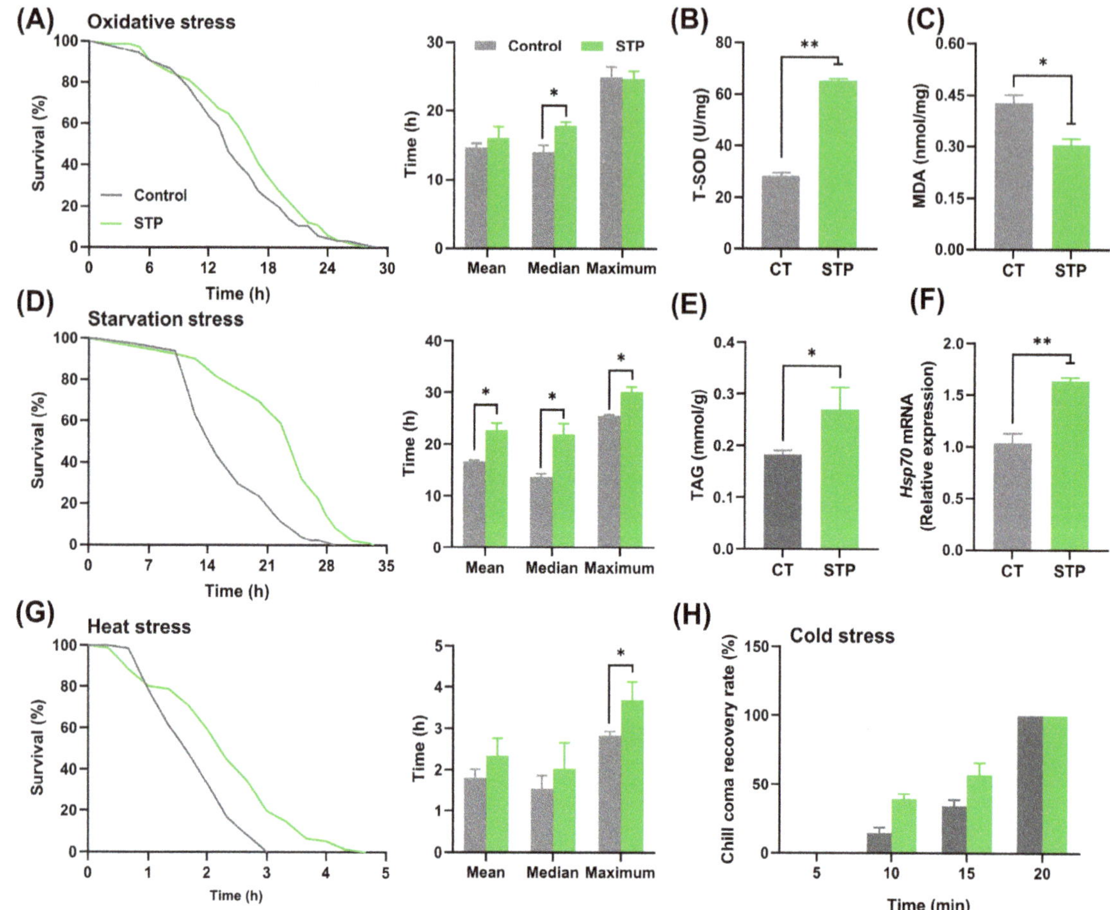

Figure 2. STP improved the resistance ability to stress in male flies. (**A**) Oxidative stress; (**B**) T-SOD activity; (**C**) MDA content; (**D**) Starvation stress; (**E**) TAG content; (**F**) Gene expression level of *Hsp70*; (**G**) Heat stress; (**H**) Cold stress. Data are shown as mean ± SEM. Statistical test: two-tailed unpaired *t*-test (* $p < 0.05$ and ** $p < 0.01$ vs. CT group).

3.4. STP Regulated TOR Signaling-Related Genes in Drosophila

As presented in Figure 4A, compared to the CT group, STP has substantially significantly upregulated the relative mRNA expression of nuclear factor-erythroid-2-like 2 (*Nrf2*) and heme oxygenase-1 (*Ho-1*) ($p < 0.001$), whereas downregulated kelch-like ECH-associated protein 1 (*Keap1*) ($p < 0.05$) in flies. Also, STP significantly downregulated the relative mRNA expression of *TORC* and upregulated the expression of autophagy-related gene1 (*Atg1*) ($p = 0.0714$) and autophagy-related gene 8a (*Atg8a*) ($p < 0.05$) in male flies compared to the CT group (Figure 4B). Taken together, the current data suggested that STP could enhance autophagy and inhibit oxidative stress by inhibiting TOR signaling pathway in flies.

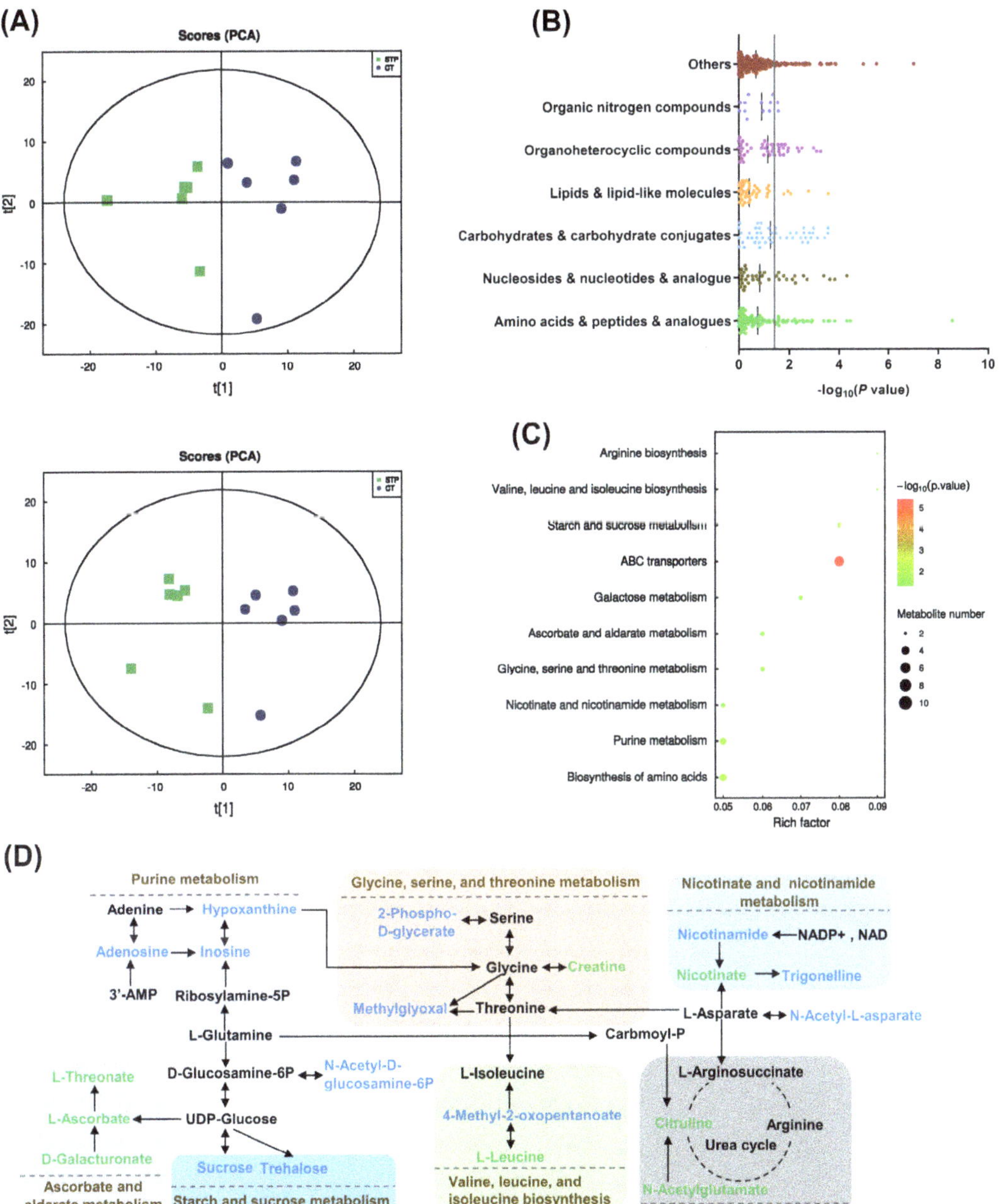

Figure 3. STP modulated metabolome in male flies. (**A**) The PCA scores plot under positive and negative ion modes; (**B**) Significant altered metabolites between CT and STP groups; (**C**) KEGG pathway resulting from the differential metabolites; (**D**) Metabolic pathways affected by STP supplementation in male flies. Metabolites colored in green/blue represent metabolites with increased/decreased levels after STP supplementation.

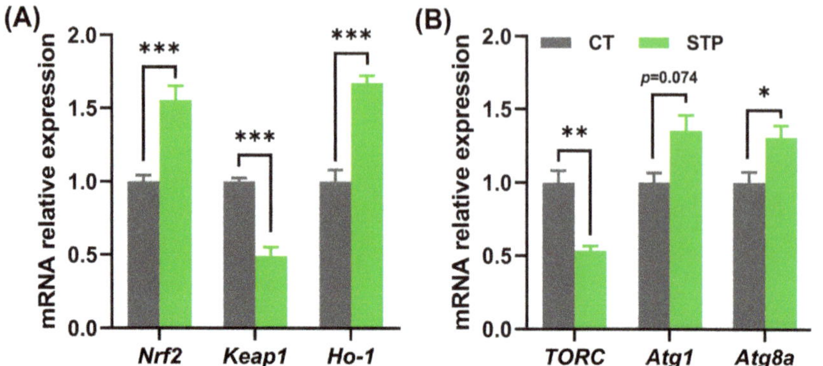

Figure 4. STP may inhibit the TOR signaling pathway to prolong the lifespan of flies. The gene expression level of (**A**) *Nrf2*, *Keap1*, and *Ho-1*; (**B**) *TORC*, *Atg1*, and *Atg8a*. Data are shown as mean $\pm$ SEM. Statistical test: two-tailed unpaired *t*-test (* $p < 0.05$, ** $p < 0.01$, and *** $p < 0.001$ vs. CT group).

3.5. Prediction of Peptides from STP with Anti-Aging Activity by Molecular Docking

In total, 187 peptide sequences were identified from STP. Table S2 revealed that the molecular weight of peptide sequences ranged between 330.1903 and 1901.9041 Da in STP, and the sequences were majorly constituted by 4–16 amino acid residues. Next, the molecular docking of these sequences with FKBP12-FRB was performed to predict potential peptides with anti-aging activity. A total of 178 peptides, accounting for 95.18% of CPTP, were successfully docked (Table S2). Among them, the #91 peptide sequence ADLETYLLEKSRVT displayed the highest (-) CDOCKER Energy value of 225.071, indicating that this peptide may have the highest docking potential for the target. Moreover, the molecular interaction between ADLETYLLEKSRVT and FKBP12-FRB is shown in Figure 5. Four weak interactions, i.e., hydrogen bonds, van der Waals, hydrophobic, and electrostatic interactions, were observed from their simulated docking with FKBP12-FRB. To be more specific, ADLETYLLEKSRVT comprising 8 residues, had 15 hydrogen bonds with Ser38, Asp37, Arg42, Tyr42, and Glu54 on the A chain and Lys184, Gln188, and Glu121 on the B chain. There were 14 van der Walls with Asp41, Phe99, Trp59, Tyr26, Phe46, Gln53, and Lys52 on the A chain and Thr187, Gly129, Tyr194, Trp190, His117, Ser124, and Phe197 on the B chain. As for the hydrophobic and electrostatic interactions, ADLETYLLEKSRVT had seven hydrophobic interactions with Phe36, His87, Ile91, Ile90, Ile56, and Val55 on the A chain, Arg125 on the B chain, while there were six electrostatic interactions with Lys35 and Arg42 on the A chain, and Asp191, Phe128, Glu121, and Glu122 on the B chain. In short, hydrogen bonds, van der Waals, and hydrophobic and electrostatic interactions may help the peptide form a more stable complex with FKBP12-FRB.

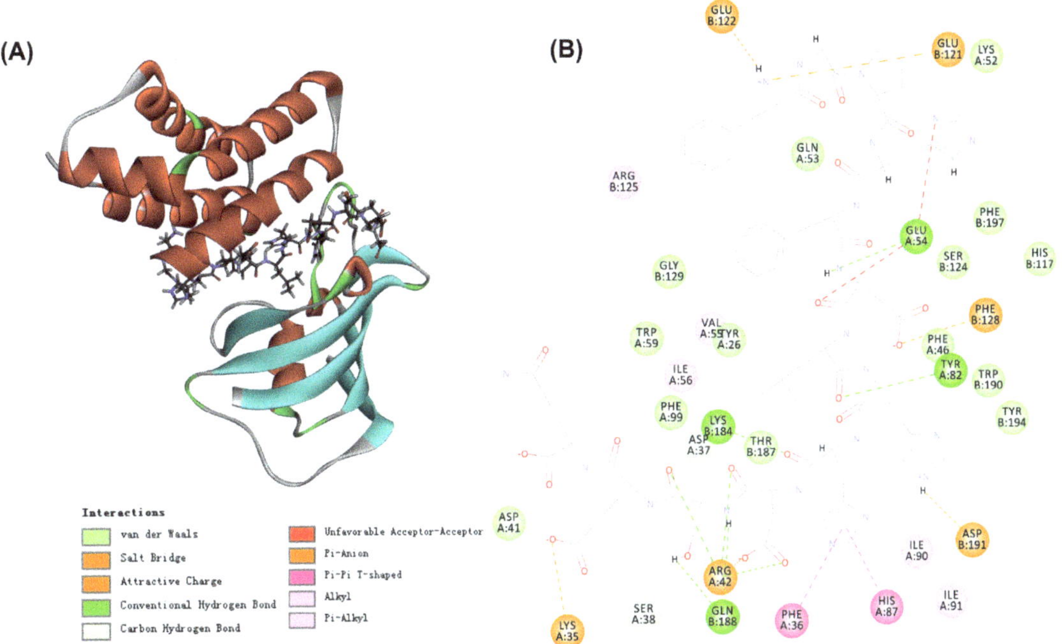

Figure 5. The 3D (**A**) and 2D (**B**) diagrams of the interactions between ADLETYLLEKSRVT and FKBP12-FRB.

4. Discussion

Bioactive peptides like cultured crocodile meat hydrolysates, sea cucumber hydrolysate, and crimson snapper peptides could increase the survival curve (indicative of increased lifespan) [5,28,29]. Similarly, STP supplementation prolonged the mean and median lifespan in both male and female flies, suggesting that STP was effective in the anti-aging of flies. Nevertheless, aging not only increases the chance of mortality, but also reduces physiological, physical, cognitive, and reproductive functions [30]. In the study, the age-related changes in muscle/neuronal function (climbing ability) and intestinal function (smurf flies) were improved by STP supplementation, which suggested that STP-treated flies were in better health than the CT group at the same age. It is reported that the incidence of intestinal dysfunction increases with age, which leads to an increased intestinal microbial load and a reduction in the host lifespan [31]. The lifespan of flies with intestinal barrier dysfunction increased with the removal of the microbes [31,32]. Therefore, STP may have reduced the microbial load to maintain the intestinal barrier function in flies, thus prolonging the lifespan of flies. Notably, the current data showed that the effect of STP on the lifespan and healthspan was more obvious in male flies than females, which suggested that the beneficial effects of STP are related to sex. Likely, Chen et al. found that crimson snapper peptides could significantly increase the healthspan in male flies but not in females [29]. This difference may be due to the higher energy required by female flies to reproduce as the possible occurrence of mating behavior. Intriguingly, it has been reported that such differences in responses to bioactive peptides between gender also exists in rodents subjected to caloric restriction [33]. One hypothesis to crack this puzzle is that the gender-biased physiologic utilization and allocation of energy difference will affect the response to daily energy alterations in males and females [34]. However, further studies on this sexual difference are greatly needed.

Generally, the phenotype alterations of longevity are accompanied by the altered resistance ability to various environmental stresses in flies. Since STP increased the lifespan and healthspan of flies, we subsequently tested whether it would enhance resistance to stress and thus lead to an increased survival rate in flies. As expected, STP supplementation could increase the survival rate in flies under oxidative stress. Correspondently, the increased antioxidant enzyme activity and decreased MDA content in the STP group further corroborated that STP could enhance the resistance to oxidative stress in flies, which was also observed in mice [10]. Similar results were found in starvation stress, and significant increases in TAG content further confirmed the finding. TAG, one of the most crucial lipid-storage molecules in insects, has been reported to affect the survival rate under starvation conditions to a great extent [35]. Additionally, STP increased the resistance to heat stress and was accompanied by significantly increased *Hsp70* gene expression, which was in agreement with the investigations of Su et al. [36]. Nevertheless, STP supplementation failed to increase the resistance to cold stress, which might be caused by the short time in the ice bath (flies in the CT and STP groups were woken up within twenty minutes). Taken together, these findings imply that STP could improve the stress resistance of flies to oxidation, heat, and starvation.

The mechanism that regulates the lifespan of an organism involves numerous metabolic changes. For a better understanding of the impact of STP on the metabolite composition, the untargeted metabolomics technique was further utilized to analyze the metabolic profiles of the whole flies in the CT and STP groups. In flies, purine metabolites such as adenosine and inosine were upregulated when the lifespan was shortened by dietary high-purine, high-sugar, or high-yeast [37] but decreased after STP treatment. Purine metabolism was used to maintain an optimal level of nucleotides in the tissues, which play a crucial role in the biochemical processes of energy metabolism. Additionally, the levels of creatine, citrulline, and L-ascorbate were upregulated after STP treatment. Aon et al. discovered that glycine–serine–threonine metabolism was one of the primary pathways of mouse longevity [38]. Supplementation with creatine, which is involved in glycine–serine–threonine metabolism, can reduce brain energy expenditure, oxidative stress, and mitochondrial dysfunction, thus ultimately improving cognitive performance in the elderly [39]. Citrulline, a component of arginine biosynthesis, has a powerful antioxidant capability to enhance long-term potentiation in aged rats [40]. L-ascorbate could ameliorate brain aging through antioxidative and anti-inflammatory effects [41]. Metabolomics findings suggested that the anti-aging activity may be linked to inhibited oxidative stress.

In the study, the molecular weight of <1000 Da of STP was 93.23%. It is reported that small peptides could interact with free radicals more efficiently through their ability to touch the intestinal barrier easily in vivo [42]. Additionally, seventeen amino acids were found in STP with a total content of 761.17 mg per g. Tyr, Cys, Met, His, Asp, Glu, Ala, Val, Pro, Phe, and Leu may contribute to the antioxidant activity of peptides [43]. In the current study, the content of the above antioxidative amino acids was 462.14 mg per g (accounting for 60.71% of STP). The results were consistent with the increased resistance to oxidative stress. Interestingly, the tolerance to oxidative stress and lifespan in flies = regulated the Nrf2/Keap1 signaling pathway [44]. Under normal conditions, Nrf2 (a stress-responsive transcription factor) is sequestered in the cytosol by Keap1 to maintain Nrf2 in an inactive state. However, the binding state is disrupted when exposed to oxidative stress, and then Nrf2 translocates to the nucleus to regulate the expression of more than a hundred genes [45]. Compared to the CT group, STP treatment significantly upregulated and downregulated the *Nrf2* and *Keap1* gene expression levels. Also, the downstream target gene (*Ho-1* and *Gclc*) expression was markedly upregulated after STP treatment. Those results implied that STP regulated the Nrf2/Keap1 signaling pathway to inhibit oxidative stress, which was consistent with the findings of increased tolerance to oxidative stress and metabolomics. Wu et al. reviewed that the homeostasis of intestinal microbiota and the balance of reactive oxygen species in the gut can be altered by bioactive peptides [46]. Therefore, STP may play a critical role in maintaining gut health and function by inhibiting oxidative stress. It is interesting to note that studies have shown that the inhibition of TOR activation was essential for the lifespan-promoting effect of flies against natural aging-induced oxidative stress. For example, rice protein hydrolysates increased flies' longevity by boosting the gene expression in Nrf2 and TOR signaling pathways [14]. Peptides derived from crimson snapper scales extended the lifespan of flies and harmful environmental exposure triggered oxidative stress by inhibiting the TOR activation [13]. Moreover, it has been demonstrated that rapamycin, a TORC1 inhibitor, activates SKN-1 (the ortholog of mammalian Nrf2) to prolong the longevity of nematodes [47]. STP significantly downregulated the *TORC* expression, suggesting that STP regulated the TOR signaling pathway to inhibit oxidative stress in flies. Additionally, inhibition of TORC1 could regulate the process of autophagy, which also has a central role in promoting longevity [15]. During the process of autophagy, lysosomes will degrade the damaged proteins, lipids, nucleic acids, and sugars, which gives cell energy and eliminates damaged cell components to exert a protective function for the cell [48]. The lifespan of flies was extended by over-expressing the autophagy kinase *Atg1* and *Atg8* [49,50]. As expected, the expression of the *Atg1* and *Atg8a* were upregulated in the current study, implying that STP promoted autophagy. During starvation, autophagy in animals mainly progresses for optimal survival [49]. STP improved the flies' tolerance to starvation stress and supported enhanced autophagy. In addition, oxidative stress and autophagy are linked via the TORC1 [15]. Namely, STP may inhibit the TOR signaling pathway to prolong the lifespan of flies. Similarly, walnut-derived peptides protected PC12 cells from oxidative stress by promoting autophagy through the Akt/TOR signaling pathway [9].

Rapamycin, an inhibitor of TORC1, could bind to the domain of FKBP-rapamycin binding (FRB) of TOR by forming a complex with the 12-kDa FK506-binding protein FKBP12, thereby inhibiting the physiological activity of TOR [15]. Therefore, the identified peptide sequences from STP were combined with FKBP12-FRB by molecular docking, and their interactions were analyzed; the responsible potential anti-aging peptides in STP were further determined in this study. As a result, 95.18% of peptides from the identified sequences in STP could successfully dock with FKBP12-FRB. The representative peptide sequences ADLETYLLEKSRVT could form a stable ternary complex with FKBP12-FRB through hydrogen bonds, van der Walls, hydrophobic interactions, and electrostatic interactions. Therefore, the peptide sequences in STP could form a stable ternary complex with FKBP12-FRB, thereby inhibiting the expression of the *TOR* gene from exerting anti-aging activity.

5. Conclusions

In this study, the supplementation of diets with STP effectively extended lifespan, improved healthspan, and enhanced stress resistances of oxidation, heat, and starvation in flies. Furthermore, STP decreased oxidative stress and enhanced autophagy by downregulating the expression of TOR, thus prolonging the lifespan of flies. Based on the molecular docking results, 95.18% of peptides with a potential inhibitory effect on TOR were identified from STP, and ADLETYLLEKSRVT could form a stable ternary complex with FKBP12-FRB by hydrogen bonds, van der Walls, hydrophobic interactions, and electrostatic interactions. Overall, this study has provided direct evidence of a longevity-promoting effect of soft-shell turtle peptides and revealed it could serve as a healthy supplementation in enhancing lifespan and healthspan. Importantly, the relationship between STP and gut microbiota in flies will be investigated in future studies.

Supplementary Materials: The following supporting information can be downloaded at: https://www.mdpi.com/article/10.3390/nu14245205/s1. Table S1: Sequences of primers, Table S2: Peptide sequences identified from STP and their docking energy to FKBP12-FRB.

Author Contributions: Conceptualization, Q.W. and F.F.; methodology, Q.W., J.Z. (Junhui Zhang), J.Z. (Jiachen Zhuang), F.S. and M.Z.; software, Q.W.; validation, Q.W.; resources, J.D. and P.Y.; writing—original draft preparation, Q.W.; writing—review and editing, Q.W., H.Z. and F.F.; funding acquisition, F.F. All authors have read and agreed to the published version of the manuscript.

Funding: This research was funded by the Ningbo Science and Technology Innovation 2025 Major Special Project, grant number 2019B10060, and the Agricultural and social development program of Yuyao, grant number 2020NS03.

Institutional Review Board Statement: Not applicable.

Informed Consent Statement: Not applicable.

Data Availability Statement: Data are contained within the article.

Conflicts of Interest: The authors declare no conflict of interest.

References

1. López-Otín, C.; Blasco, M.A.; Partridge, L.; Serrano, M.; Kroemer, G. The hallmarks of aging. *Cell* **2013**, *153*, 1194–1217. [CrossRef] [PubMed]
2. Tchkonia, T.; Kirkland, J.L. Aging, cell senescence, and chronic disease: Emerging therapeutic strategies. *JAMA-J. Am. Med. Assoc.* **2018**, *320*, 1319–1320. [CrossRef] [PubMed]
3. Hansen, M.; Kennedy, B.K. Does longer lifespan mean longer healthspan? *Trends Cell Biol.* **2016**, *26*, 565–568. [CrossRef] [PubMed]
4. Wong, R.Y. A new strategic approach to successful aging and healthy aging. *Geriatrics* **2018**, *3*, 86. [CrossRef]
5. Li, Y.; Hu, D.Y.; Huang, J.J.; Wang, S.Y. Glycated peptides obtained from cultured crocodile meat hydrolysates via maillard reaction and the anti-aging effects on *Drosophila* in vivo. *Food Chem. Toxicol.* **2021**, *155*, 112376. [CrossRef] [PubMed]
6. Qiu, W.J.; Chen, X.; Tian, Y.Q.; Wu, D.P.; Du, M.; Wang, S.Y. Protection against oxidative stress and anti-aging effect in *Drosophila* of royal jelly-collagen peptide. *Food Chem. Toxicol.* **2020**, *135*, 110881. [CrossRef]
7. Mao, J.; Zhang, Z.C.; Chen, Y.D.; Wu, T.; Fersht, V.; Jin, Y.; Meng, J.; Zhang, M. Sea cucumber peptides inhibit the malignancy of NSCLC by regulating miR-378a-5p targeted TUSC2. *Food Funct.* **2021**, *12*, 12362–12371. [CrossRef]
8. Wang, S.G.; Zheng, L.; Zhao, T.T.; Zhang, Q.; Liu, Y.; Sun, B.G.; Su, G.W.; Zhao, M.M. Inhibitory effects of walnut (*Juglans regia*) peptides on neuroinflammation and oxidative stress in lipopolysaccharide-induced cognitive impairment mice. *J. Agric. Food Chem.* **2020**, *68*, 2381–2392. [CrossRef]
9. Zhao, F.R.; Wang, J.; Lu, H.Y.; Fang, L.; Qin, H.X.; Liu, C.L.; Min, W.H. Neuroprotection by walnut-derived peptides through autophagy promotion via Akt/mTOR signaling pathway against oxidative stress in PC12 cells. *J. Agric. Food Chem.* **2020**, *68*, 3638–3648. [CrossRef]
10. Zhong, H.; Shi, J.Y.; Zhang, J.H.; Wang, Q.Q.; Zhang, Y.P.; Yu, P.; Guan, R.F.; Feng, F.Q. Soft-shelled turtle peptide supplementation modifies energy metabolism and oxidative stress, enhances exercise endurance, and decreases physical fatigue in mice. *Foods* **2022**, *11*, 600. [CrossRef]
11. Ma, M.J. Preparation of Antioxidant Peptides from Chinese Soft-Shelled Turtle and Its Anti-Aging Activity. Master's Thesis, Jiangnan University, Wuxi, China, 2020.
12. Wu, Y.C.; Liu, X.; Wang, J.L.; Chen, X.L.; Lei, L.; Han, J.; Jiang, Y.S.; Ling, Z.Q. Soft-shelled turtle peptide modulates microRNA profile in human gastric cancer AGS cells. *Oncol. Lett.* **2018**, *15*, 3109–3120. [CrossRef]
13. Cai, X.X.; Chen, S.Y.; Liang, J.P.; Tang, M.Y.; Wang, S.Y. Protective effects of crimson snapper scales peptides against oxidative stress on *Drosophila melanogaster* and the action mechanism. *Food Chem. Toxicol.* **2021**, *148*, 111965. [CrossRef]
14. Yue, Y.; Wang, M.T.; Feng, Z.P.; Zhu, Y.Y.; Chen, J.C. Antiaging effects of rice protein hydrolysates on *Drosophila melanogaster*. *J. Food Biochem.* **2021**, *45*, e13602. [CrossRef]
15. Johnson, S.C.; Rabinovitch, P.S.; Kaeberlein, M. mTOR is a key modulator of ageing and age-related disease. *Nature* **2013**, *493*, 338–345. [CrossRef]
16. Liu, G.Y.; Sabatini, D.M. mTOR at the nexus of nutrition, growth, ageing and disease. *Nat. Rev. Mol. Cell Biol.* **2020**, *21*, 183–203. [CrossRef]
17. Guimaraes, D.A.; dos Passos, M.A.; Rizzi, E.; Pinheiro, L.C.; Amaral, J.H.; Gerlach, R.F.; Castro, M.M.; Tanus-Santos, J.E. Nitrite exerts antioxidant effects, inhibits the mTOR pathway and reverses hypertension-induced cardiac hypertrophy. *Free Radic. Biol. Med.* **2018**, *120*, 25–32. [CrossRef]
18. Gurel, C.; Kuscu, G.C.; Buhur, A.; Dagdeviren, M.; Oltulu, F.; Karabay Yavasoglu, N.U.; Yavasoglu, A. Fluvastatin attenuates doxorubicin-induced testicular toxicity in rats by reducing oxidative stress and regulating the blood-testis barrier via mTOR signaling pathway. *Hum. Exp. Toxicol.* **2019**, *38*, 1329–1343. [CrossRef]
19. Xin, X.X.; Chen, Y.; Chen, D.; Xiao, F.; Parnell, L.D.; Zhao, J.; Liu, L.; Ordovas, J.M.; Lai, C.Q.; Shen, L.R. Supplementation with major royal-jelly proteins increases lifespan, feeding, and fecundity in *Drosophila*. *J. Agric. Food Chem.* **2016**, *64*, 5803–5812. [CrossRef]
20. Ulgherait, M.; Midoun, A.M.; Park, S.J.; Gatto, J.A.; Tener, S.J.; Siewert, J.; Klickstein, N.; Canman, J.C.; Ja, W.W.; Shirasu-Hiza, M. Circadian autophagy drives iTRF-mediated longevity. *Nature* **2021**, *598*, 353–358. [CrossRef]
21. Rera, M.; Clark, R.I.; Walker, D.W. Intestinal barrier dysfunction links metabolic and inflammatory markers of aging to death in *Drosophila*. *Proc. Natl. Acad. Sci. USA* **2012**, *9*, 21528–21533. [CrossRef]

22. Rana, A.; Oliveira, M.P.; Khamoui, A.V.; Aparicio, R.; Rera, M.; Rossiter, H.B.; Walker, D.W. Promoting Drp1-mediated mitochondrial fission in midlife prolongs healthy lifespan of *Drosophila melanogaster*. *Nat. Commun.* **2017**, *8*, 448. [CrossRef] [PubMed]

23. Li, J.Q.; Fang, J.S.; Qin, X.M.; Gao, L. Metabolomics profiling reveals the mechanism of caffeic acid in extending lifespan in *Drosophila melanogaster*. *Food Funct.* **2020**, *1*, 8202–8213. [CrossRef]

24. Liao, S.; Amcoff, M.; Nässel, D.R. Impact of high-fat diet on lifespan, metabolism, fecundity and behavioral senescence in *Drosophila*. *Insect Biochem. Mol. Biol.* **2021**, *133*, 103495. [CrossRef]

25. Liang, L.; Rasmussen, M.H.; Piening, B.; Shen, X.; Chen, S.; Röst, H.; Snyder, J.K.; Tibshirani, R.; Skotte, L.; Lee, N.C.; et al. Metabolic dynamics and prediction of gestational age and time to delivery in pregnant women. *Cell* **2020**, *181*, 1680–1692.e15. [CrossRef]

26. Hu, Y.M.; Lu, S.Z.; Li, Y.S.; Wang, H.; Shi, Y.; Zhang, L.; Tu, Z.C. Protective effect of antioxidant peptides from grass carp scale gelatin on the H_2O_2-mediated oxidative injured HepG2 cells. *Food Chem.* **2022**, *373*, 131539. [CrossRef]

27. Wang, Q.Q.; Yang, Z.R.; Zhuang, J.C.; Zhang, J.H.; Shen, F.; Yu, P.; Zhong, H.; Feng, F.Q. Antiaging function of Chinese pond turtle (*Chinemys reevesii*) peptide through activation of the Nrf2/Keap1 signaling pathway and its structure-activity relationship. *Front. Nutr.* **2022**, *9*, 961922. [CrossRef]

28. Lin, L.Z.; Yang, K.; Zheng, L.; Zhao, M.M.; Sun, W.Z.; Zhu, Q.Y.; Liu, S.J. Anti-aging effect of sea cucumber (*Cucumaria frondosa*) hydrolysate on fruit flies and d-galactose-induced aging mice. *J. Funct. Foods* **2018**, *47*, 11–18. [CrossRef]

29. Chen, S.Y.; Yang, Q.; Chen, X.; Tian, Y.Q.; Liu, Z.Y.; Wang, S.Y. Bioactive peptides derived from crimson snapper and in vivo anti-aging effects on fat diet-induced high fat *Drosophila melanogaster*. *Food Funct.* **2020**, *11*, 524–533. [CrossRef]

30. Fuellen, G.; Jansen, L.; Cohen, A.A.; Luyten, W.; Gogol, M.; Simm, A.; Saul, N.; Cirulli, F.; Berry, A.; Antal, P.; et al. Health and aging: Unifying concepts, scores, biomarkers and pathways. *Aging Dis.* **2019**, *10*, 883–900. [CrossRef]

31. Lee, H.Y.; Lee, S.H.; Lee, J.H.; Lee, W.J.; Min, K.J. The role of commensal microbes in the lifespan of Drosophila melanogaster. *Aging* **2019**, *11*, 4611–4640. [CrossRef]

32. Lee, H.Y.; Lee, S.H.; Min, K.J. The increased abundance of commensal microbes decreases drosophila melanogaster lifespan through an age-related intestinal barrier dysfunction. *Insects* **2022**, *13*, 219. [CrossRef]

33. Mitchell, S.J.; Madrigal-Matute, J.; Scheibye-Knudsen, M.; Fang, E.; Aon, M.; González-Reyes, J.A.; Cortassa, S.; Kaushik, S.; Gonzalez-Freire, M.; Patel, B.; et al. Effects of sex, strain, and energy intake on hallmarks of aging in mice. *Cell Metab.* **2016**, *23*, 1093–1112. [CrossRef]

34. Yamamoto, R.; Bai, H.; Dolezal, A.G.; Amdam, G.; Tatar, M. Juvenile hormone regulation of *Drosophila* aging. *BMC Biol.* **2013**, *11*, 85. [CrossRef]

35. Wei, Y.H.; Zhang, Y.J.; Cai, Y.; Xu, M.H. The role of mitochondria in mTOR-regulated longevity. *Biol. Rev. Camb. Philos. Soc.* **2015**, *90*, 167–181. [CrossRef] [PubMed]

36. Su, Y.; Wang, T.; Wu, N.; Li, D.Y.; Fan, X.L.; Xu, Z.X.; Mishra, S.K.; Yang, M.Y. Alpha-ketoglutarate extends *Drosophila* lifespan by inhibiting mTOR and activating AMPK. *Aging* **2019**, *11*, 4183–4197. [CrossRef]

37. Yamauchi, T.; Oi, A.; Kosakamoto, H.; Akuzawa-Tokita, Y.; Murakami, T.; Mori, H.; Miura, M.; Obata, F. Gut bacterial species distinctively impact host purine metabolites during aging in *Drosophila*. *iScience* **2020**, *23*, 101477. [CrossRef]

38. Aon, M.A.; Bernier, M.; Mitchell, S.J.; Di Germanio, C.; Mattison, J.A.; Ehrlich, M.R.; Colman, R.J.; Anderson, R.M.; de Cabo, R. Untangling determinants of enhanced health and lifespan through a multi-omics approach in mice. *Cell Metab.* **2020**, *32*, 100–116.e4. [CrossRef]

39. Andres, R.H.; Ducray, A.D.; Schlattner, U.; Wallimann, T.; Widmer, H.R. Functions and effects of creatine in the central nervous system. *Brain Res. Bull.* **2008**, *76*, 329–343. [CrossRef]

40. Ginguay, A.; Regazzetti, A.; Laprevote, O.; Moinard, C.; De Bandt, J.P.; Cynober, L.; Billard, J.M.; Allinquant, B.; Dutar, P. Citrulline prevents age-related LTP decline in old rats. *Sci. Rep.* **2019**, *9*, 20138. [CrossRef]

41. Nam, S.M.; Seo, M.; Seo, J.S.; Rhim, H.; Nahm, S.S.; Cho, I.H.; Chang, B.J.; Kim, H.J.; Choi, S.H.; Nah, S.Y. Ascorbic acid mitigates D-galactose-induced brain aging by increasing hippocampal neurogenesis and improving memory function. *Nutrients* **2019**, *11*, 176. [CrossRef]

42. Pan, M.F.; Liu, K.X.; Yang, J.Y.; Liu, S.M.; Wang, S.; Wang, S. Advances on food-derived peptidic antioxidants-a review. *Antioxidants* **2020**, *9*, 799. [CrossRef] [PubMed]

43. Wen, C.T.; Zhang, J.X.; Zhang, H.H.; Duan, Y.Q.; Ma, H.L. Plant protein-derived antioxidant peptides: Isolation, identification, mechanism of action and application in food systems: A review. *Trends Food Sci. Technol.* **2020**, *105*, 308–322. [CrossRef]

44. Sykiotis, G.P.; Bohmann, D. Keap1/Nrf2 signaling regulates oxidative stress tolerance and lifespan in *Drosophila*. *Dev. Cell* **2008**, *14*, 76–85. [CrossRef] [PubMed]

45. Steinbaugh, M.J.; Sun, L.Y.; Bartke, A.; Miller, R.A. Activation of genes involved in xenobiotic metabolism is a shared signature of mouse models with extended lifespan. *Am. J. Physiol. Endocrinol. Metab.* **2012**, *303*, E488–E495. [CrossRef]

46. Wu, S.J.; Bekhit, A.E.A.; Wu, Q.P.; Chen, M.F.; Liao, X.Y.; Wang, J.; Ding, Y. Bioactive peptides and gut microbiota: Candidates for a novel strategy for reduction and control of neurodegenerative diseases. *Trends Food Sci. Technol.* **2021**, *108*, 164–176. [CrossRef]

47. Robida-Stubbs, S.; Glover-Cutter, K.; Lamming, D.W.; Mizunuma, M.; Narasimhan, S.D.; Neumann-Haefelin, E.; Sabatini, D.M.; Blackwell, T.K. TOR signaling and rapamycin influence longevity by regulating SKN-1/Nrf and DAF-16/FoxO. *Cell Metab.* **2012**, *15*, 713–724. [CrossRef]

48. Aparicio, R.; Rana, A.; Walker, D.W. Upregulation of the autophagy adaptor p62/SQSTM1 prolongs health and lifespan in middle-aged *Drosophila*. *Cell Rep.* **2019**, *28*, 1029–1040.e5. [CrossRef]
49. Bjedov, I.; Cochemé, H.M.; Foley, A.; Wieser, D.; Woodling, N.S.; Castillo-Quan, J.I.; Norvaisas, P.; Lujan, C.; Regan, J.C.; Toivonen, J.M.; et al. Fine-tuning autophagy maximises lifespan and is associated with changes in mitochondrial gene expression in *Drosophila*. *PLoS Genet.* **2020**, *16*, e1009083. [CrossRef]
50. Simonsen, A.; Cumming, R.C.; Brech, A.; Isakson, P.; Schubert, D.R.; Finley, K.D. Promoting basal levels of autophagy in the nervous system enhances longevity and oxidant resistance in adult *Drosophila*. *Autophagy* **2008**, *4*, 176–184. [CrossRef]

Article

Fermented Supernatants of *Lactobacillus plantarum* GKM3 and *Bifidobacterium lactis* GKK2 Protect against Protein Glycation and Inhibit Glycated Protein Ligation

Shih-Wei Lin [1], Chi-Hao Wu [2], Ya-Chien Jao [3], You-Shan Tsai [4], Yen-Lien Chen [4], Chin-Chu Chen [5,6,7], Tony J. Fang [1] and Chi-Fai Chau [1,*]

[1] Department of Food Science and Biotechnology, National Chung Hsing University, Taichung 40227, Taiwan
[2] Graduate Programs of Nutrition Science, School of Life Science, National Taiwan Normal University, Taipei 106209, Taiwan
[3] Product and Process Research Center, Food Industry Research and Development Institute, Hsinchu 30062, Taiwan
[4] Biotech Research Institute, Grape King Bio Ltd., Taoyuan 32542, Taiwan
[5] Institute of Food Science and Technology, National Taiwan University, Taipei 10617, Taiwan
[6] Department of Food Science, Nutrition and Nutraceutical Biotechnology, Shih Chien University, Taipei 10462, Taiwan
[7] Department of Bioscience Technology, Chung Yuan Christian University, Taoyuan 32023, Taiwan
* Correspondence: chaucf@dragon.nchu.edu.tw; Tel.: +886-4-22840385 (ext. 4010)

Citation: Lin, S.-W.; Wu, C.-H.; Jao, Y.-C.; Tsai, Y.-S.; Chen, Y.-L.; Chen, C.-C.; Fang, T.J.; Chau, C.-F. Fermented Supernatants of *Lactobacillus plantarum* GKM3 and *Bifidobacterium lactis* GKK2 Protect against Protein Glycation and Inhibit Glycated Protein Ligation. *Nutrients* 2023, 15, 277. https://doi.org/10.3390/nu15020277

Academic Editor: Maria Luz Fernandez

Received: 16 November 2022
Revised: 19 December 2022
Accepted: 31 December 2022
Published: 5 January 2023

Abstract: With age, protein glycation in organisms increases continuously. Evidence from many studies shows that the accumulation of glycated protein is highly correlated with biological aging and the development of aging-related diseases, so developing a dietary agent to attenuate protein glycation is very meaningful. Previous studies have indicated that lactic acid bacteria-fermented products have diverse biological activities especially in anti-aging, so this study was aimed to investigate the inhibitory effect of the fermented supernatants of *Lactobacillus plantarum* GKM3 (GKM3) and *Bifidobacterium lactis* GKK2 (GKK2) on protein glycation. The results show that GKM3- and GKK2-fermented supernatants can significantly inhibit protein glycation by capturing a glycation agent (methylglyoxal) and/or protecting functional groups in protein against methylglyoxal-induced responses. GKM3- and GKK2-fermented supernatants can also significantly inhibit the binding of glycated proteins to the receptor for advanced glycation end products (RAGE). In conclusion, lactic acid bacteria fermentation products have the potential to attenuate biological aging by inhibiting protein glycation.

Keywords: aging; protein glycation; *Lactobacillus plantarum* GKM3; *Bifidobacterium lactis* GKK2; fermented supernatants; the receptor for advanced glycation end products (RAGE)

1. Introduction

Aging is often recognized as a progressive degeneration in physiological capacity, which may be a consequence of reactive oxygen species (ROS)-/reactive carbonyl species (RCS)-mediated biological events [1,2]. In this regard, RCS-induced protein glycation and the derived accumulation of advanced glycation end products (AGEs) during the aging process have received great attention, because these events may be the key bridge connecting aging with different human diseases such as atherosclerosis, diabetes, or neurodegenerative symptoms [3]. The content of AGEs mainly comes from the glycation reaction in the body, followed by diet, and is highly correlated with the concentration of glucose in the body and the metabolic response of glucose; therefore, it is interesting to investigate health-promoting dietary intervention that can disrupt RCS-induced glycation and AGE formation to decrease the onset and progression of aging-related diseases.

The glycation reaction can be divided into three main stages. The first stage is the interaction between carbohydrate molecules and substances with amine groups to form an unstable Schiff base, which further convert into the enaminol intermediate and then rearrange to the more stable Amadori products (e.g., ε-amino-lysine, ketoamine) [4]. The reactions in this phase are all reversible, but the formation of the Amadori products from the enaminol intermediate is faster than the opposite reaction. Therefore, the Amadori products are accumulated. The second stage is the degradation of Amadori products to produce highly active carbonyl compounds such as glyoxal (GO), methylglyoxal (MGO), or 3-deoxyglucosone (3-DG). During the production of these carbonyl compounds, the metal ions or oxygen are involved as catalysts. The third stage is the process in which active molecules interact with proteins or amino acids in the middle stage of glycation to generate stable substances (so-called glycated proteins or glycated molecules). Generally, the production of AGEs is irreversible. These AGEs usually contain arginine and lysine residues; therefore, they bind to the receptors for highly glycated end products (RAGE) located on the cell membrane triggering the cell physiological mechanism such as pro-inflammatory cytokines in the pathogenesis of chronic diseases [5].

The use of food with anti-glycation activity represents a potential strategy to treat and prevent AGE-related diseases. For example, the phenolic compounds from citrus fruits have been reported with an ability to inhibit AGE formation [6]. The isoflavones from legumes also presented with anti-glycation capacity in an in vitro experiment [7]. These anti-glycation compounds from food material usually come from enzymatic or microbial fermentation in the human gut. Therefore, focusing on gut microbiota or specific microbial supplements with anti-glycation effects might be one of the solutions for disease prevention.

Lactic acid bacteria (LAB) have been widely demonstrated as probiotics, which have several health benefits such as assistance for body weight control, the homeostasis of gut microbiota, immune modulation, and the prevention of metabolic diseases and cancer [8,9]. Our research group has demonstrated that LAB, including *Lactobacillus plantarum* GKM3 (GKM3) and *Bifidobacterium lactis* GKK2 (GKK2), also have great anti-aging properties [10,11]. In the senescence-accelerated prone mice P8 (SAMP8) animal model, the administration of GKM3 (1.0×10^9 cfu/kg B.W./day) effectively decreased TBARS and 8-OHdG levels in mice brains as well as the related features of cognitive impairment from the passive and active avoidance test, suggesting that GKM3 has great potential for delaying the oxidative-stress-related aging process [11]. The administration of GKK2 was also found to delay the progression of aging in SAMP8 mice as evidenced by the score of senescence from Takeda's grading method. Furthermore, decreased age-related muscle loss was found in GKK2-treated mice when compared to the vehicle controls [10]. Because anti-glycation characteristics such as anti-oxidative and anti-inflammatory activities were also found in other GKM3 and GKK2 studies [12–14], the present study developed an in vitro methylglyoxal-induced bovine serum albumin (MG-BSA) model mimicking protein glycation to clarify whether the anti-aging actions of GKM3 and GKK2 fermentate were partially inhibiting RCS-induced glycation.

2. Materials and Methods

2.1. Fermented Supernatant Samples

Lactobacillus plantarum GKM3 (BCRC 910787) and *Bifidobacterium lactis* GKK2 (BCRC 910826) were isolated from fresh vegetable and infant feces, respectively. Fermented supernatants of *L. plantarum* GKM3 (GKM3) and *B. lactis* GKK2 (GKK2) were provided by Grape King Bio Ltd. (Taoyuan, Taiwan). In general, bacteria strains GKM3 and GKK2 were cultured in a 15-tons fermenter, respectively, at 37 °C for 16 h with containing an 80% medium (5% glucose, 2.0% yeast extract, 0.05% $MgSO_4$, 0.1% K_2HPO_4, and 0.1% Tween 80) under pH control at 6.0. During the fermentation of the strain GKK2, the addition of CO_2 gas was needed. After incubation, the fermented supernatants were collected and centrifuged. The pH values of the fermented supernatants were then adjusted to 6.8–7.2 and these supernatants were heated at 121 °C for 1 min. Finally, GKM3- and GKK2-fermented supernatants were stored at -30 °C for study. The protein contents of the GKM3- and GKK2-fermented supernatants were 7 ng/mL and 28 ng/mL, respectively.

2.2. Protein Glycation Assays

Protein glycation was carried out by the reaction of a highly reactive dicarbonyl-methylglyoxal (MG; 25 mM) and bovine serum albumin (BSA, the most abundant albumin in the mammalian circulatory system; 5 mg/mL) at 37 °C for 7 days. GKM3- and GKK2-fermented supernatants (50 μL/mL) were added to the in vitro system of protein glycation as mentioned above to investigate their inhibitory effects on protein glycation. The control groups are BSA incubated with the vehicle, GKM3- or GKK2-fermented supernatants without MG addition. At the end of the first day of the reaction, an aliquot of the reaction solution was subjected to an analysis of the Amadori products that were the main products in the early protein glycation stage using a nitroblue tetrazolium (NBT) reagent as previously described [15]. In the NBT assay, the basic carbonate buffer containing NBT was mixed with the sample for 15 min. Since the glucose reduced NBT, the colorless NBT became to a deep blue, and the absorbance at 530 nm was then determined.

At the end of the reaction, 10% sodium dodecyl sulfate polyacrylamide gel electrophoresis (SDS-PAGE) and 3-D fluorescence spectroscopy were used to analyze the molecular characteristics of the glycated protein in each group according to the previous methods [16–18]. Specific functional groups, including free ε-amino groups measured by trinitrobenzenesulfonic acid (TNBS) [19] and carbonyl content measured by 2,4-dinitrophenylhydrazine (DNPH) [20] were further determined in these glycated proteins in order to clarify the underlying mechanism of the actions of GKM3- and GKK2-fermented supernatants in protein glycation. According to the previous study [21], an aliquot of the reaction solution was also taken for the determination of the content of MG-derived glycation products (argpyrimidine) in each group using a fluorescence spectrophotometer (F-7000, HITACHI, Tokyo, Japan) at an excitation wavelength (Ex) of 335 nm and emission wavelength (Em) of 400 nm.

2.3. MG-Trapping Assay

In general, GKM3- and GKK2-fermented supernatants (1 mL) were reacted with 2 mM MG (1 mL) in a glass vial and were incubated at 37 °C for 1 h. After the reaction, 1 mL of 12 mM OPD was added to the sample vial, and the mixtures were incubated at 37 °C for 15 min to derivatize MG into 2-MQ. Finally, the mixtures were centrifuged $(14,000 \times g,$ 5 min) and then subjected to high-performance liquid chromatography (HPLC) analysis to determine the amount of MG in each group [22]. The MG-trapping ability of the samples was determined based on the percent changes in the intensity of the MG peak of HPLC chromatograms in the sample groups compared to those in the control group.

2.4. Glycated Protein–RAGE Binding Assay

A CircuLex AGE-RAGE in vitro Binding Assay Kit (No. CY-8151, MBL Medical & Biological Laboratories Co. Ltd., Nagoya, Japan) was used to determine the inhibitory effects of GKM3- and GKK2-fermented supernatants on RAGE ligation, which is an important biological response in the onset and progression of various diseases [23]. The assay was carried out according to the manufacturer's instructions. In brief, the GKM3- or GKK2-fermented supernatants was added to the diluted recombinant His-tagged sRAGE in the microplate pre-coated with AGE2-BSA for an hour. After the wells were washed, the horse radish peroxidase (HRP)-conjugated anti-His-tag antibody was added for another hour. The wells were washed out again, then the tetra methylbenzidine (TMB), a chromogenic subtract which could be catalyzed by an anti-His-tag antibody, was incubated for 15 min at room temperature. With the addition of a stop solution, the inhibitory effect on the AGE-RAGE interaction was evaluated by measuring the amount of His-tagged sRAGE2 on the wells under 450 nm absorbance.

2.5. Statistical Analysis

The data shown are the means of the results obtained from three independent experiments, and the statistical differences between the data of the sample groups and control group were evaluated by a Student's *t* test using SPSS 18.0 software (IBM, Armonk, NY, USA). Significant differences were accepted at $p < 0.05$.

3. Results

3.1. Effects of GKM3- and GKK2-Fermented Supernatants on Early and Late Stages of Protein Glycation

One of the Amadori products, ketoamine, was detected and is shown in Figure 1 as the description of the early stages of protein glycation. The ketoamine content from the glycated BSA was 10.72 nmole/mg. With the addition of the GKM3-fermented supernatant, the concentration of ketoamine was decreased to 2.82 nmole/mg. A similar result was observed in the addition of the GKK2-fermented supernatant with a decrease in ketoamine content to 3.47 nmole/mg. These results suggested that both GKM3- and GKK2-fermented supernatants could inhibit the early stages of protein glycation by reducing the production of the Amadori product.

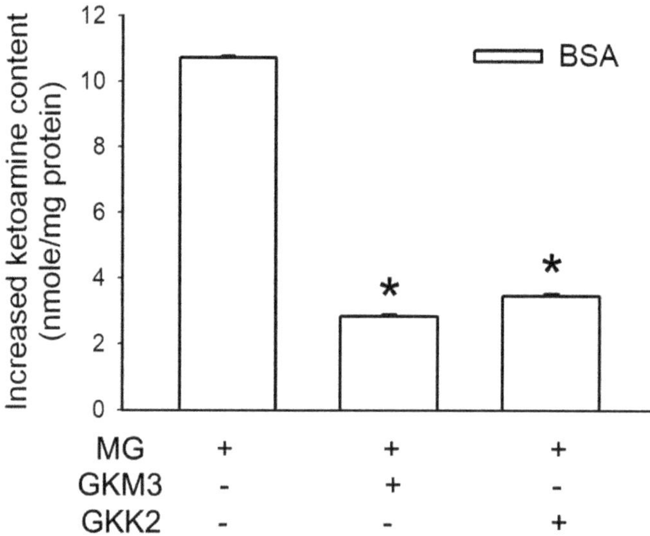

Figure 1. Effect of GKM3 and GKK2 on the formation of Amadori products in the early stage of protein glycation. Data unit was expressed as the increased nmole of ketoamines per mg of protein of samples. Data shown are means of results calculated by values of glycated BSA groups minus values of native BSA groups obtained from three independent experiments. * means statistical differences vs. data of the group of BSA treated with MG only. BSA: bovine serum albumin; MG: dicarbonyl-methylglyoxal.

Furthermore, SDS-PAGE revealed the MG-trapping ability of the probiotic fermented supernatant (Figure 2). The protein cross-link aggregation occurred in the glycated BSA, which could be reduced by the presence of either GKM3- or GKK2-fermented supernatants. This suggests that the second stage of glycation reaction was also inhibited by the fermented supernatants from probiotics GKM3 or GKK2.

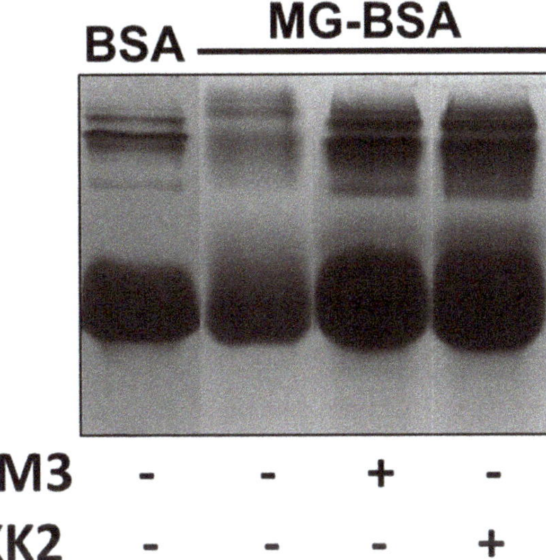

Figure 2. Effect of GKM3 and GKK2 on the SDS-PAGE results of products in the final stage of protein glycation. SDS-PAGE: sodium dodecyl sulfate polyacrylamide gel electrophoresis.

The 3-D fluorescent intensity described the late stage of the protein glycation (Figure 3). There were significant differences between the fermented supernatants and the glycated BSA from the principle component analysis (PCA) (data not shown). This indicates that GKM3- and GKK2-fermented supernatants possessed abilities in inhibiting glycation.

(a)   (b)   (c)   (d)

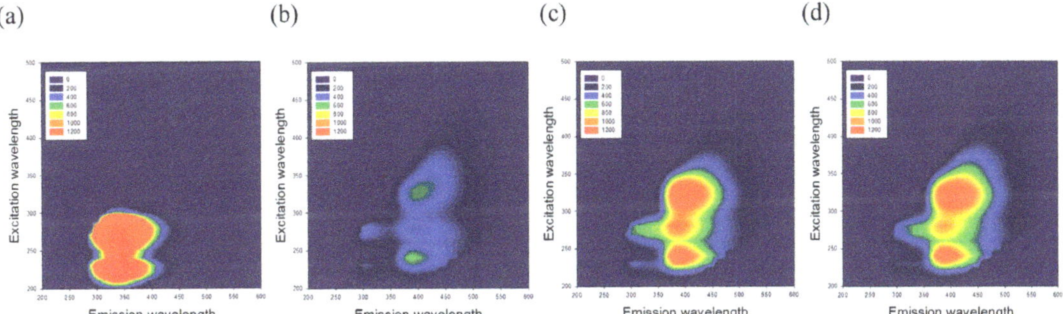

Figure 3. Effect of GKM3 and GKK2 on the 3-D fluorescent intensity of products in the final stage of protein glycation: (**a**) native BSA; (**b**) MG-BSA; (**c**) MG-BSA with GKM3; (**d**) MG-BSA with GKK2. Results shown are excitation–emission matrices (EEMs) of each group (excitation range: 200–500 nm; emission range: 200–600 nm).

An analysis of changes in the specific functional groups on glycated BSA found that the carbonyl content increased 69.74 nmoles/mg (Figure 4) and the content of the ε-amino groups decreased 37.28% (Figure 5). The addition of the fermented supernatants from the probiotics GKM3 or GKK2 showed decreases in the carbonyl contents ($p < 0.05$). It is interesting that the addition of GKM3-fermented supernatants decreases the ε-amino groups but the addition of GKK2-fermented supernatants did not affect the decrease in the ε-amino groups a lot when compared to the glycated BSA. It is assumed that the supernatants from GKM3 and GKK2 possessed different function in changing the polypeptide structures. Furthermore, both GKM3- and GKK2-fermented supernatants could increase the free sulfhydryl content (data not shown) when compared to the glycated BSA.

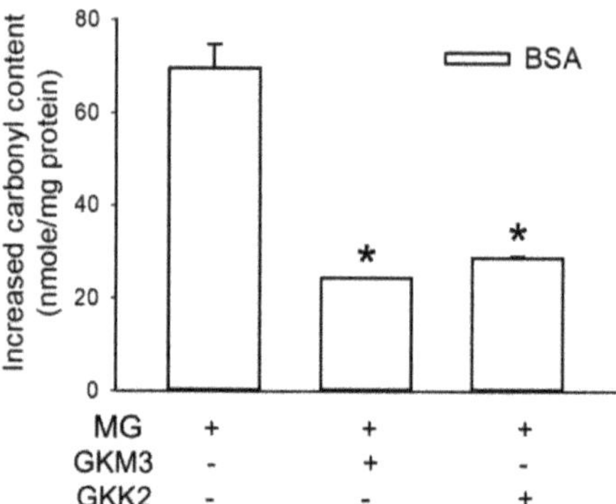

Figure 4. Effect of GKM3 and GKK2 on the content of carbonyl groups of products in the final stage of protein glycation. Data unit was expressed as the increased nmole of carbonyl groups per mg of protein of samples. Data shown are means of results calculated by values of glycated BSA groups minus values of native BSA groups obtained from three independent experiments. * means statistical differences vs. data of the group of BSA treated with MG only.

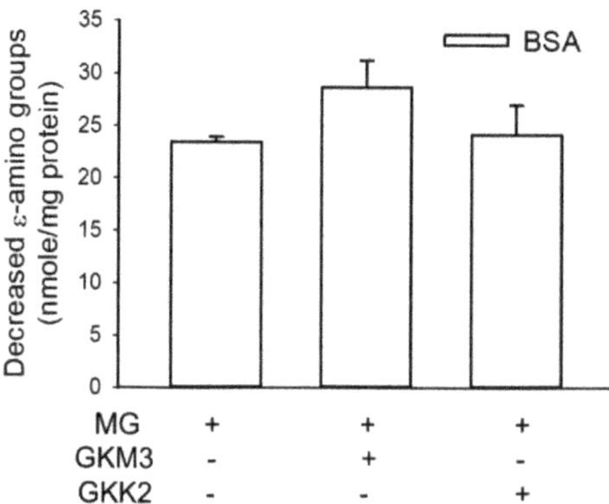

Figure 5. Effect of GKM3 and GKK2 on the content of ε-amino groups of products in the final stage of protein glycation. Data unit was expressed as the decreased nmole of ε-amino groups per mg of protein of samples. Data shown are means of results calculated by values of native BSA groups minus values of glycated BSA groups obtained from three independent experiments.

3.2. Effects of GKM3- and GKK2-Fermented Supernatants on Formation of MG-Derived Advanced Glycation end Products and Related RAGE Ligation

Argpyrimidine is one of the typical AGEs derived from MG. After the reaction among the BSA (5 mg/mL), glucose (10 mM), and MG (25 mM), the argpyrimidine was increased 84.32-fold. In the groups of GKM3- or GKK2-fermented supernatants, the contents of the argpyrimidine were significantly lower than the group of MG treatment only ($p < 0.05$) (Figure 6). Moreover, GKM3- or GKK2-fermented supernatants can decrease the MG intensity detected from HPLC analysis (Figure 7). This provided evidence that probiotic fermented supernatants suppressed the MG-derived AGEs.

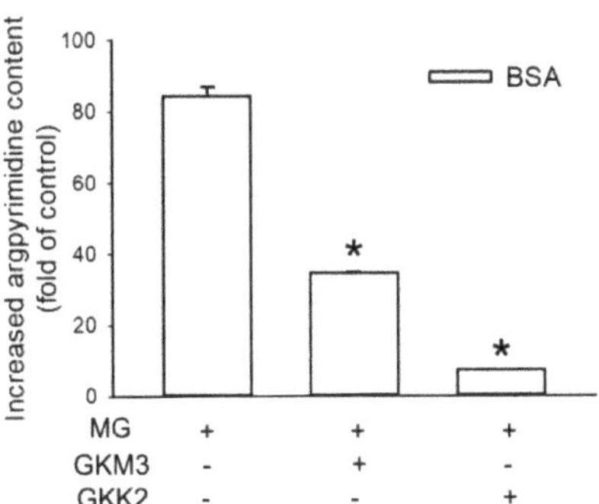

Figure 6. Effect of GKM3 and GKK2 on the formation of advanced glycation end products (AGEs) in the final stage of protein glycation. Data shown are means of ratio of values of glycated BSA groups versus values of native BSA groups obtained from three independent experiments. * means statistical differences vs. data of the group of BSA treated with MG only.

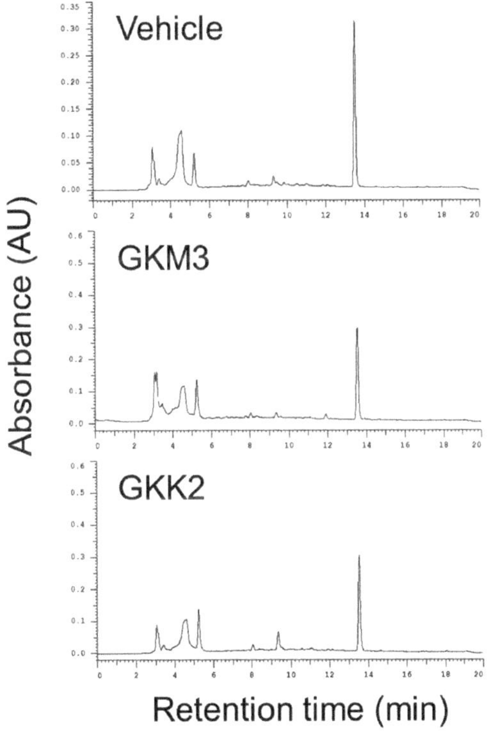

Figure 7. Methylglyoxal-trapping ability of GKM3 and GKK2. Data were expressed as the % of trapped methylglyoxal per 20 µL of samples. Data shown are means of results obtained from two independent experiments.

Figure 8 shows the inhibitory efficiency of the AGE-RAGE interaction. Both probiotic fermented supernatants from GKM3 and GKK2 showed inhibitory rates for AGE-RAGE interaction of about 77% and 85%, respectively (Figure 8b). This suggests that probiotic fermented supernatants not only attenuated the glycation by inhibiting the production of Amadori products, MG derivatives, and glycated residues, but also by blocking the combination of the AGEs and RAGE to not trigger the downstream aging-related physiological cell pathways (Figure 8a).

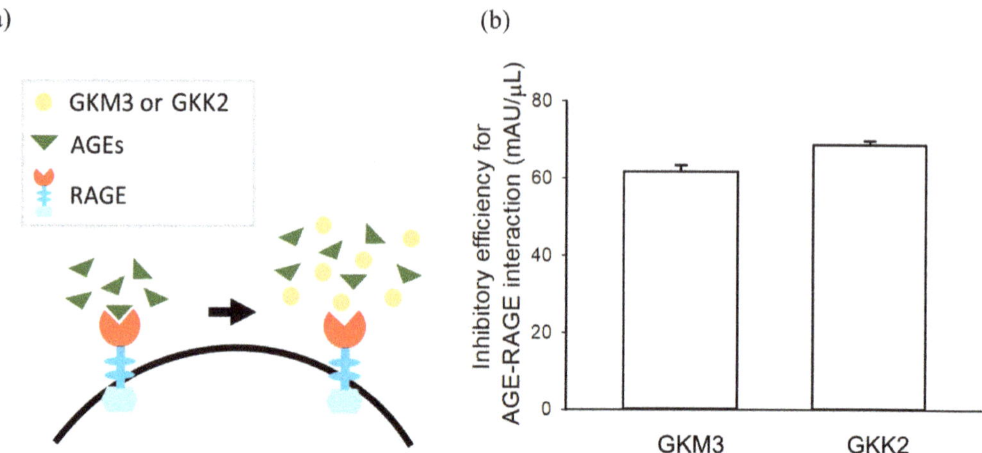

Figure 8. Effect of GKM3 and GKK2 on the ligation of RAGE with glycated proteins. (**a**) Scheme of inhibition of glycated proteins–RAGE ligation by GKM3 and GKK2 and (**b**) the experimental data. Data shown are means of results obtained from three independent experiments. RAGE: receptor for advanced glycation end products; AGE: advanced glycation end products.

4. Discussion

Ketoamine, also known as an Amadori product, is the main product in the early protein glycation stage [24]. The results indicated that GKM3- and GKK2-fermented supernatants significantly inhibited MG-induced ketoamine production in the early stage of the in vitro MG-BSA glycation model ($p < 0.05$) (Figure 1). MG can modify amino residues, specifically lysine and arginine, generating adducted amino residues, which can further result in protein cross-linking and the formation of macromolecules in the late stage of protein glycation [2]. In the present study, MG-induced protein cross-links can be found in the late stage of the in vitro MG-BSA glycation model as evidenced by the graphs of SDS-PAGE (Figure 2). According to the three-dimensional fluorescence spectrum, there are two major fluorophores in BSA (Figure 3a). One (Ex 275 nm/Em 340 nm) mainly results from the fluorescence of tryptophan and tyrosine residues, and another (Ex 225 nm/Em 340 nm) may be due to the π–π* transition of the polypeptide structures [17]. The BSA fluorescence was strongly modified after incubation with MG in the late stage of the in vitro MG-BSA glycation model (Figure 3b). In this regard, the MG-driven fluorescence quenching of tryptophan and tyrosine may result from an electron transfer or energy transfer reaction [25]. The MG-modified protein structure (Figure 2) and functional groups present in a protein including carbonyl groups and ε-amino groups (Figures 4 and 5) can also affect the π–π* transition of the polypeptide structures. It was first demonstrated that GKM3- and GKK2-fermented supernatants can abrogate MG-induced protein cross-links (Figure 2) and changes in the carbonyl content of BSA ($p < 0.05$) (Figure 4), as well as MG-driven changes in the fluorescent characteristics of BSA (Figure 3c,d). This indicates that GKM3- and GKK2-fermented supernatants have great abilities on the prevention of protein glycation.

Similar results were also found in the previous study on *Lactococcus lactis* fermentates of cyanobacteria and microalgae [26]. Kaga et al. performed several bacteria-fermented edible algae with different anti-glycation activities, which was assumed from the different chemical composition of the algae species. In our pre-screened experiment of the lactic acid bacteria strains in the MG-BSA glycation model (data not shown), supernatants from different bacterial strains also performed different anti-glycation activities even though the fermentation process was conducted by using the same medium. That is, the bacteria itself possess different utilizations of the nutrition leading to diverse health benefits. As glycation includes the production of Amadori and an increase in the methylglyoxal concentration, some compounds from the food materials such as aloe (*Aloe sinkatana*), dill (*Anethum graveolens*), and pineapple guava (*Acca sellowiana*) have been reported to reduce the glycation process [27–29]. The GKM3- and GKK2-fermented supernatants could also play a role as inhibitors during the process. This provided the possibility of applying GKM3- and GKK2-fermented supernatants as health-promoting nutritional food materials.

Argpyrimidine is an MG-derived advanced glycation end product (AGE), which is implicated in aging-related cognitive dysfunction [30] and other disorders [2]. Gomes et al. found a high concentration of argpyrimidine in the patients with familial amyloidotic polyneuropathy, a neurologic disease [31]. This indicates that the potential suppression of the production of argpyrimidine might alleviate nerves' relative symptoms. According to Figure 6, argpyrimidine residues significantly increased in MG-BSA (over 80 times to the native BSA) as evidenced by fluorescence analysis at Ex 335 nm/Em 400 nm; however, GKM3- and GKK2-fermented supernatants can significantly inhibit MG-driven increases in argpyrimidine. This may be related to the specific interaction of GKM3- and GKK2-fermented supernatants and the MG compound as evidenced by the changes in the intensity of the MG peak in the HPLC chromatogram of each group (Figure 7). Moreover, the inhibition of the content of argpyrimidine by the GKM3-fermented supernatant could explain the phenomenon of the memory retention that we observed in the age-accelerated mouse model in our previous publication [11]. In addition, the species *B. lactis* fermented supernatant reduced MG-induced argpyrimidine, which might lead to the improvement of the nerve conduction accompanied by a decreased expression of 8-OHdG or other oxidative markers in nerve cells, which is consistent with our observation in the SAMP8 mouse brain [10,32].

The receptor for AGEs (RAGE) is a multi-ligand receptor that has been demonstrated as the important role in the onset and progression of many diseases, such as diabetic complications, cardiovascular diseases, neuropathy, nephropathy, and cancer [23]. Therefore, AGE-RAGE ligation is the critical event in the progression of aging-related disorders [33]. Our study first demonstrated that GKM3- and GKK2-fermented supernatants could significantly inhibit AGE-RAGE interaction (Figure 8). This may be one of the mechanisms underlying the biological actions of the GKM3- and GKK2-fermented supernatants described in our previous studies [10,11]. The ligand–RAGE interaction activates NF-kappaB (NK-kB) and increased the expression of cytokines, adhesion molecules, and oxidative stress [34]. Consistent with other studies in probiotic fermented supernatants, liquid from fermented *Lactobacillus rhamnosus* showed an inhibitory effect on the expression of NK-kB in renal epithelial cells as well as other oxidative markers such as ERK, p53, and Caspase-3 [35]. We therefore assumed the fermented GKM3- and GKK2-supernatants presented the anti-aging effect by suppressing the relative oxidative markers involved with AGE-RAGE interaction.

5. Conclusions

Protein glycation is an important event during aging, which may increase AGE-RAGE interaction, leading to the occurrence of various diseases. In this study, GKM3- and GKK2-fermented supernatants were first found to significantly inhibit RCS-induced protein glycation, decrease the levels of argpyrimidine (the biomarker of aging diseases), and interrupt AGE-RAGE interaction, suggesting that lactic acid bacteria fermentates have great potential for the prevention of aging-related diseases.

Author Contributions: Conceptualization, Y.-L.C., C.-C.C. and C.-H.W.; methodology, Y.-C.J.; validation, S.-W.L. and Y.-C.J.; formal analysis, Y.-C.J.; investigation, S.-W.L. and Y.-C.J.; resources, S.-W.L.; data curation, Y.-S.T.; writing—original draft preparation, S.-W.L.; writing—review and editing, Y.-S.T.; visualization, Y.-S.T.; supervision, T.J.F. and C.-F.C.; project administration, S.-W.L. All authors have read and agreed to the published version of the manuscript.

Funding: This research received no external funding.

Institutional Review Board Statement: Not applicable.

Informed Consent Statement: Not applicable.

Data Availability Statement: The data presented in this study are available on request from the corresponding authors.

Acknowledgments: Special thanks to Jer-An Lin from National Chung Hsing University Graduate Institute of Food Safety (Taichung, Taiwan) for his technical support on glycation measurements.

Conflicts of Interest: The authors declare no conflict of interest.

References

1. Hohn, A.; Konig, J.; Grune, T. Protein oxidation in aging and the removal of oxidized proteins. *J. Proteom.* **2013**, *92*, 132–159. [CrossRef]
2. Lai, S.W.T.; Lopez Gonzalez, E.J.; Zoukari, T.; Ki, P.; Shuck, S.C. Methylglyoxal and its adducts: Induction, repair, and association with disease. *Chem. Res. Toxicol.* **2022**, *35*, 1720–1746. [CrossRef]
3. Chaudhuri, J.; Bains, Y.; Guha, S.; Kahn, A.; Hall, D.; Bose, N.; Gugliucci, A.; Kapahi, P. The role of advanced glycation end products in aging and metabolic diseases: Bridging association and causality. *Cell Metab.* **2018**, *28*, 337–352. [CrossRef]
4. Laga, M. Linking TNF-Induced Dysfunction and Formation of Methylglyoxal Derived AGEs in Endothelial Cells. Ph.D. Thesis, Ghent University, Ghent, Belgium, 2018.
5. Zong, H.; Ward, M.; Stitt, A.W. AGEs, RAGE, and diabetic retinopathy. *Curr. Diabetes Rep.* **2011**, *11*, 244–252. [CrossRef]
6. Fernandes, A.C.F.; Santana, A.L.; Martins, I.M.; Moreira, D.K.T.; Macedo, J.A.; Macedo, G.A. Anti-glycation effect and the α-amylase, lipase, and α-glycosidase inhibition properties of a polyphenolic fraction derived from citrus wastes. *Prep. Biochem. Biotechnol.* **2020**, *5*, 794–802. [CrossRef]
7. Genova, V.M.; Fernandes, A.C.F.; Hiramatsu, E.; Queirós, L.D.; Macedo, J.A.; Macedo, G.A. Biotransformed antioxidant isoflavone extracts present high-capacity to attenuate the in vitro formation of advanced glycation end products. *Food Biotechnol.* **2021**, *35*, 50–66.
8. Ayivi, R.D.; Gyawali, R.; Krastanov, A.; Aljaloud, S.O.; Worku, M.; Tahergorabi, R.; Silva, R.C.; Ibrahim, S.A. Lactic acid bacteria: Food safety and human health applications. *Dairy* **2020**, *1*, 202–232. [CrossRef]
9. Das, T.K.; Pradhan, S.; Chakrabarti, S.; Mondal, K.C.; Ghosh, K. Current status of probiotic and related health benefits. *Appl. Food Biotechnol.* **2022**, *2*, 100185. [CrossRef]
10. Lin, S.W.; Tsai, Y.S.; Chen, Y.L.; Wang, M.F.; Chen, C.C.; Lin, W.H.; Fang, T.J. An examination of *Lactobacillus paracasei* GKS6 and *Bifidobacterium lactis* GKK2 isolated from infant feces in an aged mouse model. *Evid.-Based Complement Alternat. Med.* **2021**, *2021*, 6692363. [CrossRef]
11. Lin, S.W.; Tsai, Y.S.; Chen, Y.L.; Wang, M.F.; Chen, C.C.; Lin, W.H.; Fang, T.J. *Lactobacillus plantarum* GKM3 promotes longevity, memory retention, and reduces brain oxidation stress in SAMP8 mice. *Nutrients* **2021**, *13*, 2860. [CrossRef]
12. Hou, Y.H.; Lin, S.W.; Chang, Z.; Lu, H.C.; Chen, Y.L.; Lin, W.H.; Chen, C.C. Effect of *Bifidobacterium lactis* GKK2 on OVA-induced asthmatic mice. *Hans J. Biomed.* **2019**, *9*, 70–80. [CrossRef]
13. Hsu, C.L.; Hou, Y.H.; Wang, C.S.; Lin, S.W.; Jhou, B.Y.; Chen, C.C.; Chen, Y.L. Antiobesity and uric acid-lowering effect of *Lactobacillus plantarum* GKM3 in high-fat-diet-induced obese rats. *J. Am. Coll. Nutr.* **2019**, *38*, 623–632. [CrossRef]
14. Tsai, Y.S.; Lin, S.W.; Chen, Y.L.; Chen, C.C. Effect of probiotics *Lactobacillus paracasei* GKS6, *L. plantarum* GKM3, and *L. rhamnosus* GKLC1 on alleviating alcohol-induced alcoholic liver disease in a mouse model. *Nutr. Res. Pract.* **2020**, *14*, 299–308. [CrossRef]
15. Armbruster, D.A. Fructosamine: Structure, analysis, and clinical usefulness. *Clin. Chem.* **1987**, *33*, 2153–2163. [CrossRef]
16. Khan, M.A.; Arif, Z.; Khan, M.A.; Moinuddin; Alam, K. Methylglyoxal produces more changes in biochemical and biophysical properties of human IgG under high glucose compared to normal glucose level. *PLoS ONE* **2018**, *13*, e0191014. [CrossRef] [PubMed]
17. Wani, T.A.; AlRabiah, H.; Bakheit, A.H.; Kalam, M.A.; Zargar, S. Study of binding interaction of rivaroxaban with bovine serum albumin using multi-spectroscopic and molecular docking approach. *Chem. Cent. J.* **2017**, *11*, 134. [CrossRef]
18. Wu, C.H.; Huang, H.W.; Lin, J.A.; Huang, S.M.; Yen, G.C. The proglycation effect of caffeic acid leads to the elevation of oxidative stress and inflammation in monocytes, macrophages and vascular endothelial cells. *J. Nutr. Biochem.* **2011**, *22*, 585–594. [CrossRef]
19. Hashimoto, C.; Iwaihara, Y.; Chen, S.J.; Tanaka, M.; Watanabe, T.; Matsui, T. Highly-sensitive detection of free advanced glycation end-products by liquid chromatography-electrospray ionization-tandem mass spectrometry with 2,4,6-trinitrobenzene sulfonate derivatization. *Anal. Chem.* **2013**, *85*, 4289–4295. [CrossRef]
20. Levine, R.L.; Garland, D.; Oliver, C.N.; Amici, A.; Climent, I.; Lenz, A.G.; Ahn, B.W.; Shaltiel, S.; Stadtman, E.R. Determination of carbonyl content in oxidatively modified proteins. *Methods Enzymol.* **1990**, *186*, 464–478.
21. Sero, L.; Sanguinet, L.; Blanchard, P.; Dang, B.T.; Morel, S.; Richomme, P.; Seraphin, D.; Derbre, S. Tuning a 96-well microtiter plate fluorescence-based assay to identify AGE inhibitors in crude plant extracts. *Molecules* **2013**, *18*, 14320–14339. [CrossRef]
22. Lo, C.Y.; Hsiao, W.T.; Chen, X.Y. Efficiency of trapping methylglyoxal by phenols and phenolic acids. *J. Food Sci.* **2011**, *76*, H90–H96. [CrossRef] [PubMed]
23. Dong, H.; Zhang, Y.; Huang, Y.; Deng, H. Pathophysiology of RAGE in inflammatory diseases. *Front. Immunol.* **2022**, *13*, 931473. [CrossRef] [PubMed]

24. Alenazi, F.; Saleem, M.; Syed Khaja, A.S.; Zafar, M.; Alharbi, M.S.; Al Hagbani, T.; Khan, M.Y.; Ahmad, W.; Ahmad, S. Antiglycation potential of plant based TiO2 nanoparticle in D-ribose glycated BSA in vitro. *Cell Biochem. Funct.* **2022**, *40*, 784–796. [CrossRef]
25. Banerjee, D.; Mandal, A.; Mukherjee, S. Excited state complex formation between methyl glyoxal and some aromatic bio-molecules: A fluorescence quenching study. *Spectrochim. Acta A Mol. Biomol. Spectrosc.* **2003**, *59*, 103–109. [CrossRef]
26. Kaga, Y.; Kuda, T.; Taniguchi, M.; Yamaguchi, Y.; Takenaka, H.; Takahashi, H.; Kimura, B. The effects of fermentation with lactic acid bacteria on the antioxidant and anti-glycation properties of edible cyanobacteria and microalgae. *LWT* **2021**, *135*, 110029. [CrossRef]
27. Elhassan, G.O.M.; Adhikari, A.; Yousuf, S.; Rahman, M.H.; Khalid, A.; Omer, H.; Fun, H.K.; Jahan, H.; Choudhary, M.I.; Yagi, S. Phytochemistry and antiglycation activity of *Aloe sinkatana* Reynolds. *Phytochem. Lett.* **2012**, *5*, 725–728. [CrossRef]
28. Oshaghi, E.A.; Khodadadi, I.; Tavilani, H.; Goodarzi, M.T. Aqueous extract of *Anethum Graveolens* L. has potential antioxidant and antiglycation effects. *Iran. J. Basic Med. Sci.* **2016**, *41*, 328–333.
29. Muniz, A.; Garcia, A.H.; Pérez, R.M.; García, E.V.; González, D.E. In vitro inhibitory activity of *Acca sellowiana* fruit extract on end products of advanced glycation. *Diabetes Ther.* **2018**, *9*, 67–74. [CrossRef]
30. Ahmed, N.; Ahmed, U.; Thornalley, P.J.; Hager, K.; Fleischer, G.; Münch, G. Protein glycation, oxidation and nitration adduct residues and free adducts of cerebrospinal fluid in Alzheimer's disease and link to cognitive impairment. *J. Neurochem.* **2005**, *92*, 255–263. [CrossRef]
31. Gomes, R.; Sousa Silva, M.; Quintas, A.; Cordeiro, C.; Freire, A.; Pereira, P.; Martins, A.; Monteiro, E.; Barroso, E.; Ponces Freire, A. Argpyrimidine, a methylglyoxal-derived advanced glycation end-product in familial amyloidotic polyneuropathy. *Biochem. J.* **2005**, *385*, 339–345. [CrossRef]
32. Nishizawa, Y.; Wada, R.; Baba, M.; Takeuchi, M.; Hanyu-Itabashi, C.; Yagihashi, S. Neuropathy induced by exogenously administered advanced glycation end-products in rats. *J. Diabetes Investig.* **2010**, *1*, 40–49. [CrossRef] [PubMed]
33. Senatus, L.M.; Schmidt, A.M. The AGE-RAGE Axis: Implications for age-associated arterial diseases. *Front. Genet.* **2017**, *8*, 187. [CrossRef]
34. Isermann, B.; Bierhaus, A.; Humpert, P.M.; Rudofsky, G.; Chavakis, T.; Ritzel, R.; Wendt, T.; Morcos, M.; Kasperk, C.; Hamann, A.; et al. AGE-RAGE: A hypothesis or a mechanism? *Herz* **2004**, *29*, 504–509. [CrossRef] [PubMed]
35. Tsai, Y.S.; Chen, Y.P.; Lin, S.W.; Chen, Y.L.; Chen, C.C.; Huang, G.J. *Lactobacillus rhamnosus* GKLC1 ameliorates cisplatin-induced chronic nephrotoxicity by inhibiting cell inflammation and apoptosis. *Biomed. Pharmacother.* **2022**, *147*, 112701. [CrossRef] [PubMed]

Article

Anti-Hyperuricemic Effect of Anserine Based on the Gut–Kidney Axis: Integrated Analysis of Metagenomics and Metabolomics

Mairepaiti Halimulati [1], Ruoyu Wang [1], Sumiya Aihemaitijiang [1], Xiaojie Huang [1], Chen Ye [1], Zongfeng Zhang [1], Lutong Li [1], Wenli Zhu [1,2,*], Zhaofeng Zhang [1,2,*] and Lixia He [3]

1   Department of Nutrition and Food Hygiene, School of Public Health, Peking University, Beijing 100871, China
2   Beijing's Key Laboratory of Food Safety Toxicology Research and Evaluation, Beijing 100871, China
3   Division of Molecular and Cellular Oncology, Dana-Farber Cancer Institute, Brigham and Women's Hospital, Harvard Medical School, Boston, MA 02115, USA
*   Correspondence: zhuwenli@bjmu.edu.cn (W.Z.); zhangzhaofeng@bjmu.edu.cn (Z.Z.); Tel.: +86-13661306973 (W.Z.); +86-18614033568 (Z.Z.)

Abstract: Nowadays, developing effective intervention substances for hyperuricemia has become a public health issue. Herein, the therapeutic ability of anserine, a bioactive peptide, was validated through a comprehensive multiomics analysis of a rat model of hyperuricemia. Anserine was observed to improve liver and kidney function and modulate urate-related transporter expressions in the kidneys. Urine metabolomics showed that 15 and 9 metabolites were significantly increased and decreased, respectively, in hyperuricemic rats after the anserine intervention. Key metabolites such as fructose, xylose, methionine, erythronic acid, glucaric acid, pipecolic acid and trans-ferulic acid were associated with ameliorating kidney injury. Additionally, anserine regularly changed the gut microbiota, thereby ameliorating purine metabolism abnormalities and alleviating inflammatory responses. The integrated multiomics analysis indicated that *Saccharomyces*, *Parasutterella excrementihominis* and *Emergencia timonensis* were strongly associated with key differential metabolites. Therefore, we propose that anserine improved hyperuricemia via the gut–kidney axis, highlighting its potential in preventing and treating hyperuricemia.

Keywords: hyperuricemia; anserine; metagenomics; urine metabolomics; integrated analysis; gut–kidney axis

Citation: Halimulati, M.; Wang, R.; Aihemaitijiang, S.; Huang, X.; Ye, C.; Zhang, Z.; Li, L.; Zhu, W.; Zhang, Z.; He, L. Anti-Hyperuricemic Effect of Anserine Based on the Gut–Kidney Axis: Integrated Analysis of Metagenomics and Metabolomics. *Nutrients* **2023**, *15*, 969. https://doi.org/10.3390/nu15040969

Academic Editor: Jaime Uribarri

Received: 26 December 2022
Revised: 9 February 2023
Accepted: 11 February 2023
Published: 15 February 2023

1. Introduction

With advancements in economic growth and changes in lifestyles, the number of people suffering from hyperuricemia is increasing annually worldwide. Approximately 20.2% and 20.0% of American males and females, respectively, are estimated to suffer from hyperuricemia [1]. As the largest developing country, the incidence of hyperuricemia in China has been reported to be up to 17.4%, with the age of onset gradually decreasing [2]. Regardless of the presence of symptoms, hyperuricemia has tremendous harmful effects on the body, posing a challenge to patients and clinicians. In addition to being the main hazard for gout, hyperuricemia also increases the risk of other diseases [3], such as diabetes and chronic kidney disease. Furthermore, a recent Irish study illustrated that compared to the normal population, blood uric acid of males and females above 535 μM/L and 416 μM/L, respectively, decreased their median survival years by 11.7 and 6 years, respectively [4]. Therefore, there is an urgent need to identify effective measures to prevent and treat hyperuricemia.

As an end product of purine metabolism, uric acid plays an important antioxidative role similar to Vitamin C [5]. Hyperuricemia occurs as a result of purine metabolism dysregulation and has a complex pathogenesis. The risk factors of hyperuricemia include genetics, nutritional status, sleep and stress, which consequently induce inflammation, oxidative stress and insulin resistance in the body. Moreover, it is accompanied by liver and kidney damage and toxic epidermal necrolysis [6]. Under these circumstances, the body compensates by increasing uric acid levels to counteract the damage caused by the metabolic disorder.

Current treatments for hyperuricemia include xanthine oxidase inhibitors (allopurinol or febuxostat), which decrease uric acid production, and uricosurics (probenecid), which increase uric acid excretion. Additionally, a low-purine diet has been recommended for lowering blood uric acid levels. Although this diet is safe, the effects of dietary interventions on hyperuricemia are limited. A recent meta-analysis on the effects of diet and genetics on blood uric acid concentrations in Caucasians in New Zealand demonstrated that 63 types of food combinations accounted for only 4.29% of the variation in blood uric acid concentrations. Contrastingly, 23.8% and 40.3% variations in blood uric acid concentrations in males and females, respectively, were due to genetic factors [7]. Therefore, simply reducing uric acid concentrations does not deter the development of hyperuricemia. Furthermore, there is a need to elucidate the pathogenesis of hyperuricemia and identify an efficient dietary intervention using natural products to improve overall body metabolism.

Overproduced uric acid or reduced uric acid excretion are the two primary causes of hyperuricemia. The kidneys excrete two-thirds of the uric acid, while the intestines excrete a third [8]. Excretory disorders related to the kidney are usually associated with the regulation of molecular signals, such as insulin resistance, inflammation, oxidative stress and cell damage [5]. Low-grade chronic systemic inflammation can directly lead to kidney damage and then affect the kidney's uric acid-related transporters expression, ultimately affecting the excretion of uric acid [9]. Additionally, studies show that high uric acid impairs mitochondrial function and produces reactive oxygen species, which activates the inflammatory bodies of the NOD-like receptor (NLR) family containing pyrimidine domain 3 (NLRP3). This cascade further aggravates kidney injury by secreting interleukin-1β (IL-1β) [10]. Moreover, excess uric acid enters the cells and becomes pro-oxidative, subsequently causing oxidative stress, aggravating kidney dysfunction and improving insulin resistance, thus forming a vicious circle. Therefore, reducing oxidative stress and low-grade chronic inflammation in the kidney is vital for the excretion of uric acid via the kidney. Various metabolites produced during this pathophysiological process are ultimately excreted by the kidneys through urine. Therefore, exploring urinary metabolites in hyperuricemia could provide a more comprehensive elucidation of its pathogenesis.

Recently, the gut microbiota has been reported to influence hyperuricemia, especially via the gut–kidney axis [11]. Then, the kidneys are damaged due to hyperuricemia, and uric acid and uremic toxins accumulate in the blood [12]. Moreover, low-grade inflammation produces proinflammatory cytokines. Elevated uric acid levels, uremic toxins and inflammatory cytokines negatively affect gut microbiota homeostasis. Dysbiosis of the gut microbiota increases intestinal permeability, which allows bacteria and intestinal metabolites, such as lipopolysaccharide (LPS), to be transported out of the intestine. LPS forms a complex with its CD14 receptor and is detected by Toll-like receptor 4 (TLR4), which induces chronic low-grade inflammation and thereby aggravates kidney injury [13]. The entire process forms a vicious cycle and increases the risk of hyperuricemia. Furthermore, gut dysbacteriosis increases the abundance of the xanthine oxidase gene-related microbiota and decreases the abundance of allantoinase gene-related flora, resulting in elevated uric acid levels in the intestinal tract. Therefore, this study aims to explore intervening substances that act on the gut–kidney axis.

The ocean is a uniquely rich source of bioactive peptides. In studies on peptides associated with hyperuricemia, pelagic fishes, such as tuna and bonito, have been found to migrate tens of thousands of kilometres at high speeds without causing an acid build-up in the body. This phenomenon was attributed to the presence of an important dipeptide, anserine, in the body, which has currently become a new target for dietary intervention in hyperuricemia [14]. Anserine is a multifunctional and highly stable histidine carnosine-like dipeptide found in fish skeletal muscles. In human clinical trials, anserine has been shown to reduce blood glucose and inflammation and elevate kidney functions [15]. Additionally, it improved gut microbiota disorders caused by hyperuricemia in mice [16]. However, the effect of anserine on host metabolism remains unclear and the alleviation of hyperuricemia via the gut–kidney axis remains to be investigated.

Thus, this study establishes a rat model for hyperuricemia using potassium oxycyanate and yeast and verifies the ameliorating effect of anserine on hyperuricemia. It also explores the mechanism of anserine-related amelioration of hyperuricemia through gut microbiota and metabolites using a combined ultraperformance liquid chromatography–tandem mass spectrometry (UPLC–MS) and macrogenomic analysis. Thus, this study aims to explore novel ideas for preventing and treating hyperuricemia.

2. Materials and Methods

2.1. Materials and Reagents

A rodent regular diet was obtained from Beijing Weitong Lihua Laboratory Animal Technology Co., Housed in a clean-grade animal house at the Peking University Health Science Center and the provided nutrients met the needs for rodent growth and development per GB 14924.3-2010. The diet for hyperuricemic rats comprised a normal diet combined with 4% potassium oxonate and 20% yeast, which was obtained from Beijing BotaiHongda Biotechnology Co., Ltd. Anserine was obtained from Sinopharm Holding Starshark Pharmaceutical (Xiamen) Co., Ltd. Allopurinol was obtained from Beijing Balinwei Technology Co., Ltd.

Primary antibodies against TLR4, myeloid differentiation factor88 (MyD88), nuclear factor E2-related factor 2 (Nrf2) antibody, NLRP3 inflammation, ATP-binding cassette, subfamily G, member 2 (ABCG2), matrix metallopeptidase 2 (MMP2), matrix metallopeptidase 2 (MMP9), β-actin and secondary antibodies were obtained from Abcam Co., Ltd. The primary antibody nuclear factor-κB (NF-kB) was obtained from CST Co., Ltd., whereas that against urate transporter 1 (URAT1) was obtained from Proteintech Biotech Co., Ltd. The primary antibodies against glucose transporter 9 (GLUT9) and the tissue inhibitor of metalloproteinase1 (TIMP-1) were obtained from Novus Biologicals Co., Ltd. and BOSTER Co., Ltd., respectively.

2.2. Animal Treatment

A total of 60 male Sprague–Dawley rats (180–220 g) were obtained from the Beijing Vital River Laboratory Animal Technology Co., Ltd. The animals were housed in a climate-controlled room at 25 ± 1 °C with a 12/12 h dark/light cycle and free access to sterile water and standard feed. All laboratory procedures were reviewed and approved by the Institutional Animal Care and Use Committee of Peking University (Approved No. 2019PHE017) on the 5th May 2019 and conformed to the Guidelines for the Care and Use of Laboratory Animals (NIH Publication No. 85-23, Revised 1996).

After four days of acclimatization, the rats were randomly divided into six groups (n = 10): normal control group (NC group), hyperuricemia group (HUA group), allopurinol group (Allo group, 10 mg/kg·bw allopurinol), three anserine groups (Ans1, Ans10 and Ans100 groups were treated with 1 mg/kg·bw, 10 mg/kg·bw and 100 mg/kg·bw anserine, respectively). During the six weeks of intervention, the general conditions of experimental animals, including health and behaviour, were observed daily, and food intake, water intake, urine volume and weight gain were monitored weekly.

2.3. Plasma and Urine Biochemical Analysis

Urine was collected for 24 h after six weeks of intervention and the volume was measured. Following centrifugation at 3000 r/min for 10 min, the supernatant was collected and stored at −80 °C. Subsequently, blood samples were collected and centrifuged at 3000 r/min for 10 min at 4 °C after the rats were sacrificed. The concentrations of serum uric acid (SUA), serum creatinine (SCr), blood urea nitrogen (BUN), urinary uric acid (UUA), urinary creatinine (UCr), alanine aminotransferase (ALT), aspartate aminotransferase (AST), albumin (ALB), globulin (GLB), albumin/globulin (A/G) and total protein (TP) were measured using an automatic biochemical instrument (AU480, Japan Olympus Corporation). Uric acid clearance (CUA) and creatinine clearance (CCr) were calculated using the following formulas:

$$\text{CUA} = \text{UUA}/\text{SUA} \times \text{Urine volume (mL/min)}; \quad \text{CCr} = \text{UCr}/\text{SCr} \times \text{Urine volume (mL/min)}.$$

2.4. Serum and Liver Xanthine Oxidase (XOD) and Adenosine Deaminase (ADA) Assay

Serum XOD activity was measured using the colourimetric method. After the liver was excised, 1 g of liver tissue was added to 9 mL of 0.9% normal saline to prepare 10% liver homogenate using mechanical homogenizers. The homogenate was centrifuged at 3000 r/min for 10 min and then 100 μL of the 10% liver homogenate supernatant was collected to evaluate liver XOD activity using colourimetry. A total of 50 μL of the liver homogenate supernatant was used to determine ADA activity. ADA generates ammonia by hydrolysing adenosine, which is then used to calculate ADA activity.

2.5. Histopathological Evaluation

The rats were sacrificed after the experiment. Then, their right kidneys were fixed with 4% paraformaldehyde, and paraffin sections were made by a series of techniques such as dewaxing, hematoxylin staining, eosin staining, dehydration and sealing. Finally, the tissue was examined under an optical microscope (Olympus BX43) and the image was recorded. The remaining half of the right kidney was fixed with 4% glutaraldehyde at a size of 1 mm × 1 mm × 1 mm, and then underwent a series of processes such as dehydration, transparency, wax immersion, embedding and sectioning to obtain electron microscopic sections for observation under an electron microscope (JEM1400).

2.6. Western Blotting

The expressions of GLUT9, URAT1, ABCG2, TLR4, MyD88, NF-κB, NLRP3, Nrf2, MMP-1, MMP-2 and TIMP-1 proteins in the renal tissue were detected using Western blotting. Commercial kits (Beijing Prily Gene Technology Co., Ltd. C1053) were used to extract total protein from kidney tissue, and BCA protein detection kits (Beijing Prily Gene Technology Co., Ltd. P1511) were used to quantify protein concentrations. After protein separation using 12% sodium dodecyl-sulphate polyacrylamide gel electrophoresis, an Immobilon-P transfer membrane (Millipore, Germany, IPVH00010) was used for transferring the protein. Following this, the membrane was sealed with 5% nonfat milk, incubated overnight at 4 °C with primary antibodies, and then incubated with secondary antibodies conjugated with horseradish peroxidase. Colour development using ECL Chemiluminescence Chromogenic solution (PerkinElmer, America, NEL105001EA) was performed. The grayscale was calculated and normalised to the values of β-actin.

2.7. Untargeted Urine Metabolomic Analysis

To 100 μL of urine sample in an Eppendorf tube, 400 μL of 80% methanol aqueous solution (Thermo Fisher, Waltham, MA, USA, 67-56-1) was added. The sample was vortexed, placed in an ice bath for 5 min and centrifuged at 15,000 r/min for 20 min at 4 °C. A quantity of the supernatant was diluted with MS-grade water to obtain a methanol content of 53%. This sample was then centrifuged at 15,000 r/min for 20 min at 4 °C and the supernatant was analysed using a UPLC–MS. Metabolomics workflow was based on the R language MetaboAnalystR package. First, the sample data were subjected to quality control and batch correction, then the sample data were standardized, and finally, the metabolite content was counted. Metabolites that differed between groups were identified using an orthogonal partial least squares discriminant method.

2.8. Metagenomic Sequencing of Gut Microbiota

DNA was extracted from stool samples using the CTAB method. After the DNA samples were qualified, the library was constructed. Qubit 2.0 was used for preliminary quantification, and then, an Agilent 2100 was used for the detection of inserted fragments from the library. Using real-time polymerase chain reaction, the effective concentration of the library could be accurately quantified after the insert size reached the expected size. Following library detection, different libraries were pooled into a flow-through pool based on effective concentration and target data volume. After cBOT clustering, the Illumina PE150 (2 × 150) high-throughput sequencing platform was used for sequencing. KneadData software was used for raw data quality control (Trimmom-based) and de-hosting (Bowtie2-2-based). Before and after KneadData, FastQC was used to verify the justification and effectiveness of the quality control. Furthermore, Kraken2 was used to analyse the species composition and diversity information of samples, whereas Bracken was used for predicting species relative abundances. The identification of microbial genera and species that differed between the six groups was performed using LEfSe biomarker mining analysis (microorganisms with LDA > 2 by default). To functionally annotate genes, we used HUMAnN2 software to compare the sequences after quality control (based on DIAMOND) and host removal with the protein database (UniRef90), further filtering out reads that failed to be compared. Based on the mapping between the ID of UniRef90 and the ID of the KEGG and EC functional databases, the relative abundance of the corresponding functions of each functional database was calculated.

2.9. Statistical Analysis

2.9.1. Statistical Analysis of Non-Omics Data

One-way analysis of variance (ANOVA) was calculated using SPSS 25.0 software; variables were transformed if they did not meet the requirements of normality for variance. If the transformed variables still did not meet the requirements, a nonparametric test was used for statistical analysis. The experimental and control groups were compared using the least significant difference (LSD) method, and a difference of $p < 0.05$ was considered to be statistically significant.

2.9.2. Statistical Analysis of Omics Data

The omics data were analysed using R software (version 4.1.1). Principal coordinates analysis (PcoA) and principal components analysis (PCA) were used for dimension reduction analysis to show the degree of variation among the samples. Binary unpaired samples were tested for significance using the rank-sum test, and multiple groups were compared using the Kruskal–Wallis rank-sum test.

2.9.3. Multiomics Association Analysis

The correlation between gut microbiota and metabolites was analysed using the Pearson correlation coefficient method, with a significance threshold of $p < 0.05$.

3. Results

3.1. Anserine Reduced Uric Acid Levels and Improved Kidney Damage

The HUA group had a significantly higher SUA than the NC group after 6 weeks of potassium oxalate and yeast feeding ($p < 0.05$), indicating successful modelling. After the anserine intervention, the SUA level was significantly decreased ($p < 0.05$). Moreover, BUN was significantly elevated in the HUA group but tended to decrease after the anserine intervention. SCr level was not significantly elevated in the HUA group; however, it was notably decreased on the high-dose anserine intervention compared to the HUA group (Figure 1A–C).

Hyperuricemia is often related to impaired kidney function, which can be reflected by CUA and CCr levels. CUA and CCr levels were significantly decreased in the HUA group but elevated on anserine intervention. Notably, the levels of CCr were elevated in both the HUA group and the anserine intervention groups; however, the anserine groups exhibited higher levels. In the Allo group, there were no significant differences in SUA, BUN or SCr levels compared with the HUA group; however, CUA and CCr levels were elevated (Figure 1D,E).

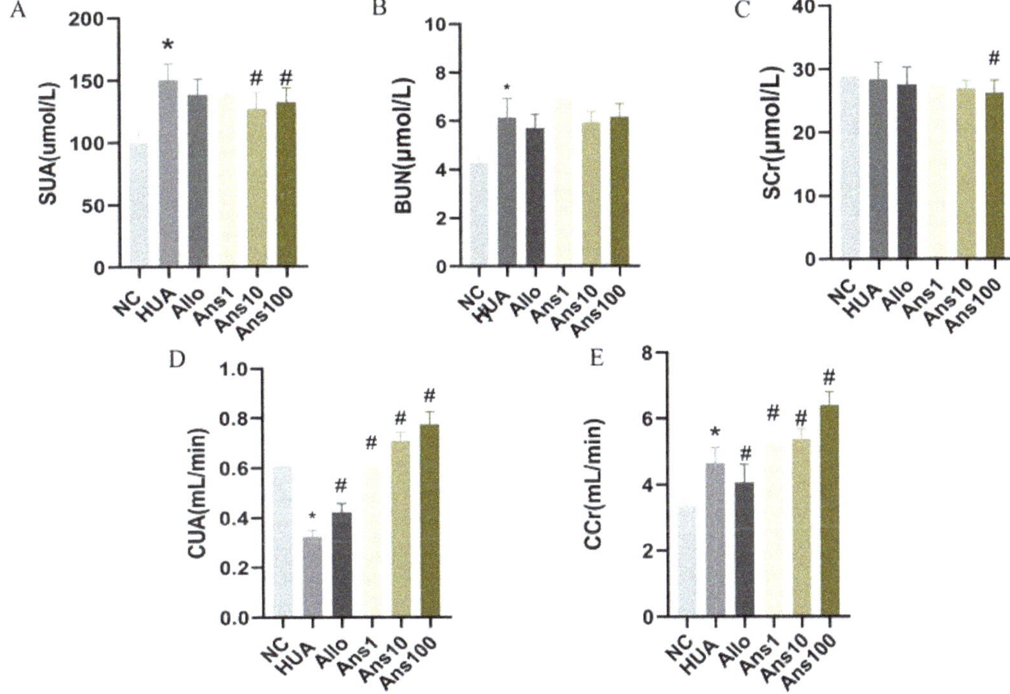

Figure 1. Effect of anserine on hyperuricemia-related indicators. The level of serum (**A**) SUA (serum uric acid); (**B**) BUN (blood urea nitrogen); (**C**) SCr (serum creatinine); (**D**) CUA (uric acid clearance); and (**E**) CCr(creatinine clearance); * $p < 0.05$ indicates that the difference is significant when compared with the NC group. # $p < 0.05$ indicates that the difference is significant when compared with the HUA group.

3.2. Anserine Inhibited Uric Acid Production and Improved Liver Function

Serum XOD and ADA in the HUA group were higher than in the NC group; however, serum ADA was decreased in the Ans100 group and Allo group ($p < 0.05$) (Supplementary Figure S1A,B). Additionally, liver XOD levels did not significantly differ among the groups (Supplementary Figure S1C). In the HUA group, serum ALT and AST levels were increased and A/G levels were decreased compared with the NC group; however, anserine intervention altered these changes and showed similar levels to that of the NC group. Furthermore, allopurinol intervention showed similar effects as those of anserine ($p < 0.05$) (Supplementary Figure S1C–E).

3.3. Anserine Promoted Uric Acid Excretion by Regulating Kidney Urate Transport-Related Proteins

We further investigated the excretion-related proteins of the kidney. URAT1 and GLUT9 mainly mediate uric acid reabsorption, while ABCG2 primarily mediates uric acid excretion. In this study, URAT1 and GLUT9 expressions showed no significant difference in the NC and HUA groups; however, they were notably decreased by anserine intervention in the Ans group ($p < 0.05$). Moreover, ABCG2 expression in the HUA group was lower than that in the NC group but was subsequently reversed by anserine, ($p < 0.05$). However, there was no significant difference between the Allo and HUA groups (Figure 2A–C).

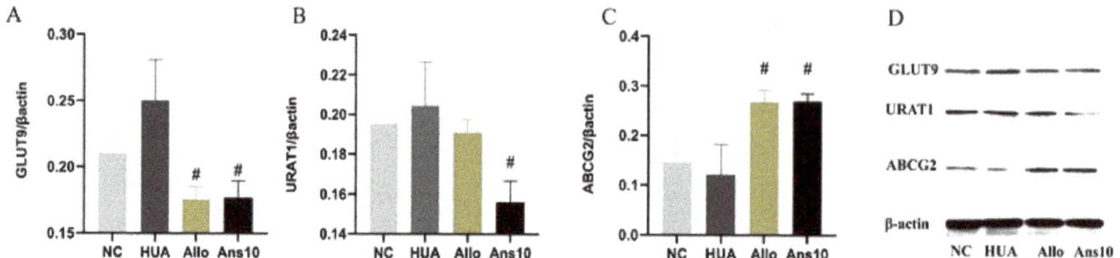

Figure 2. Effect of anserine on kidney urate transport-related proteins. (**A**) GLUT9; (**B**) URAT1; (**C**) ABCG2; and (**D**) Western blot result of GLUT9, URAT1 and ABCG2 protein in each group. $^{\#} p < 0.05$ indicates that the difference is significant when compared to the HUA group.

3.4. Anserine Ameliorated Pathological Changes in the Kidneys in Hyperuricemia Rats

To investigate kidney damage in hyperuricemic rats, a histological evaluation was performed (Figure 3). The kidney surface of the HUA rats was rough and vacuolated; however, after the anserine intervention, the surface became smooth and the vacuole disappeared. Furthermore, in HUA rats, the tubular lumen was dilated and inflammatory cell infiltration was observed in the kidneys, which was greatly improved on anserine intervention. Electron microscopy further showed that the epithelial cells were swollen and kidney podocyte fusion was observed in the HUA group; however, these changes were also reversed on anserine intervention. Although the kidney surface roughness was improved in the Allo group compared to the HUA group, the phenomenon of enlargement and vacuolation along with podocyte fusion remained.

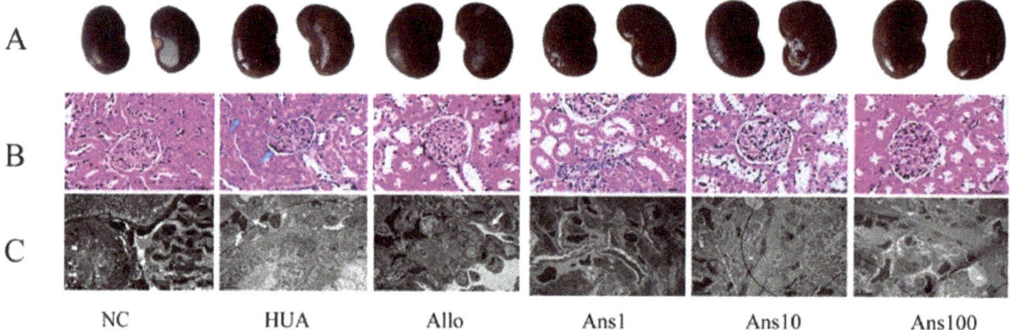

Figure 3. Morphological observation of anserine on kidney pathological damage caused by hyperuricemia. Morphological pictures of different groups of kidneys were observed: (**A**) naked eyes, (**B**) light microscopy and (**C**) electron microscopy.

3.5. Effects of Anserine on the Kidney TLR4/MyD88/NF-κB Pathway, NLRP3 Inflammasome, Nrf2 and Cell Damage-Related Proteins

In our study, compared to the NC group, the expressions of TLR4, MyD88 and NF-κBp65 were enhanced slightly in the HUA group but were decreased on anserine intervention ($p < 0.05$) (Figure 4A–C). Compared to the NC group, NLRP3 expression was increased in the HUA group but decreased on the anserine and allopurinol interventions ($p < 0.05$) (Figure 4D). Moreover, the Nrf2 expression in the NC and HUA groups did not differ significantly; however, the anserine intervention enhanced the Nrf2 expression, while the allopurinol intervention reduced it ($p < 0.05$) (Figure 4E). Furthermore, neither the MMP-2 nor the MMP-9 expression significantly differed among the groups (Figure 4F–H). However, compared to the NC group, the TIMP-1 level in the HUA group decreased but was reversed on the anserine intervention ($p < 0.05$).

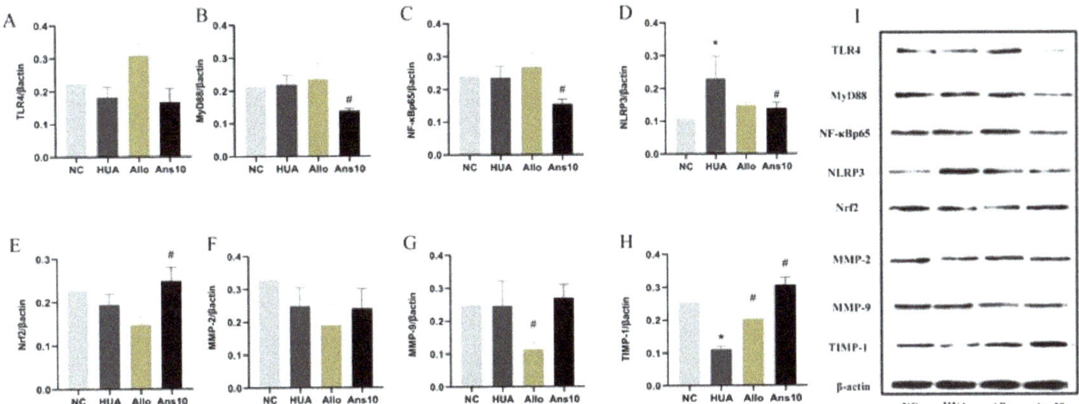

Figure 4. Effect of anserine on the TLR4/MyD88/NF-κB pathway, NLRP3 inflammasome, Nrf2, and MMP-2, MMP-9, and TIMP-1. The relative expression of: (**A**) TLR4; (**B**) MyD88;(**C**) NF-κBp65; (**D**) NLRP3; (**E**) Nrf2; (**F**) MMP-2; (**G**) MMP-9; and (**H**) TIMP-1 among groups. (**I**) Western blot results of TLR4/MyD88/NF-κBp65/NLRP3/Nrf2/MMP-2/MMP-9/TIMP-1 proteins in each group. * $p < 0.05$ indicates that the difference is significant when compared to the NC group. # $p < 0.05$ indicates that the difference is significant when compared to the HUA group.

3.6. Metagenomic Sequencing of Gut Microbiota

3.6.1. Overview of the Changes in Gut Metagenomic Diversity

Significant differences were observed between the gut microbiota structures of the HUA and NC groups and between the anserine dose and HUA groups based on the PCA and PcoA results. (Figure 5A,B) ($p < 0.05$). Intestinal flora diversity was analysed using the Shannon index. The HUA group had lower diversity than the NC group, which was reversed by anserine treatment. However, allopurinol failed to elicit a significant effect (Figure 5C) ($p < 0.05$).

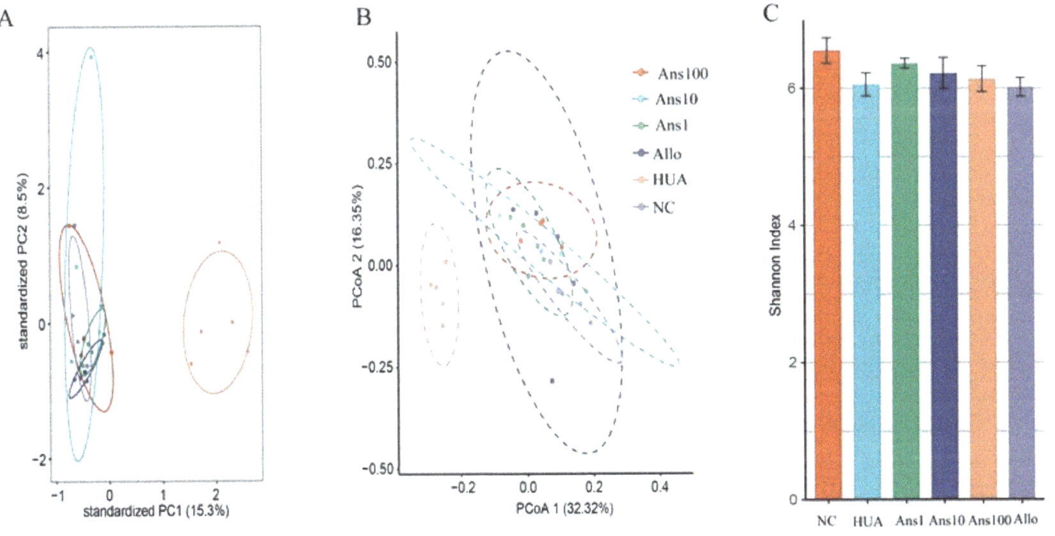

Figure 5. Effect of anserine on the gut microbiota structure and diversity: (**A**) PCA, (**B**) PcoA, and (**C**) Shannon index ($p < 0.05$). ("Ans100" refers to "Anserine 100 mg group"; "Ans10" refers to "Anserine 10 mg group"; "Ans1" refers to "Anserine 1 mg group"; "Allo" refers to "Allopurinol group"; "HUA" refers to "hyperuricemic group" "NC" refers to "Normal control group").

3.6.2. Changes in Gut Microbiota at Different Levels

At the phylum level, compared to the NC group, the abundance of *Bacteroidetes* and *Proteobacteria* was increased and that of *Firmicutes* was decreased in the HUA group; however, anserine and allopurinol treatment reversed these changes (Figure 6A) ($p < 0.05$). Seven different families were identified by LDA analysis. At the family level, the abundance of *Lactobacillaceae* in the HUA group was the lowest, while that of *Lachnospiraceae* and *Clostridiales* was the highest (Figure 6B) ($p < 0.05$).

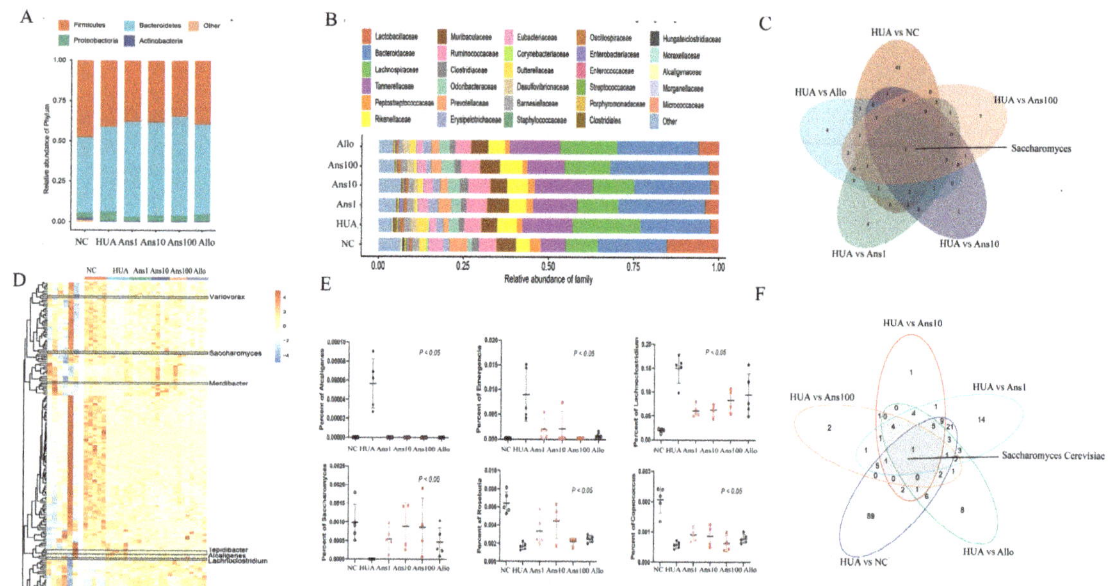

Figure 6. Alterations in the gut microbiota of each group. Abundance distribution: (**A**) phylum level; (**B**) genus level; (**D**) species level; (**E**) The three differential genus with the highest and lowest abundance in the HUA group compared to the Anserine and allopurinol intervention groups. (**C**,**F**) Common genus and species shared with five compared groups ($p < 0.05$).

Furthermore, the lesser method was used to obtain 79 different genera between the five comparison groups (NC vs. HUA, HUA vs. Ans100, HUA vs. Ans 10, HUA vs. Ans1, HUA vs. Allo) (Supplementary Figure S2), with *Saccharomyces* being the common differential genus (Figure 6C). Heatmap analysis revealed the relative abundances of these 79 differential genera among the six groups (Figure 6D). *Alcaligenes*, *Emergencia* and *Lachnoclostridium* were the highest while *Saccharomyces*, *Roseburia* and *Coprococcus* were the lowest in the HUA group (Figure 6E) ($p < 0.05$).

At the species level, based on the results of the OUT comparison, a total of 2568 microbial species were identified. The lesser analysis yielded a total of 189 differential species between the five comparison groups (Supplementary Figure S3), with *Saccharomyces cerevisiae* being the common differential species (Figure 6F) ($p < 0.05$).

3.6.3. Changes in the Metabolic Function of Gut Microbiota

Furthermore, the intestinal flora-related function changes after anserine intervention were examined. Based on the KEGG database, at level 1, the relative abundance of metabolic pathways was the highest in each group, followed by cellular processes (Supplementary Figure S4A). Furthermore, KEGG included 42 subdivided pathways, and differential analysis yielded 25 differential pathways among the groups (Supplementary Figure S4B). The comparison of the abundance of the KEGG map and KO enrichment revealed a heatmap that showed the relative abundance of the top 20 differential maps and the top 30 differential KOs among the six groups (Supplementary Figure S5A,B). Among them, three metabolic pathways in the HUA group were enriched compared to other groups, including D-Arginine and D-ornithine metabolism and Lipoarabinomannan (LAM) biosynthesis (Figure 7A). An in-depth analysis of the D-Arginine and D-ornithine metabolism pathways revealed that the genes 2,4-diaminopentanoate dehydrogenase and D-ornithine 4,5-aminomutase subunit beta corresponding to two proteins ord and oraE were more abundant in the HUA group than other groups but were significantly reduced on anserine and allopurinol interventions (Figure 7B).

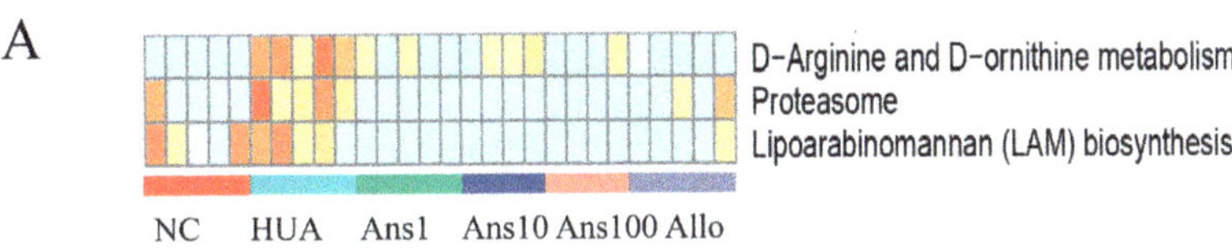

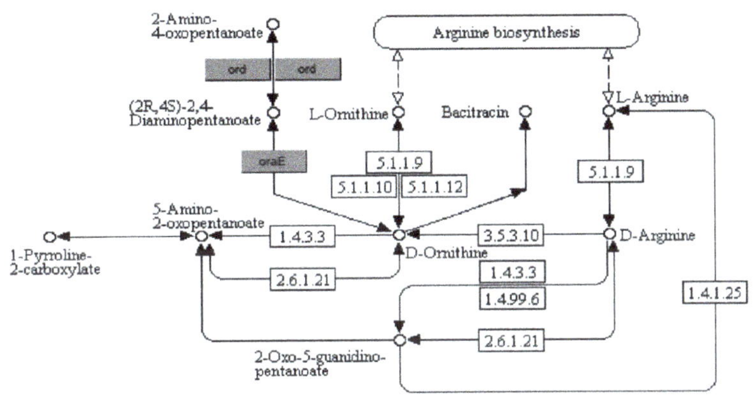

Figure 7. Changes in the metabolic function of gut microbiota: (**A**) three metabolic pathways in the HUA group are richer than other groups, including D-Arginine and D-ornithine metabolism, proteasome and Lipoarabinomannan (LAM) biosynthesis pathways; (**B**) altered genes in the D-arginine and D-ornithine metabolism pathways ($p < 0.05$).

3.7. Urine Metabolomic Analysis

The HUA group was compared with five other groups to form five comparison groups. There were 38 differential metabolites in these five comparison groups (Figure 8A). Figure 5B demonstrates the changing trends in differential metabolites for the five comparison groups. Compared to the NC group, the HUA group had 21 different metabolites, with 11 showing a significant decrease, including galactose, mannose, threonine and D-glucuronic acid. However, inositol, uracil, γ-amino, valine and another 10 metabolites increased significantly in the HUA group. Compared to the HUA group, 15 metabolites were significantly increased and 9 metabolites were significantly decreased in the anserine intervention groups, with fructose, xylose, threonine D-glucuronide and methionine showing an increasing trend, and glycine and pipecolic acid showing a decreasing trend (Figure 8B). Moreover, four metabolites, namely erythronic acid, glucaric acid, pipecolic acid and trans-ferulic acid, were the common differential metabolites of the five comparison groups. Compared with the other groups, the contents of erythronic acid, glucaric acid and trans-ferulic acid in the HUA group were lower, while that of pipecolic acid was higher (Figure 8C).

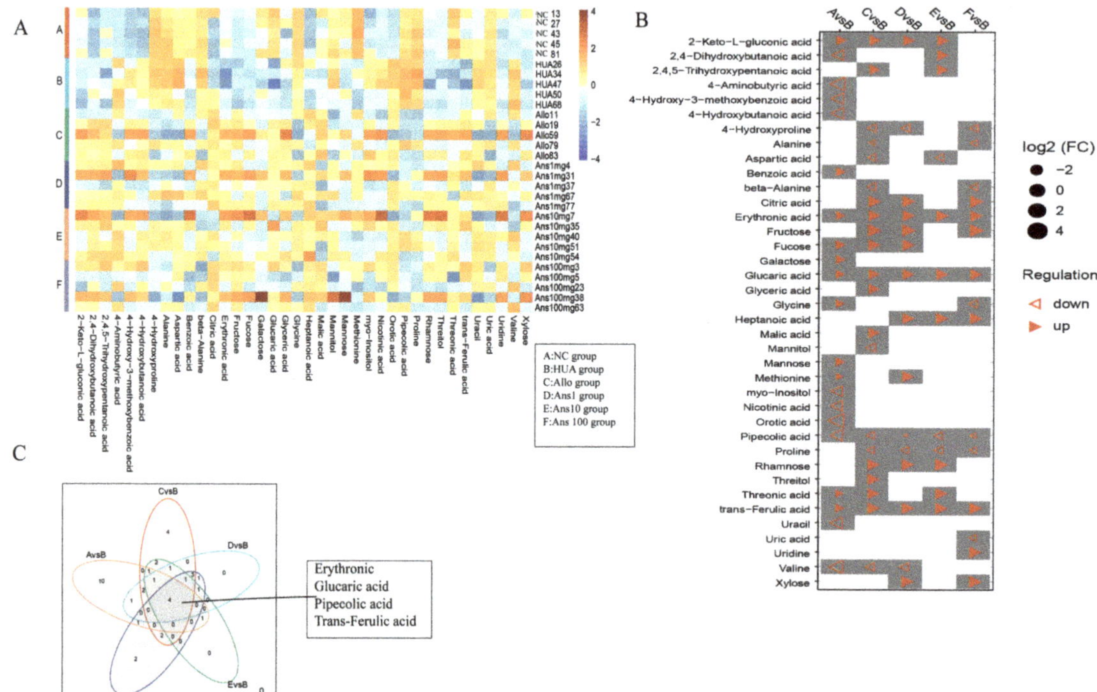

Figure 8. Differential urine metabolites among groups: (**A**) the heatmap of the relative level of 38 differential metabolites in each group; (**B**) the trend changes for 38 differential metabolites in five comparison groups; (**C**) the Venn diagram shows common metabolites for the five comparison groups ($p < 0.05$).

3.8. Association Analysis of Differential Gut Microbiota and Differential Urinary Metabolites

We explored the correlation of 38 differential metabolites with 79 differential genera. The results showed that methionine was significantly correlated with 39 genera, which was the highest association (Supplementary Figure S6). Among all genera, *Parasutterella*, *Oligella*, *Catabacter*, *Emergencia* and *Bacteroides* were significantly associated with 18, 15, 13, 13 and 13 metabolites, respectively, and formed the top five genera associated with most differential metabolites. Among them, *Saccharomyces* was only related to methionine (Supplementary

Figure S6). At the species level, we analysed 189 differences in species and approximately 38 differences in metabolites (Supplementary Figure S7). Different species that were significantly associated with erythronic acid, glucaric acid, pipecolic acid and trans-ferulic acid were extracted, revealing three species, *Parasutterella excrementihominis*, *Emergencia timonensis* and *Bacteroides uniformis*. Considering that pipecolic acid and differential species *Parasutterella excrementihominis* and *Emergencia timonensis* were more abundant in the HUA group than in the other five groups, we speculated that there was a positive correlation between them. Conversely, erythronic acid, glucaric acid and trans-ferulic acid were negatively correlated with the abundance of *Parasutterella excrementihominis* and *Emergencia timonensis* (Figure 9). Furthermore, glucaric acid was associated with the greatest number of differential species (Supplementary Figure S7).

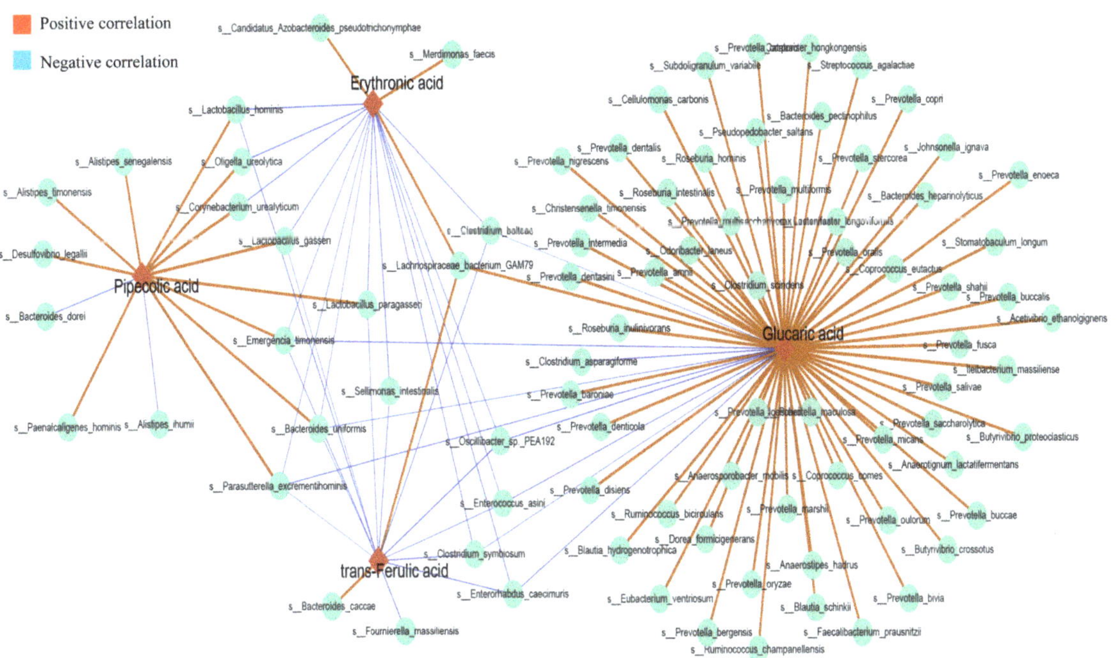

Figure 9. Species that are significantly associated with erythroid acid, glucaric acid, pipecolic acid, and trans-ferulic acid. The orange and blue lines represent positive and negative correlations, respectively ($p < 0.05$).

4. Discussion

This study used yeast combined with potassium oxalate to establish a rat model with hyperuricemia. Anserine was found to significantly reduce blood uric acid levels and improve liver and kidney damage, indicating that anserine had a preventative effect on hyperuricemia. Additionally, the changes in the gut microbiota structure and function as well as host metabolism, which were induced by hyperuricemia, were partially reversed by anserine. Moreover, the underlying mechanism of anserine ameliorating hyperuricemia was explored using integrated macrogenomics and metabolomics for the first time, to the best of our knowledge, in this study.

It is reported that, in general, bonito, bigeye tuna, and southern tuna contain about 1070 mg, 1260 mg and 636 mg of anserine per 100 g of fish flesh, respectively [17]. We learned that the doses of anserine used in previous animal experiments ranged from 2 mg/kg·bw to 80 mg/kg·bw [18,19], so we chose the 1 mg/10 mg/100 mg kg·bw as the low/medium/high doses of anserine intervention in our experiment. Our results did show that medium doses of anserine were the most effective in improving hyperuricemia.

Consequently, we speculate that the effect of anserine in improving hyperuricemia is dose-dependent. Therefore, in subsequent experiments using Western blotting to detect the expression of uric acid-related transporter proteins and proteins associated with renal inflammation, oxidative stress, and cellular damage in the kidney, we only examined kidney samples from the mid-dose group of the goose-peptide intervention.

Initially, the effect of anserine on uric acid production and excretion was investigated. Anserine showed no obvious effects on ADA and XOD enzymes, which are closely related to uric acid production. However, anserine ameliorated liver damage induced by hyperuricemia. Then, the effect of anserine on kidney uric acid excretion was investigated. URAT1, GLUT9 and ABCG2 expression changes indicated that anserine reduced serum uric acid levels by inhibiting uric acid reabsorption and promoting uric acid excretion. The changes in CUA and CCr, two indicators of kidney function, also implied that anserine improved kidney function, which is consistent with the morphological changes observed in the kidneys. It is noteworthy that, in our experiments, the levels of CCr were elevated in the HUA group. We assume this is very likely because when the blood creatinine in hyperuricemic rats started to rise, the body compensated by increasing the renal creatinine clearance to keep the body's creatinine level at a steady state. Additionally, the timing of the measurement also has some influence on the results. In short, this phenomenon is very interesting and deserves further in-depth study.

We further investigated the effect of anserine on kidney injury caused by hyperuricemia by evaluating the effect of anserine on kidney inflammation, oxidative stress and cellular injuries. The TLR4/MyD88/NF-κB and NLRP3 proteins are reported to be the major signalling pathways closely associated with renal inflammation caused by hyperuricemia [20]. Contrarily, Nrf-2 neutralises the activation of cellular oxidative stress and ameliorates kidney injury by inhibiting NF-κB expression [21]. Moreover, MMP2 and MMP9 are reported to cleave collagen IV in the basement membrane of cell bands and have activity in kidney tissue [22]. Furthermore, TIMP-1 inhibits the activities of most MMPs, thus improving cell damage [23]. Given that anserine decreased the expression of TLR4/MyD88/NF-κB and NLRP3 and elevated Nrf2 and TIMP1 in hyperuricemic rats, we speculate that anserine improves overall kidney function by decreasing inflammation, oxidative stress and cellular damage in hyperuricemia. Anserine also regulates the expressions of uric acid-related transporters, thereby reducing serum uric acid levels.

Increasing evidence suggests that gut microbiota balance is closely related to metabolic disorders, and patients with hyperuricemia have a different intestinal microbiota structure from normal individuals [24]. Abnormally high levels of uric acid in the blood enter the intestine and affect the steady state of the intestinal flora, thereby affecting the intestinal metabolism of uric acid and consequently aggravating hyperuricemia. Therefore, we analysed the effect of anserine on the changes in intestinal flora structure caused by hyperuricemia. Our study showed a decrease in the intestinal microbial diversity of hyperuricemic rats, which is consistent with previous studies [24]; however, this was reversed on anserine intervention.

Uric acid is the end product of purine metabolism; moreover, intestinal flora has also been shown to play an essential role in purine oxidative metabolism. For example, *Escherichia coli* in the human gut produces XOD to influence the production of uric acid [25]. The *Lactobacillaceae* family inhibits the growth of *E. coli* by secreting reuterin [26], indirectly inhibiting uric acid accumulation. Additionally, *Lactobacillus* can synthesize various UA metabolic enzymes, such as uricase, allantoinase and allantoicase, which can decompose uric acid into 5-hydroxyisothreonate, allantoin, allantoate and finally degrade it into urea [27]. Similarly, the *Clostridiaceae* family also degrades uric acid [28]. *Saccharomyces cerevisiae* is a fungus that secretes urate oxidase, which can catalyse uric acid oxidation and plays an essential role in the purine degradation pathway, thereby preventing uric acid accumulation [29]. This study showed that the abundance of *Lactobacillaceae*, *Clostridiaceae* family and *Saccharomyces cerevisiae* was reduced in hyperuricemic rats but elevated after the anserine intervention, suggesting the preventive effect of anserine was partially due to the changes in some specific microbiota.

A dysregulated gut microbiota is accompanied by imbalanced intestinal metabolites, such as trimethylamine, short-chain fatty acids (SCFAs) and LPS, which are considered mediators between the intestinal microbiome and their human hosts [30]. Anserine increases the abundance of *Roseburia* and *Coprococcus* in hyperuricemic rats, which are crucial in SCFA generation. SCFAs regulate gut microbiota homeostasis, repair intestinal permeability and are beneficial to kidney function [31]. Moreover, butyrate, a major SCFA in the intestine, is reported to be increased in a *Lactobacillaceae*-enriched environment [32]. Therefore, *Lactobacillaceae* is speculated to not only participate in purine metabolism but also play a role in increasing butyrate levels in the intestinal tract. Additionally, gut microbiome dysbiosis can cause the excretion of LPS from the cell walls of Gram-negative bacteria, and the inflammation in the liver and kidney is further activated by the excreted LPS entering the bloodstream through a disrupted gut barrier [33]. Furthermore, members of the Gram-negative *Proteobacteria phylum*, *Alcaligenes* genus and *Lachnoclostridium* genus were increased in the HUA group but were reduced on anserine intervention. Additionally, a Proteobacterial strain has also been shown to enhance intestinal nitrogen fixation [34], wherein, nitrogen is converted to ammonia. Notably, excess ammonia entering the host's circulatory system through the intestinal barrier can aggravate kidney damage. Additionally, we observed the *Emergencia timonensis* genus was more enriched in the HUA group than in other groups. Furthermore, *Emergencia timonensis*, a potential key bacterium for the conversion of carnitine to trimethylamine N-oxide (TMAO), is also a toxin that can aggravate kidney damage [35]. This study indicated that anserine alleviated hyperuricemia owing to its ability to maintain the balance in the composition of the intestinal microbiota (the increase in beneficial bacteria and the decrease in pathogenic bacteria). Moreover, it also promotes purines and uric acid catabolism, regulates intestinal epithelial cell proliferation, reduces chronic inflammation and improves uric acid excretion.

The intestinal microflora greatly affects the health of the host by regulating its metabolic function. In this study, six metabolic pathways were altered in the HUA group compared to the NC group, three of which were elevated and the other three were decreased; however, anserine intervention reversed these changes. For example, D-Arginine and D-ornithine metabolism pathways were significantly enriched by hyperuricemia but anserine supplementation reversed the change, which was verified by the changes in two key proteins in this pathway, namely 4-diaminopentanoate dehydrogenase and D-ornithine 4,5-aminomutase subunit beta. The D-Arginine and D-ornithine metabolism are related to the urea cycle, indicating that more urea is metabolized in the intestine to produce ammonia in hyperuricemia. The disturbance of the gut microbiota combined with the reduction in beneficial metabolites such as SCFAs increases the permeability of the intestine, which increases the ammonia levels entering the circulation system, thereby aggravating liver and kidney function.

Metabolic profiling of host biofluids provides profound insights into the gut microbiota's impact on host health/disease status, therefore exploring differential urinary metabolites aids in identifying the causative agent rather than the presence of the metabolite [36]. By comparing urinary metabolites in the HUA group with the different doses of the anserine group, Allo and NC groups, we identified erythronic acid, glucaric acid, pipecolic acid and trans-ferulic acid as the four common differential metabolites. This suggests that these four metabolites and their associated metabolic pathways play a critical role in the pathogenesis and amelioration of hyperuricemia. Erythronic acid is related to mitochondrial dysfunction in transaldolase deficiency [37], highlighting its role in mediating energy metabolism in humans. D-gluconic acid has toxin-reducing and antioxidant abilities, wherein it can improve diabetic kidney tubular damage by inhibiting inositol oxygenase, preventing mitochondrial damage and apoptosis and reducing oxidative stress through the ascorbic acid and aldehyde metabolic pathway, thereby improving kidney function [38]. Moreover, ferulic acid has been shown to lighten oxidative stress through the activation of the AMPK signalling pathway in vitro [39]. Pipecolic acid is an intermediate in the lysine degradation pathway, with an enhanced lysine degradation pathway indicating enhanced levels of oxidative stress in the host [40]. Thus, by enhancing the levels of erythronic acid, glucaric acid and trans−ferulic acid and decreasing the levels of pipecolic acid in hyperuricemic rats, anserine exerts an anti-hyperuricemia effect by improving energy metabolism and reducing oxidative stress and inflammation. Notably, *Parasutterella excrementihominis*, *Emergencia timonensis* and *Bacteroides uniformis* were associated with these four metabolites. As *Parasutterella excrementihominis* and *Emergencia timonensis* are positively associated with pipecolic acid but negatively associated with erythronic acid, glucaric acid and trans-ferulic acid, we speculated that anserine primarily reduced the abundance of *Parasutterella excrementihominis* and *Emergencia timonensis* to exert an ameliorating effect on kidney injuries.

Notably, methionine was associated with the highest number of differential genera and species; however, the *Saccharomyces* genus was only correlated with methionine. Methionine produces strong antioxidative metabolites such as glutathione, cysteine and sulphate through the trans-sulphuration pathway. Previous studies have demonstrated the ameliorative effect of methyl and S-adenosylmethionine produced by the methionine cycle on systemic inflammation and liver damage [41]. The methionine cycle is widely active in *Saccharomyces cerevisiae* [42]. This suggests that *Saccharomyces cerevisiae* could be a target probiotic for anserine to improve hyperuricemia. Additionally, the anserine group exhibited significantly higher levels of two differential metabolites (fructose and xylose) in the starch and sucrose metabolism pathway than the HUA group. Starch and sucrose metabolic pathways are directly related to the development of diseases involved in energy metabolism and insulin resistance, such as obesity, diabetes and kidney tubular dysfunction [43].

5. Conclusions

This study reveals the beneficial effects of anserine on the reversal of hyperuricemia. It exerts a beneficial effect by regulating intestinal microbiota and host metabolites. Additionally, the key differential metabolites that improve anserine-associated hyperuricemia were identified to be fructose, xylose, methionine, erythronic acid, glucaric acid, pipecolic acid and trans-ferulic acid, and key differential gut microbiota were identified to be *Saccharomyces*, *Parasutterella excrementihominis* and *Emergencia timonensis*, which are involved in the gut–kidney axis. These key microbiota and metabolites have the potential as disease markers to predict the onset of disease, thereby improving the efficiency and accuracy of early clinical diagnosis. However, this study also has some limitations. This study lacked gut metabolite and enzyme data related to uric acid metabolism, which requires further study. Furthermore, this study provides insight into the pathogenesis of hyperuricemia and highlights the anti-hyperuricemic properties of anserine.

Supplementary Materials: The following Supplementary Materials can be downloaded at: https://www.mdpi.com/article/10.3390/nu15040969/s1. Figure S1. Effect of anserine on uric acid production and liver function-related indicators. Figure S2. Using the Lefser method obtained differential genus (LDA > 2) identification between the HUA group and the NC group, Allo group, Ans 1 mg, Ans 10 mg and Ans 100 mg group. Figure S3. Using the Lefser method obtained differential species (LDA > 2) identification between the HUA group and the NC group, Allo group, Ans 1 mg, Ans 10 mg, and Ans 100 mg group. Figure S4. Annotation of genes in different groups in the KEGG database. Figure S5. The enrichment of genes in different groups in the KEGG map and KO. Figure S6. Association analysis between metabolites and differential genera. Figure S7. Association analysis between metabolites and differential species.

Author Contributions: Formal analysis, M.H. and R.W.; investigation, S.A., X.H., C.Y., Z.Z. (Zongfeng Zhang) and W.Z.; methodology, M.H. and Z.Z. (Zhaofeng Zhang); visualization, L.L.; writing—original draft, M.H.; writing—review and editing, L.H. All authors have read and agreed to the published version of the manuscript.

Funding: This research is supported by National Key R&D Program of China, the project number is 2022YFF1100104.

Institutional Review Board Statement: All laboratory procedures were reviewed and approved by the Institutional Animal Care and Use Committee of Peking University (Approved No. 2019PHE017) and conformed to the Guidelines for the Care and Use of Laboratory Animals (NIH Publication No. 85-23, Revised 1996).

Informed Consent Statement: Not applicable.

Data Availability Statement: The data that supports the findings of this study are available from the corresponding author upon reasonable request.

Acknowledgments: This study did not receive any specific grant from funding agencies in the public, commercial or not-for-profit sectors.

Conflicts of Interest: The authors declare that they have no competing interest.

Abbreviations

NOD-like receptor (NLR), NOD-like receptor family containing pyrimidine domain 3 (NLRP3), interleukin-1β (IL-1β), lipopolysaccharide (LPS), Toll-like receptor 4 (TLR4), ultraperformance liquid chromatography–tandem mass spectrometry (UPLC–MS), myeloid differentiation factor88 (MyD88), nuclear factor E2-related factor 2 (Nrf2), ATP-binding cassette, subfamily G, member 2 (ABCG2), matrix metallopeptidase 2 (MMP2), matrix metallopeptidase 2 (MMP9), nuclear factor-κB (NF-kB), urate transporter 1 (URAT1), glucose transporter 9 (GLUT9), tissue inhibitor of metalloproteinase1(TIMP-1), normal control group (NC group), serum uric acid (SUA), serum creatinine (SCr), blood urea nitrogen (BUN), urinary uric acid (UUA), urinary creatinine (UCr), alanine aminotransferase (ALT), aspartate aminotransferase (AST), albumin (ALB), globulin (GLB), albumin/globulin (A/G) and total protein (TP), Uric acid clearance (CUA), creatinine clearance (CCr), analysis of variance (ANOVA), least significant difference (LSD), short-chain fatty acids (SCFAs), trimethylamine N-oxide (TMAO).

References

1. Chen-Xu, M.; Yokose, C.; Rai, S.K.; Pillinger, M.H.; Choi, H.K. Contemporary Prevalence of Gout and Hyperuricemia in the United States and Decadal Trends: The National Health and Nutrition Examination Survey, 2007–2016. *Arthritis Rheumatol.* **2019**, *71*, 991–999. [CrossRef] [PubMed]
2. Huang, J.; Ma, Z.F.; Zhang, Y.; Wan, Z.; Li, Y.; Zhou, H.; Chu, A.; Lee, Y.Y. Geographical distribution of hyperuricemia in mainland China: A comprehensive systematic review and meta-analysis. *Glob. Health Res. Policy* **2020**, *5*, 52. [CrossRef] [PubMed]
3. Wei, C.Y.; Sun, C.C.; Wei, J.C.; Tai, H.C.; Sun, C.A.; Chung, C.F.; Chou, Y.C.; Lin, P.L.; Yang, T. Association between Hyperuricemia and Metabolic Syndrome: An Epidemiological Study of a Labor Force Population in Taiwan. *BioMed Res. Int.* **2015**, *2015*, 369179. [CrossRef] [PubMed]
4. Browne, L.D.; Jaouimaa, F.Z.; Walsh, C.; Perez-Ruiz, F.; Richette, P.; Burke, K.; Stack, A.G. Serum uric acid and mortality thresholds among men and women in the Irish health system: A cohort study. *Eur. J. Intern. Med.* **2021**, *84*, 46–55. [CrossRef] [PubMed]
5. El Ridi, R.; Tallima, H. Physiological functions and pathogenic potential of uric acid: A review. *J. Adv. Res.* **2017**, *8*, 487–493. [CrossRef] [PubMed]
6. Yang, C.Y.; Chen, C.H.; Deng, S.T.; Huang, C.S.; Lin, Y.J.; Chen, Y.J.; Wu, C.Y.; Hung, S.I.; Chung, W.H. Allopurinol Use and Risk of Fatal Hypersensitivity Reactions: A Nationwide Population-Based Study in Taiwan. *JAMA Intern. Med.* **2015**, *175*, 1550–1557. [CrossRef] [PubMed]
7. Major, T.J.; Topless, R.K.; Dalbeth, N.; Merriman, T.R. Evaluation of the diet wide contribution to serum urate levels: Meta-analysis of population based cohorts. *BMJ Clin. Res. Ed.* **2018**, *363*, k3951. [CrossRef] [PubMed]
8. Mandal, A.K.; Mount, D.B. The molecular physiology of uric acid homeostasis. *Annu. Rev. Physiol.* **2015**, *77*, 323–345. [CrossRef] [PubMed]
9. Chen, M.; Lu, X.; Lu, C.; Shen, N.; Jiang, Y.; Chen, M.; Wu, H. Soluble uric acid increases PDZK1 and ABCG2 expression in human intestinal cell lines via the TLR4-NLRP3 inflammasome and PI3K/Akt signaling pathway. *Arthritis Res. Ther.* **2018**, *20*, 20. [CrossRef] [PubMed]
10. Cao, J.Y.; Zhou, L.T.; Li, Z.L.; Yang, Y.; Liu, B.C.; Liu, H. Dopamine D1 receptor agonist A68930 attenuates acute kidney injury by inhibiting NLRP3 inflammasome activation. *J. Pharmacol. Sci.* **2020**, *143*, 226–233. [CrossRef] [PubMed]

11. Monteiro, R.C.; Berthelot, L. Role of gut-kidney axis in renal diseases and IgA nephropathy. *Curr. Opin. Gastroenterol.* **2021**, *37*, 565–571. [CrossRef]

12. Yang, T.; Richards, E.M.; Pepine, C.J.; Raizada, M.K. The gut microbiota and the brain-gut-kidney axis in hypertension and chronic kidney disease. *Nat. Rev. Nephrol.* **2018**, *14*, 442–456. [CrossRef] [PubMed]

13. Cani, P.D.; Bibiloni, R.; Knauf, C.; Waget, A.; Neyrinck, A.M.; Delzenne, N.M.; Burcelin, R. Changes in gut microbiota control metabolic endotoxemia-induced inflammation in high-fat diet-induced obesity and diabetes in mice. *Diabetes* **2008**, *57*, 1470–1481. [CrossRef] [PubMed]

14. Li, Q.; Kang, X.; Shi, C.; Li, Y.; Majumder, K.; Ning, Z.; Ren, J. Moderation of hyperuricemia in rats via consuming walnut protein hydrolysate diet and identification of new antihyperuricemic peptides. *Food Funct.* **2018**, *9*, 107–116. [CrossRef] [PubMed]

15. Wu, G. Important roles of dietary taurine, creatine, carnosine, anserine and 4-hydroxyproline in human nutrition and health. *Amino Acids* **2020**, *52*, 329–360. [CrossRef] [PubMed]

16. Han, J.; Wang, Z.; Lu, C.; Zhou, J.; Li, Y.; Ming, T.; Zhang, Z.; Wang, Z.J.; Su, X. The gut microbiota mediates the protective effects of anserine supplementation on hyperuricaemia and associated renal inflammation. *Food Funct.* **2021**, *12*, 9030–9042. [CrossRef] [PubMed]

17. Boldyrev, A.A.; Aldini, G.; Derave, W. Physiology and pathophysiology of carnosine. *Physiol. Rev.* **2013**, *93*, 1803–1845. [CrossRef]

18. Chen, M.; Ji, H.; Song, W.; Zhang, D.; Su, W.; Liu, S. Anserine beneficial effects in hyperuricemic rats by inhibiting XOD, regulating uric acid transporter and repairing hepatorenal injury. *Food Funct.* **2022**, *13*, 9434–9442. [CrossRef]

19. Sakano, T.; Egusa, A.S.; Kawauchi, Y.; Wu, J.; Nishimura, T.; Nakao, N.; Kuramoto, A.; Kawashima, T.; Shiotani, S.; Okada, Y.; et al. Pharmacokinetics and tissue distribution of orally administrated imidazole dipeptides in carnosine synthase gene knockout mice. *Biosci. Biotechnol. Biochem.* **2022**, *86*, 1276–1285. [CrossRef]

20. Tan, J.; Wan, L.; Chen, X.; Li, X.; Hao, X.; Li, X.; Li, J.; Ding, H. Conjugated Linoleic Acid Ameliorates High Fructose-Induced Hyperuricemia and Renal Inflammation in Rats via NLRP3 Inflammasome and TLR4 Signaling Pathway. *Mol. Nutr. Food Res.* **2019**, *63*, e1801402. [CrossRef]

21. Hejazian, S.M.; Hosseiniyan Khatibi, S.M.; Barzegari, A.; Pavon-Djavid, G.; Razi Soofiyani, S.; Hassannejhad, S.; Ahmadian, E.; Ardalan, M.; Zununi Vahed, S. Nrf-2 as a therapeutic target in acute kidney injury. *Life Sci.* **2021**, *264*, 118581. [CrossRef] [PubMed]

22. Wang, W.; Schulze, C.J.; Suarez-Pinzon, W.L.; Dyck, J.R.; Sawicki, G.; Schulz, R. Intracellular action of matrix metalloproteinase-2 accounts for acute myocardial ischemia and reperfusion injury. *Circulation* **2002**, *106*, 1543–1549. [CrossRef] [PubMed]

23. Sharma, C.; Dobson, G.P.; Davenport, L.M.; Morris, J.L.; Letson, H.L. The role of matrix metalloproteinase-9 and its inhibitor TIMP-1 in burn injury: A systematic review. *Int. J. Burns Trauma* **2021**, *11*, 275–288. [PubMed]

24. Xi, Y.; Yan, J.; Li, M.; Ying, S.; Shi, Z. Gut microbiota dysbiosis increases the risk of visceral gout in goslings through translocation of gut-derived lipopolysaccharide. *Poult. Sci.* **2019**, *98*, 5361–5373. [CrossRef] [PubMed]

25. Crane, J.K. Role of host xanthine oxidase in infection due to enteropathogenic and Shiga-toxigenic *Escherichia coli*. *Gut Microbes* **2013**, *4*, 388–391. [CrossRef] [PubMed]

26. Azad, M.A.K.; Sarker, M.; Li, T.; Yin, J. Probiotic Species in the Modulation of Gut Microbiota: An Overview. *BioMed Res. Int.* **2018**, *2018*, 9478630. [CrossRef]

27. Hsieh, C.Y.; Lin, H.J.; Chen, C.H.; Lai, E.C.; Yang, Y.K. Chronic kidney disease and stroke. *Lancet Neurol.* **2014**, *13*, 1071. [CrossRef]

28. Hartwich, K.; Poehlein, A.; Daniel, R. The purine-utilizing bacterium *Clostridium acidurici* 9a: A genome-guided metabolic reconsideration. *PLoS ONE* **2012**, *7*, e51662. [CrossRef] [PubMed]

29. Fazel, R.; Zarei, N.; Ghaemi, N.; Namvaran, M.M.; Enayati, S.; Mirabzadeh Ardakani, E.; Azizi, M.; Khalaj, V. Cloning and expression of *Aspergillus flavus* urate oxidase in *Pichia pastoris*. *SpringerPlus* **2014**, *3*, 395. [CrossRef] [PubMed]

30. Pan, L.; Han, P.; Ma, S.; Peng, R.; Wang, C.; Kong, W.; Cong, L.; Fu, J.; Zhang, Z.; Yu, H.; et al. Abnormal metabolism of gut microbiota reveals the possible molecular mechanism of nephropathy induced by hyperuricemia. *Acta Pharm. Sin. B* **2020**, *10*, 249–261. [CrossRef] [PubMed]

31. van der Hee, B.; Wells, J.M. Microbial Regulation of Host Physiology by Short-chain Fatty Acids. *Trends Microbiol.* **2021**, *29*, 700–712. [CrossRef] [PubMed]

32. Zhu, L.B.; Zhang, Y.C.; Huang, H.H.; Lin, J. Prospects for clinical applications of butyrate-producing bacteria. *World J. Clin. Pediatr.* **2021**, *10*, 84–92. [CrossRef]

33. Bian, M.; Wang, J.; Wang, Y.; Nie, A.; Zhu, C.; Sun, Z.; Zhou, Z.; Zhang, B. Chicory ameliorates hyperuricemia via modulating gut microbiota and alleviating LPS/TLR4 axis in quail. *Biomed. Pharmacother.* **2020**, *131*, 110719. [CrossRef] [PubMed]

34. Ni, J.; Shen, T.D.; Chen, E.Z.; Bittinger, K.; Bailey, A.; Roggiani, M.; Sirota-Madi, A.; Friedman, E.S.; Chau, L.; Lin, A.; et al. A role for bacterial urease in gut dysbiosis and Crohn's disease. *Sci. Transl. Med.* **2017**, *9*, eaah6888. [CrossRef] [PubMed]

35. Wu, W.K.; Panyod, S.; Liu, P.Y.; Chen, C.C.; Kao, H.L.; Chuang, H.L.; Chen, Y.H.; Zou, H.B.; Kuo, H.C.; Kuo, C.H.; et al. Characterization of TMAO productivity from carnitine challenge facilitates personalized nutrition and microbiome signatures discovery. *Microbiome* **2020**, *8*, 162. [CrossRef]

36. Holmes, E.; Li, J.V.; Marchesi, J.R.; Nicholson, J.K. Gut microbiota composition and activity in relation to host metabolic phenotype and disease risk. *Cell Metab.* **2012**, *16*, 559–564. [CrossRef]

37. Engelke, U.F.; Zijlstra, F.S.; Mochel, F.; Valayannopoulos, V.; Rabier, D.; Kluijtmans, L.A.; Perl, A.; Verhoeven-Duif, N.M.; de Lonlay, P.; Wamelink, M.M.; et al. Mitochondrial involvement and erythronic acid as a novel biomarker in transaldolase deficiency. *Biochim. Biophys. Acta* **2010**, *1802*, 1028–1035. [CrossRef]
38. Zhan, M.; Usman, I.M.; Sun, L.; Kanwar, Y.S. Disruption of renal tubular mitochondrial quality control by Myo-inositol oxygenase in diabetic kidney disease. *J. Am. Soc. Nephrol. JASN* **2015**, *26*, 1304–1321. [CrossRef]
39. Singh, S.S.B.; Patil, K.N. trans-ferulic acid attenuates hyperglycemia-induced oxidative stress and modulates glucose metabolism by activating AMPK signaling pathway in vitro. *J. Food Biochem.* **2022**, *46*, e14038. [CrossRef]
40. Fujimoto, K.; Kishino, H.; Hashimoto, K.; Watanabe, K.; Yamoto, T.; Mori, K. Biochemical profiles of rat primary cultured hepatocytes following treatment with rotenone, FCCP, or (+)-usnic acid. *J. Toxicol. Sci.* **2020**, *45*, 339–347. [CrossRef]
41. Vaccaro, J.A.; Naser, S.A. The Role of Methyl Donors of the Methionine Cycle in Gastrointestinal Infection and Inflammation. *Healthcare* **2021**, *10*, 61. [CrossRef] [PubMed]
42. Chen, H.; Wang, Z.; Cai, H.; Zhou, C. Progress in the microbial production of S-adenosyl-L-methionine. *World J. Microbiol. Biotechnol.* **2016**, *32*, 153. [CrossRef] [PubMed]
43. Rani, L.; Saini, S.; Shukla, N.; Chowdhuri, D.K.; Gautam, N.K. High sucrose diet induces morphological, structural and functional impairments in the renal tubules of *Drosophila melanogaster*: A model for studying type-2 diabetes mediated renal tubular dysfunction. *Insect Biochem. Mol. Biol.* **2020**, *125*, 103441. [CrossRef] [PubMed]

 **nutrients**

MDPI

Article

Dietary Protein Intake and Physical Function in Māori and Non-Māori Adults of Advanced Age in New Zealand: LiLACS NZ

Maia Lingman [1], Ngaire Kerse [2], Marama Muru-Lanning [3] and Ruth Teh [2,*]

[1] Te Whatu Ora, Waitematā, Auckland 0622, New Zealand
[2] School of Population Health, The University of Auckland, Auckland 1023, New Zealand
[3] James Henare Māori Research Centre, The University of Auckland, Auckland 1023, New Zealand
* Correspondence: r.teh@auckland.ac.nz

Abstract: The population of older adults is growing exponentially. Research shows that current protein intake recommendations are unlikely to meet the ageing requirements and may be linked to reduced physical function. Ensuring optimal function levels is crucial for independence and quality of life in older age. This study aims to quantify the protein intake in those over 90 years of age and determine the association between historical protein intake (2011) and subsequent physical function at ten years follow-up (2021). Eighty-one participants (23 Māori and 54 non-Māori) undertook dietary assessment 24 h multiple-pass recall (MPR) and a standardised health and social questionnaire with physical assessment in 2011 and 2021. Intake24, a virtual 24 h MPR, was utilised to analyse dietary intake. Functional status was measured using the Nottingham Extended Activities of Daily Living Scale (NEADL), and physical performance was the Short Physical Performance Battery (SPPB). Māori men and women consumed less protein (g/day) in 2021 than in 2011 (P = 0.043 in men), but weight-adjusted protein intake in Māori participants over the ten years was not significantly reduced. Both non-Māori men and women consumed significantly less protein (g/day) between 2011 and 2021 ($p = 0.006$ and $p = 0.001$, respectively), which was also significant when protein intake was adjusted for weight in non-Māori women ($p = 0.01$). Weight-adjusted protein intake in 2011 was independently associated with functional status (NEADL score) in 2021 ($p =< 0.001$). There was no association between past protein intake and SPPB score ($p = 0.993$). Animal protein was replaced with plant-based protein over time. In conclusion, a reduction in protein intake was seen in all participants. The independent association between past protein intake and future functional status supports recommendations to keep protein intake high in advanced age.

Keywords: protein; physical performance; older adults; nonagenarians; indigenous

Citation: Lingman, M.; Kerse, N.; Muru-Lanning, M.; Teh, R. Dietary Protein Intake and Physical Function in Māori and Non-Māori Adults of Advanced Age in New Zealand: LiLACS NZ. *Nutrients* **2023**, *15*, 1664. https://doi.org/10.3390/nu15071664

Academic Editors: Lei Zhao and Liang Zhao

Received: 21 February 2023
Revised: 23 March 2023
Accepted: 24 March 2023
Published: 29 March 2023

1. Introduction

New Zealand has a growing population of older adults, with people aged over 85 years predicted to grow from 2% of the total population in 2020 to 4–5% in 2048 and 5–8% in 2073 [1]. The older Māori population is projected to grow more quickly [2]. Optimal quality of life is the priority as the population ages. Social relationships, environmental spaces, emotional well-being, health and functional ability are crucial factors impacting quality of life of older adults worldwide [3–5]. Both health and decline in functional ability are affected by ageing [6], and diet contributes most significantly to health losses from disease in New Zealand [7].

The New Zealand Adult Nutrition Survey 2008/09 reported that those over 71 years of age had the highest prevalence of inadequate protein intake (13.4% in men and 15.5% in women) than any other age group [8]. Adequate protein and energy intake are important for optimal health, particularly older adults. Adequate protein intake is essential in lean muscle mass maintenance [6,9,10]. Protein-energy malnutrition is prevalent in the institutionalised, hospitalised and up to 25% of community-dwelling populations of older people [11]. In community-dwelling older adults, mean age between 69 and 86 years, up to 45% of sample had protein intake below the current recommended protein intake (<0.8 g/kg BW/d) [12]. It is essential to achieve adequate protein intake, as prolonged protein-energy malnutrition can severely affect well-being, health and functional status [13].

Additionally, international researchers and nutrition agencies such as ESPEN and PROT-AGE suggest that current recommended protein intakes are insufficient to meet the requirements to maintain muscle strength and function [14–16]. Updated information on protein intake for older adults related to supporting nutritional status and health in ageing is needed.

Impaired mobility, functional decline and mortality are linked to reducing muscle mass and strength with increasing age, significantly impacting quality of life [17–20]. Adequate protein intake is associated with better maintenance of lean muscle mass and physical function in older adults [6,9,15,21]. Good physical function is essential not only on a micro-environmental level and meso-environmental level, affecting both an individual's and the broader age groups collective quality of life, but also on a macro-environmental level. The requirement for support services due to age-related reduction in functional status and independence cost New Zealand Government $983 million in 2015 [22].

Investigating elements that may impact functional decline in ageing, including diet, may lead to strategies to offset the high cost to individuals and the government of providing support. The ageing demographic, with greater proportional increases in Māori, indigenous to NZ, justifies further investigation into the relationship between physical function and dietary protein intake in New Zealand [6,23]. Whilst frailty in this population has previously been investigated [24], there is little research on protein intake and physical function longitudinally in the 85+ demographic. Focusing research efforts on New Zealand's older-aged population has the potential to positively impact the quality of life and health of this age group.

The aim of this study was to investigate the protein intake of over 90-year-olds in New Zealand and to measure the impact of dietary protein intake on functional status in Māori and non-Māori over age 90-year in New Zealand.

2. Materials and Methods

2.1. Life and Living in Advanced Age: A Cohort Study in New Zealand (LiLACS-NZ)

Te Puā waitanga o Ngā Tapuwae Kia ora Tonu, Life and Living in Advanced Age: A Cohort Study in New Zealand (LiLACS-NZ) is a cohort study with inception in 2010 investigating the predictors of successful ageing in adults of advanced age. This research was carried out within the Bay of Plenty District Health Board and Lakes District Health Board areas. The 10th year follow up was completed in 2021 with participants over 90 years of age. A detailed study protocol has been previously published [25], and data outcomes have been published over the follow-up period [26–38]. In brief, Māori, indigenous peoples of Aotearoa New Zealand, aged 80–90 years, (10-year birth cohort 1920–1930) and non-Māori aged 85 years (single birth year cohort, 1925) were identified through multiple overlapping strategies, invited, given informed consent, and then interviewed in 2010–2011. A wider age band was necessary for Māori to allow Māori-specific analyses and provide pro-equity data analysis. Participants were followed up yearly for five years with one 10-year follow-up contact in 2021 (delayed by COVID restrictions from the planned year 2020). A detailed nutrition assessment completed in 2011 forms the baseline for this paper [39]. At baseline, two dietary assessments (24 h multiple pass recall) and a health and well-being questionnaire with physical assessment were conducted by trained interviewers. One 24 h multiple pass recall and the health and well-being interview were repeated for the 10-year follow-up of the study (2021; wave seven). Written informed consent was gained from all participants in this study.

2.2. Dietary Assessment: Multiple-Pass 24-H Recall (MPR)

Trained interviewers conducted a multiple-pass 24 h recall. At 10-year follow-up, this was completed on one occasion to reduce participant burden. The MPR protocol has been proven suitable for the oldest-old population and matches the dietary assessment methods used in the LILACS-NZ and Newcastle 85+ studies [39].

This study utilised an online version of the MPR, the Intake24 virtual dietary assessment system developed by Newcastle University, UK [40,41] with incorporated data of common foods from the New Zealand FOODfiles 2016 database (e.g., mussels, pipi, pūha, silverbeet) [42]. INTAKE24 has a built-in photographic atlas and uses household measures to aid portion size estimations. Reported foods that were not already listed in the Intake24 were identified in the dataset, and the closest alternatives were found in the FOODfiles 2016 database or nutrition information panels (NIPs) by nutrition trained researchers. Nutrient analysis for these items was manually entered into the dataset. Interviewers took care to ensure accurate estimates were attained to ensure optimal accuracy in reporting nutrient intake. Missing foods were coded into the most appropriate food group categories.

For participants residing in rest homes who may have memory or cognitive decline and are not self-preparing meals, food service managers were approached for information regarding the menu and recipes used in rest homes where able; otherwise, rest home managers or nurse managers were approached. Nurses and allied health assistants gave information regarding patient intake as they were directly involved in mealtimes. They were asked to note how much participants left on their plates the day prior to interviewing. A copy of the recipes was also obtained. A short questionnaire collected demographic information from all participants, assessed whether the reported intake was usual, perceived ability to eat a healthy diet, weekly expenses on food and drink, and noted issues with the interview and whether the interviewer thought this recall reflected the participants' true intake.

Selection of Nutrients

INTAKE24 data were collated and analysed using FOODfiles 2016. Key macronutrients analysed included energy (Kcal), protein (gram), carbohydrate (g), total fat (g), saturated fat (g) and cholesterol (g).

2.3. Health and Well-Being Questionnaire and Physical Assessment

Trained interviewers conducted a standardised structured Health and Well-being Questionnaire and Physical Assessment. The development of these interviews and examinations are described [25]. Relevant to this paper, data collected were age, marital status, living situation, anthropometric data (height, weight, body mass index (BMI)), body composition using a Tanita Bioimpedance Analysis BC545 (BIA) scale (total body fat, bone mass, muscle mass), sitting and standing blood pressure, grip strength using a hand-held dynamometer-Grip D, the Physical Activity Scale (PASE) [43], Nottingham Extended Activities of Daily Living Scale (NEADL) [44] and Short Physical Performance Battery (SPPB) [45]. Demographic information (sex, ethnicity, education, and NZ Deprivation Index) was collected once at baseline (2010).

Physical assessments were not completed if it was unsafe for the participant or if they opted out of this part of the assessment. Reasons for incomplete physical assessment were recorded.

2.4. Statistical Analysis

Analyses were completed separately for Māori and non-Māori where possible. Māori have different culture-related foods [46,47] and are disproportionately socio-economically disadvantaged than other ethnic groups in New Zealand. Separate analyses were completed to inform protein intake and the potential impact on physical function in Māori living into their ninth decades.

Descriptive analyses were completed for all variables. Continuous data were presented as means and standard deviation (SD) or median, interquartile range (IQR), and categorical data such as number (*n*) and percentage (%).

Further computation on dietary protein intake data was completed to reflect current New Zealand Nutrient Reference Values: Recommended Daily Intake (RDI); inadequate protein intake (<0.8 g/kg BW/day), adequate protein intake (>0.8 g/kg BW/day) and Estimated Average Requirement (EAR) (>0.86 g/kg BW/day in men and >0.75 g/kg BW/day in women) [48]. Sources of protein were reported according the 2008/09 New Zealand Adult Nutrition Survey NZ food groups.

Paired *t*-test (or McNemar) was used to determine difference in variables of interest between 2011 and 2021 (see Tables 1 and 2 in the results section). Independent *t*-tests (or Kruskal–Wallis tests) were used to determine the association between protein intake (g/day, g/kg BW/day) and gender. Pearson's (or Spearman's) correlation test was used to determine the association between protein intake and functional status (NEADL), physical performance (SPPB), BIA measures and PASE score. All results for univariate analysis were presented in mean (SD) or median (IQR) and P-value. Variables with $p < 0.2$ were included in the multiple regression model. Generalised linear models (GLM) were completed to determine the independent association between protein intake (g/Kg BW/day) and physical function as the dependent variable, adjusting for relevant confounders (age, gender, ethnicity, education status, socio-economic status, body weight, BMI, muscle mass, energy intake, gait speed, physical activity levels and living situation). Three models were tested, each with increasing number of possible confounding variables, and the final models are presented in the main text (see Tables 3 and 4 in the results section). Models 1 and 2 can be found in the Supplementary Materials. We did not split the data by ethnicity in the final GLM because there were too few Māori men to provide stable models. Beta coefficients, 95% confidence intervals and P-values were presented. The significance level is set at $p < 0.05$. Data were analysed using the statistical package SPSS version 27.0 (IBM SPSS Statistics for Windows).

3. Results

3.1. Participant Characteristics, Physical Function and Nutrient Intake

A total of eighty-one participants were recruited into the current study (23 Māori and 58 non-Māori). This was 9% (81/937) of those still alive in original LiLACS NZ and 72% (81/112) of those invited; 5 did complete the dietary assessment interview. Reasons for nonresponse: decline or unable to consent (21), and not contactable (5). Participant characteristics, physical function measures and nutrient intake are presented for Māori and non-Māori participants separately in Tables 1 and 2, respectively, showing changes in measures from 2011 to 2021.

3.1.1. Māori Participants

There were five Māori men and 18 Māori women aged between 92 and 94 years old. Fifty-six percent of Māori women and all men lived with others. Over the ten years, SPPB and NEADL scores decreased in both men and women. A similar trend was observed in grip strength and physical activity (PASE) score, and this was only significant in women (Table 1).

Overall, the total energy intake significantly reduced to 1054 kcal/day and 1337 kcal/day in men and women, respectively, between 2011 and 2021. Protein intake almost halved in Māori men from 81.2 g/day to 40.8 g/day ($p = 0.043$) but was relatively stable in Māori women (from 56 g/day to 51.7 g/day, $p = 0.096$). The mean weight-adjusted protein intake remained stable in Māori women (0.86 g/kg BW/day) but declined in Māori men (0.69 g/kg BW/day), although it was not statistically significant. Percent of energy from protein remained relatively stable, about 16% for both men and women, meeting the lower end of the acceptable macronutrient distribution range (AMDR). However, the percentage of animal protein and plant-based protein has changed (Figure 1). There was a significant reduction in fat intake in both men ($p = 0.043$) and women ($p = 0.009$) between 2011 and 2021; a similar declining trend was also observed for cholesterol intake. However, the percentage of energy from fat did not change significantly. Interestingly, the percentage of energy from carbohydrates increased significantly in Māori women from 44.7% to 50.3% ($p = 0.022$). Figure 2 shows the primary dietary sources of protein in Māori in 2011 and 2021. A key difference between these two time periods was that poultry, beef and veal, and lamb and mutton did not make it to the ten most common dietary protein sources in 2021. Interestingly, fish and seafood are the most common source of dietary protein in Māori, which are followed closely by bread.

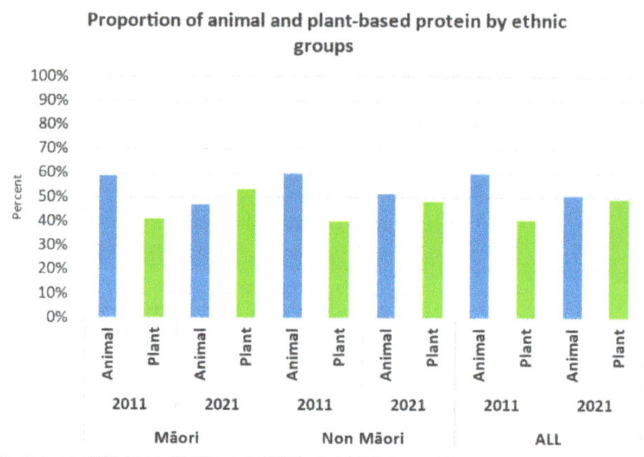

Figure 1. The proportion of animal and plant-based protein in 2011 and 2021 by ethnic groups.

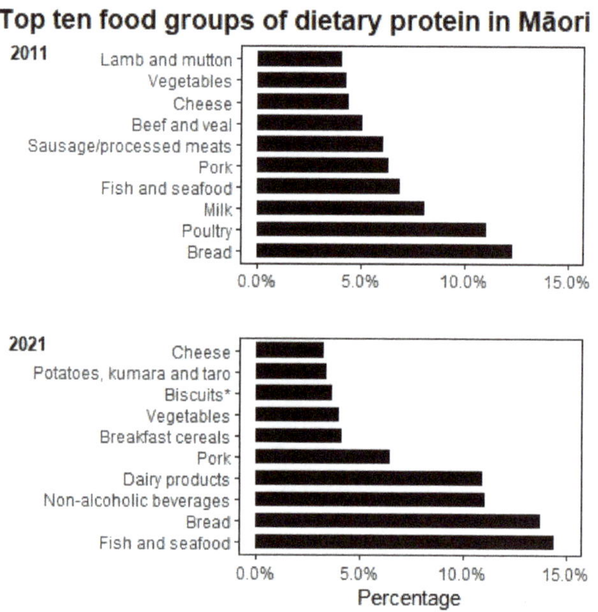

Figure 2. Top ten food groups of dietary protein intake for Māori in 2011 and 2021. Notes: Classification of food groups were based on the 2008/09 New Zealand Adult Nutrition Survey NZ food groups. * Sweet biscuits (plain, chocolate coated, fruit filled, cream filled), crackers.

3.1.2. Non-Māori Participants

The mean age of non-Māori participants was 95 years. The majority of men and women lived in private homes with others. All measures of physical function declined significantly over the ten years. The decline in mean grip strength was more notable in men (from 33.2 to 24.4 kg) than in women (17.8 to 15 kg). Notably, there was a significant decline in physical activity (PASE score from 116 to 27 and 88 to 28 in men and women, respectively) (Table 2).

The overall energy intake reduced from 2040 to 1581 kcal in men ($p = 0.003$) and 1637 to 1477.6 kcal in women between 2011 and 2021. Mean protein intake (g/day) and body weight-adjusted protein intake declined in both men and women between 2011 and 2021; body weight-adjusted protein intake in women declining significantly from 1.00 ± 0.29 g/kg BW/day to 0.84 g/kg BW/day ($p = 0.010$), but this was not significant in men (1.08 to 0.94 g/kg BW/day, $p = 0.295$). Non-Māori men had a significant reduction in carbohydrate ($p = 0.025$) and fat intake ($p = 0.004$) but not women. Non-Māori women had a significant reduction in percent energy intake from protein from 16.4% in 2011 to 14.2% in 2021 ($p = 0.001$); this was also reflected in a significant decline in the proportion of the group not attaining the AMRD for protein; this was not observed in their male counterparts (Table 2). We observed that the distribution of animal versus plant-based protein was changed from 60% animal protein to 40% plant-based protein in 2011 to 52%:48% animal: plant-based protein (Figure 1). There was a significant increase in percent energy from carbohydrate between 2011 and 2021 ($p < 0.05$) and only a modest reduction in percent energy from fat ($p > 0.05$). Figure 3 shows the primary sources of protein in non-Māori in 2011 and 2021; dairy products, non-alcoholic beverages (e.g., hot drinks and fruit juice) and biscuits replaced milk, fish and seafood, and cheese as the top ten food groups of dietary protein in 2021.

3.1.3. Relationship between Protein Intake in 2011 and Physical Function Measures in 2021

The changes in function levels (NEADL and SPPB score) between 2011 and 2021 in relation to achieving the Recommended Daily Intake (RDI) for protein (≥ 0.8 g/kg BW/day) in 2011 by sex and ethnicity are presented visually in Figure S1 (Supplementary Materials). Participants with inadequate protein intake in 2011 had a larger reduction in NEADL score over the 10-year period than those with adequate intake (38% and 43% reduction in those who did and did not meet the RDI in 2011, respectively). Participants with inadequate protein intake in 2011 had a mean reduction in SPPB score of 51% over ten years compared to 46% in those with adequate protein intake.

Regression models examined the relationship between physical function outcomes (NEADL and SPPB scores in 2021) and protein intake in 2011. The models were not split by ethnic groups due to the small sample size. Models 1 and 2 can be found in the Supplementary Materials. The multicollinearity between independent variables was examined, and we found that historical NEADL was correlated with current PASE. We chose to include PASE over NEADL in the final GLM because it fit the model better. Table 3 shows the final regression model of the relationship between weight-adjusted protein intake in 2011 and 2021 NEADL score. We observed that weight-adjusted protein intake in 2011 was independently associated with 2021 NEADL score (13.6 (95% CI 6.7–20.6), $p < 0.001$) but not SPPB score ($p = 0.99$) (Table 4). Male sex was associated with lower SPPB score at 10 years follow-up ($p = 0.012$); and those living in low NZ deprivation areas compared to high NZ deprivation areas was associated with higher a SPPB score at 10 years follow-up ($p = 0.048$).

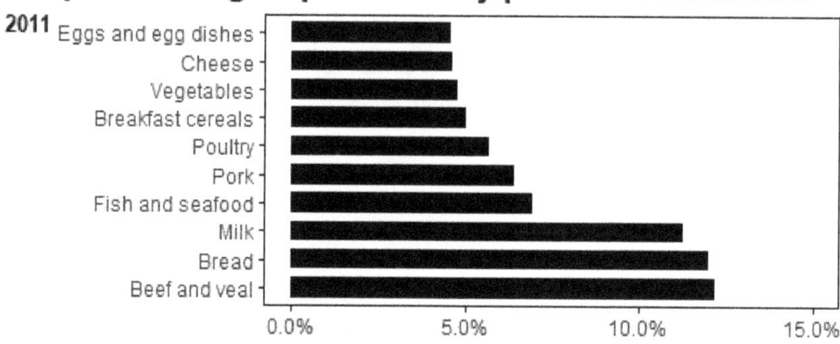

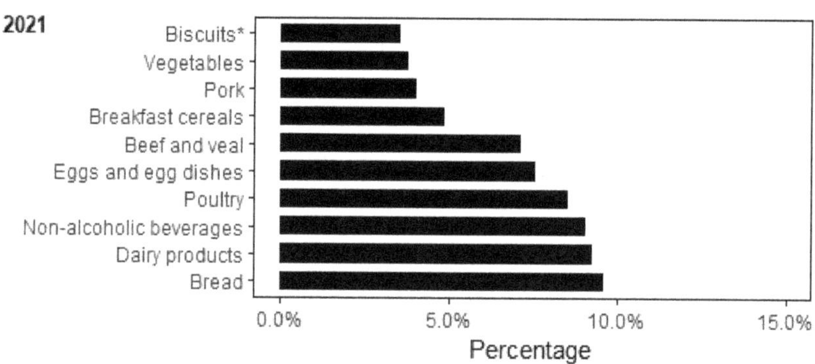

Figure 3. Top ten food groups of dietary protein intake for non-Māori in 2011 and 2021. Notes: Classification of food groups were based on the 2008/09 New Zealand Adult Nutrition Survey NZ food groups. * Sweet biscuits (plain, chocolate coated, fruit filled, cream filled), crackers.

Table 1. Demographic Characteristics, Physical Function and Nutrient Intake in Māori Participants.

		Men, *n* = 5	Women, *n* = 18
Age, years (mean ± SD)		93.5 ± 3.8	92.0 ± 2.4
Education, *n* (%)	Primary or no school	2 (40.0)	3 (16.7)
	Secondary school, no qualification	1 (20.0)	9 (50.0)
	Secondary school, qualification	2 (40.0)	4 (22.2)
	Trade/tertiary qualification	0 (0.0)	2 (11.1)
Marital status, *n* (%)	Never married	1 (33.3)	2 (18.2)
	Married	2 (66.7)	6 (54.5)
	Widow	0 (0.0)	1 (9.1)
	Divorced	0 (0.0)	2 (18.2)
NZ deprivation index, *n* (%) [a]	1–3 (low)	2 (40.0)	2 (11.1)
	4–7 (medium)	1 (20.0)	5 (27.8)
	8 and above (high)	2 (40.0)	11 (61.1)
Who do you live with? *n* (%)	Alone	0 (0.0)	8 (44.4)
	With others	5 (100.0)	10 (55.6)
Where do you live? *n* (%)	Private home	4 (100.0)	12 (70.6)
	Retirement village	0 (0.0)	2 (11.8)
	Rest home	0 (0.0)	3 (17.6)
	Other	0 (0.0)	0 (0.0)

Table 1. *Cont.*

	Men, *n* = 5			Women, *n* = 18		
	2011	2021	*p*-Value	2011	2021	*p*-Value
BMI (Kg/m^2), mean ± SD	28.9 ± 5.4	29.0 ± 4.7	0.655	28.0 ± 3.4	24.5 ± 5.0	0.161
Height (m), mean ± SD	1.65 ± 0.01	1.61 ± 0.1	0.180	1.56 ± 0.1	1.54 ± 0.1	0.017 *
Weight (Kg), mean ± SD	78.0 ± 13.4	70.9 ± 8.9	0.285	67.7 ± 7.9	58.1 ± 9.1	0.002 *
Muscle mass (Kg), mean ± SD	52.4 ± 3.4	54.9 ± 2.5	0.285	39.1 ± 3.0	35.9 ± 2.4	0.213
Fat mass (%), mean ± SD	28.1 ± 9.5	19.7 ± 7.8	0.109	38.2 ± 4.6	33.9 ± 5.7	0.025 *
Bone mass (Kg), mean ± SD	2.8 ± 0.2	2.5 ± 0.2	0.276	2.1 ± 0.2	2.0 ± 0.2	0.048 *
Systolic BP standing (mmHg), mean ± SD	147 ± 11	161 ± 12	0.285	150 ± 19	154 ± 13	0.398
Systolic BP sitting (mmHg), mean ± SD	151 ± 22	160 ± 28	0.715	148 ± 14	153 ± 17	0.477
Diastolic BP standing (mmHg), mean ± SD	85 ± 17	80 ± 9	0.109	92 ± 11	88 ± 10	0.018 *
Diastolic BP sitting (mmHg), mean ± SD	77 ± 7	72 ± 12	0.715	83 ± 10	85 ± 9	0.929
NEADL score, median (IQR)	17.5 (16.0–18.0)	16.5 (14.0–19.0)	1.000	19.5 (18.0–21.0)	15.0 (8.0–20.0)	<0.001 *
SPPB score, mean ± SD	9.8 ± 1.7	4.5 ± 5.5	0.285	9.2 ± 3.4	6.8 ± 3.6	0.006 *
Grip strength (Kg), mean ± SD	33.0 ± 5.8	22.4 ± 6.1	0.109	21.4 ± 3.3	16.7 ± 4.1	0.004 *
Total PASE score, median (IQR)	145 (107–193)	32 (0–185)	0.285	133 (83–167)	70 (18-93)	0.013 *
Energy intake (kcal), mean ± SD	1874 ± 705	1054 ± 515	0.05 *	1483 ± 331	1337 ± 473	0.026 *
Protein intake (g), mean ± SD	81.2 ± 27.5	40.8 ± 17.62	0.043 *	56.0 ± 16.5	51.7 ± 22.3	0.096
Protein intake (g/Kg BW/day), mean ± SD	1.09 ± 0.44	0.69 ± 0.07	0.109	0.86 ± 0.26	0.86 ± 0.36	0.859
Carbohydrate intake (g), mean ± SD	194.7 ± 90.4	132.5 ± 76.5	0.138	166.2 ± 46.9	166.9 ± 67.6	0.433
Fat intake (g), mean ± SD	86.0 ± 37.6	40.3 ± 18.8	0.043 *	63.9 ± 16.0	48.1 ± 17.3	0.009 *
Saturated fat intake (g), mean ± SD	34.7 ± 15.2	16.1 ± 9.6	0.080	26.6 ± 8.3	21.7 ± 9.9	0.109
Cholesterol intake (g), mean ± SD	282.4 ± 171.9	178.3 ± 171.4	0.043 *	221.1 ± 110.6	127.9 ± 64.4	0.016 *
Percent energy intake from protein (%), mean ± SD	17.7 ± 3.3	15.9 ± 4.4	0.686	15.2 ± 3.5	15.5 ± 4.8	0.826
• *n* (%) within the AMDR (15–25% of energy intake)	4 (80.0%)	3 (60.0%)	1.000	9 (64.3%)	5 (35.7%)	0.219
Percent energy intake from carbohydrate (%), mean ± SD	41.1 ± 12.4	48.6 ± 5.1	0.225	44.7 ± 6.1	50.3 ± 9.6	0.022 *
• *n* (%) within the AMDR (45–65% of energy intake)	2 (40%)	3 (60%)	1.000	7 (15%)	8 (57.1%)	1.000
Percent energy intake from fat (%), mean ± SD	41.7 ± 13.3	35.7 ± 5.5	0.225	39.0 ± 5.9	32.7 ± 8.4	0.056
• *n* (%) within the AMDR (20–35% of energy intake)	2 (40%)	2 (40%)	1.000	4 (28.6%)	7 (50.0%)	0.453

[a] NZ Deprivation Status was collected in 2010. Low: NZ Deprivation (Dep) Index 1–3, Middle: NZ Dep Index 4–7, High: NZ Dep Index 8–10. Notes: AMDR, Acceptable Macronutrient Distribution Range; BW, body weight; g, gram; Kg, kilogram; NEADL, Nottingham Extended Activities of Daily Living Scale; PASE, Physical Activity Scale SD, standard deviation; SPPB, Short Physical Performance Battery. * Paired *t*-test (or McNemar) was used to determine difference in variables of interest between 2011 and 2021, statistically significant at *p* < 0.05.

Table 2. Demographic Characteristics, Physical Function and Nutrient Intake in Non-Māori Participants.

		Men, *n* = 26	Women, *n* = 32
Age, years (mean ± SD)		95.2 ± 0.4	95.3 ± 0.4
Education, *n* (%)	Primary or no school	3 (11.5)	1 (3.1)
	Secondary school, no qualification	9 (34.6)	11 (34.4)
	Secondary school, qualification	7 (26.9)	10 (31.3)
	Trade/Tertiary qualification	7 (26.9)	10 (31.3)
Marital status, *n* (%)	Never married	0 (0.0)	0 (0.0)
	Married	6 (66.7)	1 (12.5)
	Widow	3 (33.3)	5 (62.5)
	Divorced	0 (0.0)	2 (25.0)
NZ deprivation index, *n* (%) [a]	1–3 (low)	8 (30.8)	5 (15.6)
	4–7 (medium)	11 (42.3)	11 (34.4)
	8 and above (high)	7 (26.9)	16 (50.0)
Who do you live with? *n* (%)	Alone	8 (30.8)	14 (43.8)
	With others	18 (69.2)	18 (56.3)
Where do you live? *n* (%)	Private home	12 (52.2)	13 (41.9)
	Retirement village	8 (34.8)	6 (19.4)
	Rest home	3 (13.0)	12 (38.7)
	Other	0 (0.0)	0 (0.0)

Table 2. *Cont.*

	Men, *n* = 26			Women, *n* = 32		
	2011	2021	*p*-Value	2011	2021	*p*-Value
BMI (Kg/m^2), mean $\pm$ SD	26.4 $\pm$ 3.6	24.7 $\pm$ 4.6	0.013 *	27.1 $\pm$ 3.9	26.0 $\pm$ 3.3	0.398
Height (m), mean $\pm$ SD	1.68 $\pm$ 0.1	1.64 $\pm$ 0.1	<0.001 *	1.56 $\pm$ 0.1	1.52 $\pm$ 0.1	<0.001 *
Weight (Kg), mean $\pm$ SD	74.2 $\pm$ 9.5	65.8 $\pm$ 11.4	0.001 *	65.8 $\pm$ 9.3	62.0 $\pm$ 9.4	0.002 *
Muscle mass (Kg), mean $\pm$ SD	50.96 $\pm$ 4.7	45.71 $\pm$ 3.4	0.034 *	38.03 $\pm$ 3.9	36.81 $\pm$ 3.4	0.266
Fat mass (%), mean $\pm$ SD	27.39 $\pm$ 6.5	24.97 $\pm$ 10.0	0.504	37.65 $\pm$ 7.2	35.23 $\pm$ 8.7	0.085 *
Bone mass (Kg), mean $\pm$ SD	2.71 $\pm$ 0.2	2.46 $\pm$ 0.2	0.003 *	2.06 $\pm$ 0.27	1.98 $\pm$ 0.2	0.005 *
Systolic BP standing (mmHg), mean $\pm$ SD	147 $\pm$ 19	142 $\pm$ 29	0.510	152 $\pm$ 22	148 $\pm$ 17	0.717
Systolic BP sitting (mmHg), mean $\pm$ SD	147 $\pm$ 20	139 $\pm$ 24	0.083	147 $\pm$ 21	144 $\pm$ 26	0.728
Diastolic BP standing (mmHg), mean $\pm$ SD	83 $\pm$ 10	74 $\pm$ 16	<0.001 *	87 $\pm$ 15	83 $\pm$ 1	<0.001 *
Diastolic BP sitting (mmHg), mean $\pm$ SD	79 $\pm$ 11	70 $\pm$ 13	0.011 *	77 $\pm$ 12	78 $\pm$ 15	0.687
NEADL score wave 7, median (IQR)	18.0 (15.0–19.0)	12.0 (8.0–16.0)	<0.001 *	20.0 (19.0–21.0)	10.0 (6.0–16.0)	<0.001 *
SPPB score wave 7, mean $\pm$ SD	10.4 $\pm$ 1.5	5.1 $\pm$ 3.4	<0.001 *	8.5 $\pm$ 2.5	4.7 $\pm$ 3.5	<0.001 *
Grip strength (Kg), mean $\pm$ SD	33.2 $\pm$ 5.8	24.4 $\pm$ 6.2	<0.001 *	17.8 $\pm$ 4.2	15.0 $\pm$ 3.61	<0.001 *
Total PASE score, median (IQR)	116 (86–151)	27 (0–67)	<0.001 *	88 (55–142)	28 (9–83)	<0.001 *
Energy (kcal), mean $\pm$ SD	2040 $\pm$ 636	1581 $\pm$ 464	0.003 *	1637 $\pm$ 418	1478 $\pm$ 475	0.086
Protein (g), mean $\pm$ SD	80.3 $\pm$ 20.7	61.2 $\pm$ 27.4	0.006 *	66.7 $\pm$ 18.6	51.8 $\pm$ 15.6	0.001 *
Protein (g/Kg BW/day), mean $\pm$ SD	1.08 $\pm$ 0.29	0.94 $\pm$ 0.38	0.295	1.00 $\pm$ 0.29	0.84 $\pm$ 0.26	0.010 *
Carbohydrate (g), mean $\pm$ SD	212.9 $\pm$ 61.5	187.2 $\pm$ 66.1	0.025 *	175.2 $\pm$ 44.6	180.7 $\pm$ 66.4	0.393
Fat (g), mean $\pm$ SD	86.2 $\pm$ 33.3	61.5 $\pm$ 29.9	0.004 *	71.6 $\pm$ 27.2	59.3 $\pm$ 30.9	0.106
Saturated fat (g), mean $\pm$ SD	35.5 $\pm$ 16.3	26.1 $\pm$ 14.2	0.023 *	28.3 $\pm$ 14.0	26.4 $\pm$ 13.5	0.572
Cholesterol (mg), mean $\pm$ SD	302.4 $\pm$ 169.0	222.2 $\pm$ 128.5	0.062	249.3 $\pm$ 127.9	239.9 $\pm$ 170.9	0.428
Percent energy intake from protein (%), mean $\pm$ SD						
• *n* (%) within the AMDR (15–25% of energy intake)	16.1 $\pm$ 2.7 / 16 (61.5%)	15.9 $\pm$ 5.9 / 14 (53.8%)	0.849 / 0.791	16.4 $\pm$ 3.1 / 20 (66.7%)	14.2 $\pm$ 2.8 / 11 (36.7%)	0.001 * / 0.035 *
Percent energy intake from carbohydrate (%), mean $\pm$ SD						
• *n* (%) within the AMDR (45–65% of energy intake)	42.9 $\pm$ 8.9 / 7 (26.9%)	47.6 $\pm$ 10.5 / 12 (46.2%)	0.038 * / 0.227	43.5 $\pm$ 7.6 / 14 (46.7%)	49.0 $\pm$ 10.8 / 18 (60.0%)	0.037 * / 0.424
Percent energy intake from fat (%), mean $\pm$ SD						
• *n* (%) with the AMDR (20–35% of energy intake)	37.4 $\pm$ 6.3 / 7 (26.9%)	34.0 $\pm$ 9.9 / 14 (53.8%)	0.052 / 0.118	38.6 $\pm$ 7.1 / 11 (36.7%)	35.6 $\pm$ 11.0 / 13 (43.3%)	0.271 / 0.804

[a] NZ Deprivation Status was collected at in 2010. Low: NZ Deprivation (Dep) Index 1–3, Middle: NZ Dep Index 4–7, High: NZ Dep Index 8–10. Notes: AMDR, Acceptable Macronutrient Distribution Range; BW, body weight; g, gram; Kg, kilogram; NEADL, Nottingham Extended Activities of Daily Living Scale; PASE, Physical Activity Scale SD, standard deviation; SPPB, Short Physical Performance Battery. * Paired *t*-test (or McNemar) was used to determine difference in variables of interest between 2011 and 2021, statistically significant at $p < 0.05$.

Table 3. Multivariate Regression Model Examining the Relationship between 2011 Protein Intake, as Predictor, and 2021 NEADL Score in all Participants [#].

Variables	B (95% CI), *p*-Value
2011 Protein intake (g/kg BW/day) [#]	13.59 (6.67–20.51), <0.001 *
2011 Energy intake (kcal/day) [#]	−0.009 (−0.015−−0.003), 0.004 *
Age, years	−0.32 (−0.87–0.23), 0.253
Gender (ref: Women)	0.49 (−2.77–3.76), 0.767
Ethnicity (ref: Māori)	−1.43 (−9.49–6.64), 0.729
NZ deprivation index [a]	
• low (ref: High)	0.02 (−2.31–2.35), 0.988
• medium (ref: High)	−2.01 (−3.80−−0.22), 0.028 *
Current living arrangement [#] Who do you live with? (ref: with others)	2.83 (0.89–4.77), 0.004 *
Current housing situation [#] Where do you live?	
• private house (ref: rest home)	2.76 (0.17–5.35), 0.037 *
• retirement village (ref: rest home)	3.82 (0.72–6.91), 0.016 *
Current [#] PASE [b] score	0.04 (0.03–0.05), <0.001 *
Current [#] Fat mass (%)	0.05 (−0.06–0.15), 0.385
Current [#] Grip strength (kg)	0.16 (−0.06–0.37), 0.151

[#] Current data collected in 2021. * Generalised linear models (GLM) adjusted for relevant confounder, statistically significant at $p < 0.05$. [a] NZ Deprivation index: Low-status 1–3, Middle-status 4–7, High-status 8–10. [b] PASE, Physical Activity Scale for the Elderly. Note: Interaction term Ethnicity*Wave 2 Protein intake; $p = 0.012$; Interaction term Ethnicity×Wave 2 energy intake; $p = 0.018$. There is a observed correlation between wave two NEADL and wave 7 PASE (r = 0.356, $p = 0.003$). Due to the multicollinearity of this data, we chose to include PASE in the GLM as it fit the model better.

Table 4. Multivariate Regression Model Examining the Relationship between 2011 Protein Intake, as Predictor, and 2021 SPPB Score in all Participants [#].

Variables	B (95% CI), *p*-Value
2011 Protein intake (g/kg BW/day) [#]	−0.030 (−6.858–6.798), 0.993
2011 Energy intake (kcal/day) [#]	−0.001 (−0.005–0.003), 0.597
Age, years	0.207 (−0.281–0.694), 0.406
Gender (ref: Women)	−4.675 (−8.324–−1.027), 0.012 *
Ethnicity (ref: Māori)	−5.841 (−12.703–1.021), 0.095
Education status (ref: tertiary qualification)	
• no or primary school	−1.154 (−6.238–3.931), 0.657
• secondary school, no qualification	−1.311 (−4.188–1.565), 0.371
• secondary school, qualification	1.338 (−1.274–3.950), 0.315
• Trade	1.187 (−2.176–4.550), 0.489
NZ deprivation index (ref: High) [a]	
• Low	−1.813 (−3.607–−0.018), 0.048 *
• Medium	−0.703 (−2.406–1.000), 0.418
Current living arrangement[#] Who do you live with (ref: with others)	−0.551 (−2.478–1.376), 0.575
Current housing situation[#] Where do you live?	
• private house (ref: rest home)	1.423 (−1.289–4.135), 0.304
• retirement village (ref: rest home)	0.945 (−1.873–3.763), 0.511
Current [#] PASE [b] score	0.001 (−0.001–0.014), 0.828
Current [#] Fat mass (%)	−0.086 (−0.169–−0.003), 0.043 *
Current [#] Grip strength (kg)	0.308 (0.059–0.558), 0.015 *
2011 SPPB score	0.299 (−0.122–0.719), 0.164

[#] Current data collected in 2021. * Generalised linear models (GLM) adjusted for relevant confounder, statistically significant at *p* < 0.05. [a] NZ Deprivation index: Low status 1–3, Middle status 4–7, High status 8–10. [b] PASE, Physical Activity Scale for the Elderly. Note: No interaction between ethnicity and wave 2 protein or energy intake (Interaction terms; Ethnicity × * Wave 2 Protein intake, *p* = 0.848 and Ethnicity * Wave 2 Energy intake, *p* = 0.099).

4. Discussion

This study aims to investigate protein intake in New Zealand nonagenarians and determine the impact of dietary protein intake on functional status in Māori and non-Māori at ten years follow-up. We found that intake reduced over 10 years along with physical activity, functional status and physical performance. Higher weight-adjusted protein intake (g/kg BW/day) (2011) was associated with better subsequent functional status ten years later. This finding supports the hypothesis that protein intake can impact functional status in those of advanced age.

4.1. Macronutrient Intake

Reducing dietary intake is common with ageing [16,49]. We found reductions in protein and fat intake with an increase in percentage energy from carbohydrates over ten years follow-up of octogenarians. This may be attributed to known risk factors such as oral health [29,50] and food cost [29]. The ten-year interval between two dietary assessments data points limit further inference. Other unknown factors affecting dietary intake, such as food accessibility, food preparation, and attitude towards food intake, may influence food choices. Interestingly, fish and seafood are the primary protein sources for Māori participants, which are followed closely by bread. Although the number of people in this sample of older Māori was small, it does appear that Māori were more able to maintain dietary protein intake than non-Māori. This is probably linked to the living environment, where Māori live closer to the sea and the provision of seafood from family and whānau, as was observed in 2011 [29]. For non-Māori, bread was the main source of protein. We hypothesise that the participants in their nineties may have replaced protein and fat consumption with carbohydrate foods which are more accessible and easier to prepare.

Protein Intake

Older adults in New Zealand have the highest level of inadequate protein intake compared to any other age group [8]. Protein requirements are amplified with increasing age due to age-related physiological changes, changes in dentition, food preferences, health status and social circumstances [6,15,16,49,51,52]. In the New Zealand Adult Nutrition Survey (NZ ANS) 2008/09, Māori men and women over 51 consumed a mean of 95 g/day and

68 g/day of protein, respectively [8]. In our study, Māori men in their nineties consumed about 40 g of protein per day, and Māori women consumed about 52 g/day. Protein intake, when adjusted by body weight, declined significantly in Māori men (from 1.09 ± 0.44 to 0.69 ± 0.07 g/kg BW/day) while it remained stable in Māori women (0.86 ± 0.36 g/kg BW/day). The marked difference in protein intake observed in Māori men needs to be interpreted with caution, as the small sample is likely to have skewed the value. Our study observed that while the percentage energy from protein is relatively stable over time, the protein sources switched from animal protein as the primary source in 2011 to plant-based protein in 2021. Future research is needed to confirm our findings.

Among non-Māori participants, reduction in protein intake was more significant in women than men. Protein intake adjusted by body weight for women reduced from 1.00 ± 0.29 g/kg BW/day in 2011 to 0.84 ± 0.26 in 2021, and percentage energy from protein reduced from 16% to 14%. Both measures of protein intake were relatively stable in men. In non-Māori, animal protein was a major source of protein in 2011 and 2021. However, the ratio between animal and plant-based protein was closer to 1 in 2021 than the 60:40 animal-protein ratio of 1.5 in 2011. Research that shows that plant-based proteins have lower leucine content, reduced digestibility, and are deficient in some essential amino acids which are needed for muscle anabolism compared to animal-based proteins. Although it has been postulated that this could be negated by increasing volume of plant-based protein consumed, further research is required [53,54].

Overall, the quantity and quality of protein intake changed over time and varied by sex in Māori and non-Māori. A consistent finding across both sexes in the sample, irrespectively of ethnic groups, was that in all participants, the recommended percentage energy from protein hovered around the lower range of the Acceptable Macronutrient Distribution Range (AMDR), i.e., 15–25% energy from protein [48]. Animal protein was replaced with plant-based protein over time. It is difficult to distinguish exactly what this may be related to, although it would be feasible to believe this may occur due to changes in dentition (ability to chew animal-based protein) taste change/food preference and anorexia of ageing, which has been discussed in the literature [16].

4.2. Relationship between Protein Intake and Physical Function

Physical function is associated with daily living activities [20,55]. We observed NEADL and SPPB scores decrease during the 10-year follow-up period. This is in line with previous research showing a decreasing trend in physical function in adults 65 years and older, and the annual decline rate was greater in the 85+ group compared to those aged 65 to 74 years [56]. Granic et al. reported an inverse association between age, grip strength and Timed Up and Go Test (TUG) over five years in the Newcastle 85+ study and that the declines were not different between those who had low or high protein intake at baseline (<1 g/kg aBW/d vs. $\geq$1 g/kg aBW/d) [57]. We reported previously that protein intake was not associated with grip strength over five years [30]. Our current study extends the evidence on the association between protein intake and subsequent physical function at ten years follow-up. We found a positive association between weight-adjusted protein intake at age 80-years+ with functional status ($p < 0.001$) over 10 years but not with physical performance ($p = 0.094$). Functional status is correlated with physical performance, but it is a different measure impacted by psychological, environmental and social factors [58–60]. This finding supports the hypothesis that protein intake is associated with future functional status. The potential mechanism by which of protein intake benefits function may not only be through physical performance or muscle strength. Other age-related physiological changes throughout the life course may have a role.

The physical activity level (PASE score) of participants in the current study also decreased dramatically in all participants over the ten years (from a median of 93–150 in wave two to 27–70 in wave seven).

4.3. Relationship between Socio-Demographic Characteristics, Physical Activity and Physical Function

Physical activity levels usually are associated with higher protein intake and improved physical performance and ability to perform activities of daily living [6,61,62], so it is not surprising that we saw a reduction in PASE score along with a decrease in SPPB and NEADL scores, and that physical activity was positively associated with activities of daily living. However, physical activity was not related to physical performance in this sample in advanced age. The type of physical activities (housework vs. recreational activities) could have masked this association. Interestingly, the current study found that participants with a better socio-economic status had poorer functional and physical performance than those with worse socio-economic status, challenging the wealth–health directionality [63]. Socio-economic status and food insecurity are factors affecting inadequate nutrient intake that has been discussed throughout the literature [14]. A previous study of 70-year-old Korean people reported total weight-adjusted protein intake increased proportionately with household income and education status [64]. Our study on nonagenarians found the contrary. This prompts the question of whether the socio-economic status of people in advanced age plays a role in the quality and quantity of protein consumed and the further impact on physical performance capability. Further research is needed to understand the interplay between socio-economic status, living environment, and health status impacting dietary intake in older adults.

4.4. Strengths and Limitations

This study is part of the LiLACS-NZ cohort study, allowing us to draw associations between past protein intake and functional status over ten years. Protein intake was recorded in a standardised way over two time points, ensuring continuity for comparison. This allows us to gain insight into the prevalence of inadequate protein intake and the differences in protein quantity and quality (protein sources) in this population over ten years. In addition, this prospective study highlighted the temporal sequence of events. It enabled the inclusion and examination of multiple exposure variables and potential confounding related to the outcome of physical function (NEADL score and SPPB score).

The main limitation of this study was the relatively small sample size of 81 participants, reducing the ability to produce ethnic-specific analyses. The annual mortality rate in the first four years of the follow-up for the LiLACS-NZ study was approximately 9%, and this was the main reason for attrition [65]. Due to this population being of advanced age, it is unsurprising that many participants were lost to follow-up or passed away before the 10-year study interviews. The small sample may introduce type II errors, and we advise a cautious interpretation of the results. Health conditions and the use of medication may also affect the study outcome. At baseline, 93% of the sample had two or more chronic conditions; the median of health conditions and use of prescribed medications was five, respectively [65,66]. The ubiquitous multimorbidity in the sample restricts our ability to untangle the cause–effect relationship between physical activity and chronic conditions. Considering the issue of over-adjustment, we selected physical activity level as a covariate in the model based on the notion that nutrition status and physical activity go hand in hand.

We completed Intake24 (an electronic version of the original interviewer-led 24 h MPR 24-h) once to reduce the participant burden in the current study. Completing one MPR may be less accurate, but it is common practice in research studies when balancing practicality, resources, and data [67]. MPR has been validated and shown to produce similar estimates of group intake in those who do not have extreme intake [68]. MPR can be less accurate when interviewing people with cognitive decline as it relies heavily on memory; thus, we aimed to mitigate this by having a Kaiāwhina or support person with participants to assist. Intake24, with the additions of New Zealand foods from the FOODfiles 2016 database, has not been validated in a New Zealand context. Intake24 underestimated energy intake by 1%, and mean macro- and micronutrient intake was within 4% of those recorded in interviewer-led recalls in a UK population [40].

To our knowledge, this is the first cohort study to assess protein intake in the nonagenarian age group. These findings fill a gap in the existing literature. Causality cannot be proven. Therefore, we can only show an association and recommend a cautious interpretation of the findings.

The current study's findings may improve quality of life of adults of advanced age in the future by supporting international recommendations for adequate protein intake to support physical function in advanced age.

Future research on the effect of protein from plant and animal sources and their impact on physical function over time is required to confirm these findings. Exploration of Indigenous versus Westernised protein food sources and their impact on physical function in needed. Insights from the current study will contribute to the design of future prospective studies and trials of supplements in very old adult populations.

5. Conclusions

The LiLACS-NZ cohort study highlighted a reduction in protein intake with a change in protein sources from animal to plant-based protein over ten years in octogenarians. The study found protein intake 10 years prior was associated with functional status ten years later. The novel nature of this research was that it documents change in dietary intake in nonagenarians related to physical function over time.

Supplementary Materials: The following supporting information can be downloaded at: https://www.mdpi.com/article/10.3390/nu15071664/s1, Figure S1: Trend in 2021 NEADL and SPPB scores in Relation to Adequacy of Protein Intake in 2011; Table S1: GLM Model 1—Relationship Between 2021 NEADL Score and 2011 Protein Intake in all Participants; Table S2: GLM Model 1—Relationship Between 2021 SPPB Score and 2011 Protein Intake in all Participants.

Author Contributions: Conceptualisation, R.T. and M.L.; methodology, M.L. and R.T.; formal analysis, M.L. and R.T.; investigation, M.L., R.T. and N.K.; resources, R.T. and N.K.; data curation, R.T. and M.L.; writing—original draft preparation, M.L.; writing—review and editing, M.L., R.T., N.K. and M.M.-L.; supervision, R.T.; project administration, N.K.; funding acquisition, N.K. and R.T. All authors have read and agreed to the published version of the manuscript.

Funding: This research was funded by Lottery Health Research, grant number 3720336.

Institutional Review Board Statement: The study was conducted in accordance with the Declaration of Helsinki and approved by the Institutional Review Board (or Ethics Committee) of The Northern A Health and Disability Ethics Committee of The Ministry of Health, New Zealand (Reference: NXT/10/12/127/AM15) on 27 November 2020.

Informed Consent Statement: Informed consent was obtained from all subjects involved in the study.

Data Availability Statement: We are happy to share LiLACS NZ data with interested researchers on condition of a mutually acceptable agreement in data usage. The process to apply for LiLACS NZ data access is available at [https://www.fmhs.auckland.ac.nz/en/faculty/lilacs.html] (accessed on 28 March 2023).

Acknowledgments: We thank all participants, their family and whānau (extended family) for their contribution. We also thank Karen Hayman who organised the study, and Kiri Martin and Sue MacDonell who assisted with the dietary assessment.

Conflicts of Interest: The authors declare no conflict of interest.

References

1. Stats-NZ. *National Population Projections: 2020(Base)–2073*; Stats NZ: Wellington, New Zealand, 2020.
2. Stats-NZ. Māori Population Estimates: At 30 June 2020 New Zealand: Stats NZ; 2020 [cited 2021 24/06/2021]. Available online: https://www.stats.govt.nz/information-releases/maori-population-estimates-at-30-june-2020 (accessed on 16 January 2023).
3. Borglin, G.; Edberg, A.-K.; Hallberg, I.R. The experience of quality of life among older people. *J. Aging Stud.* **2005**, *19*, 201–220. [CrossRef]
4. Gabriel, Z.; Bowling, A. Quality of life from the perspectives of older people. *Ageing Soc.* **2004**, *24*, 675–691. [CrossRef]
5. Stephens, C.; Allen, J.; Keating, N.; Szabó, Á.; Alpass, F. Neighborhood environments and intrinsic capacity interact to affect the health-related quality of life of older people in New Zealand. *Maturitas* **2020**, *139*, 1–5. [CrossRef] [PubMed]
6. Deer, R.; Volpi, E. Protein intake and muscle function in older adults. *Curr. Opin. Clin. Nutr. Metab. Care.* **2015**, *18*, 248–253. [CrossRef] [PubMed]
7. Ministry of Health. *Health Loss in New Zealand 1990-2013: A report from the New Zealand Burden of Diseases, Injuries and Risk Factors Study*; Ministry of Health: Wellington, New Zealand, 2016.
8. Parnell, W.; Wilson, N.; Thomson, C.; Mackay, S.; Stefanogiannis, N. *A Focus on Nutrition: Key Findings of the 2008/09 New Zealand Adult Nutrition Survey*; University of Otago, Ministry of Health: Wellington, New Zealand, 2011.
9. Houston, D.K.; Nicklas, B.J.; Ding, J.; Harris, T.B.; Tylavsky, F.A.; Newman, A.B.; Lee, J.S.; Sahyoun, N.R.; Visser, M.; Kritchevsky, S.B.; et al. Dietary protein intake is associated with lean mass change in older, community-dwelling adults: The Health, Aging, and Body Composition (Health ABC) Study. *Am. J. Clin. Nutr.* **2008**, *87*, 150–155. [CrossRef]
10. Mitchell, C.J.; Milan, A.M.; Mitchell, S.M.; Zeng, N.; Ramzan, F.; Sharma, P.; Knowles, S.O.; Roy, N.C.; Sjödin, A.; Wagner, K.-H.; et al. The effects of dietary protein intake on appendicular lean mass and muscle function in elderly men: A 10-wk randomized controlled trial. *Am. J. Clin. Nutr.* **2017**, *106*, 1375–1383. [CrossRef]
11. Crichton, M.; Craven, D.; Mackay, H.; Marx, W.; De Van Der Schueren, M.; Marshall, S. A systematic review, meta-analysis and meta-regression of the prevalence of protein-energy malnutrition: Associations with geographical region and sex. *Age Ageing* **2018**, *48*, 38–48. [CrossRef]
12. Hengeveld, L.M.; Boer, J.M.A.; Gaudreau, P.; Heymans, M.W.; Jagger, C.; Mendonça, N.; Ocké, M.C.; Presse, N.; Sette, S.; Simonsick, E.M.; et al. Prevalence of protein intake below recommended in community-dwelling older adults: A meta-analysis across cohorts from the PROMISS consortium. *J. Cachexia Sarcopenia Muscle* **2020**, *11*, 1212–1222. [CrossRef]
13. Saunders, J.; Smith, T. Malnutrition: Causes and consequences. *Clin. Med.* **2010**, *10*, 624–627. [CrossRef]
14. Deutz, N.E.; Bauer, J.M.; Barazzoni, R.; Biolo, G.; Boirie, Y.; Bosy-Westphal, A.; Cederholm, T.; Cruz-Jentoft, A.; Krznarić, Z.; Nair, K.S.; et al. Protein intake and exercise for optimal muscle function with aging: Recommendations from the ESPEN Expert Group. *Clin. Nutr.* **2014**, *33*, 929–936. [CrossRef]
15. Bauer, J.; Biolo, G.; Cederholm, T.; Cesari, M.; Cruz-Jentoft, A.J.; Morley, J.E.; Phillips, S.; Sieber, C.; Stehle, P.; Teta, D.; et al. Evidence-Based Recommendations for Optimal Dietary Protein Intake in Older People: A Position Paper From the PROT-AGE Study Group. *J. Am. Med. Dir. Assoc.* **2013**, *14*, 542–559. [CrossRef]
16. Volpi, E.; Campbell, W.; Dwyer, J.; Johnson, M.; Jensen, G.; Morley, J.; Wolfe, R.R. Is the Optimal Level of Protein Intake for Older Adults Greater Than the Recommended Dietary Allowance? *J. Gerontol. Ser. A Biol. Sci. Med. Sci.* **2013**, *68*, 677–681. [CrossRef]
17. Landi, F.; Cruz-Jentoft, A.J.; Liperoti, R.; Russo, A.; Giovannini, S.; Tosato, M.; Capoluongo, E.D.; Bernabei, R.; Onder, G. Sarcopenia and mortality risk in frail older persons aged 80 years and older: Results from ilSIRENTE study. *Age Ageing* **2013**, *42*, 203–209. [CrossRef]
18. Arango-Lopera, V.E.; Arroyo, P.; Gutiérrez-Robledo, L.M.; Perez-Zepeda, M.U.; Cesari, M. Mortality as an adverse outcome of sarcopenia. *J. Nutr. Health Aging* **2013**, *17*, 259–262. [CrossRef]
19. McLean, R.R.; Shardell, M.D.; Alley, D.E.; Cawthon, P.M.; Fragala, M.S.; Harris, T.B.; Kenny, A.M.; Peters, K.W.; Ferrucci, L.; Guralnik, J.M.; et al. Criteria for Clinically Relevant Weakness and Low Lean Mass and Their Longitudinal Association with Incident Mobility Impairment and Mortality: The Foundation for the National Institutes of Health (FNIH) Sarcopenia Project. *J. Gerontol. A Biol. Sci. Med. Sci.* **2014**, *69*, 576–583. [CrossRef]
20. Wang, D.X.; Yao, J.; Zirek, Y.; Reijnierse, E.M.; Maier, A.B. Muscle mass, strength, and physical performance predicting activities of daily living: A meta-analysis. *J. Cachexia Sarcopenia Muscle* **2020**, *11*, 3–25. [CrossRef]
21. Haaf, D.T.; van Dongen, E.; Nuijten, M.; Eijsvogels, T.; de Groot, L.; Hopman, M. Protein Intake and Distribution in Relation to Physical Functioning and Quality of Life in Community-Dwelling Elderly People: Acknowledging the Role of Physical Activity. *Nutrients* **2018**, *10*, 506.
22. New Zealand Productivity Commission. More Effective Social Services-Appendix E. New Zealand. 2015. Available online: https://www.productivity.govt.nz/assets/Documents/8981330814/Final-report-v2.pdf (accessed on 16 January 2023).
23. Ministry of Health. *Older People's Health Chart Book 2006*; Ministry of Health: Wellington, New Zealand, 2006.
24. Teh, R.; Mendonça, N.; Muru-Lanning, M.; MacDonell, S.; Robinson, L.; Kerse, N. Dietary Protein Intake and Transition between Frailty States in Octogenarians Living in New Zealand. *Nutrients* **2021**, *13*, 2843. [CrossRef]
25. Hayman, K.J.; Kerse, N.; Dyall, L.; Kepa, M.; The, R.; Wham, C.; Jatrana, S. Life and Living in Advanced Age: A Cohort Study in New Zealand -Te Puāwaitanga o Nga Tapuwae Kia Ora Tonu, LiLACS NZ: Study protocol. *BMC Geriatr.* **2012**, *12*, 33. [CrossRef]

26. Wham, C.A.; Teh, R.; Moyes, S.; Dyall, L.; Kepa, M.; Hayman, K.; Kerse, N. Health and social factors associated with nutrition risk: Results from life and living in advanced age: A cohort study in New Zealand (LILACS NZ). *J. Nutr. Health Aging* **2015**, *19*, 637–645. [CrossRef]

27. Pillay, D.; Wham, C.; Moyes, S.; Muru-Lanning, M.; The, R.; Kerse, N. Intakes, adequacy, and biomarker status of iron, folate, and vitamin b12 in māori and non-māori octogenarians: Life and living in advanced age: A cohort study in New Zealand (LiLACS NZ). *Nutrients* **2018**, *10*, 1090. [CrossRef] [PubMed]

28. Firebaugh, C.M.; Moyes, S.; Jatrana, S.; Rolleston, A.; Kerse, N. Physical Activity, Function, and Mortality in Advanced Age: A Longitudinal Follow-Up (LiLACS NZ). *J. Aging Phys. Act.* **2018**, *26*, 583–588. [CrossRef] [PubMed]

29. Wham, C.; The, R.; Moyes, S.; Rolleston, A.; Muru-Lanning, H.K.; Adamson, A.; Hayman, K. Macronutrient intake in advanced age: Te Puawaitanga o Nga Tapuwae Kia ora Tonu, Life and Living in Advanced Age: A Cohort Study in New Zealand (LiLACS NZ). *Br. J. Nutr.* **2016**, *116*, 1103–1115. [CrossRef] [PubMed]

30. Wham, C.; Moyes, S.A.; Rolleston, A.; Adamson, A.; Kerse, N.; Teh, R. Association between dietary protein intake and change in grip strength over time among adults of advanced age: Life and Living in Advanced Age: A Cohort Study in New Zealand (LiLACS NZ). *Australas. J. Ageing* **2021**, *40*, 430–437. [CrossRef] [PubMed]

31. Wham, C.; Baggett, F.; Teh, R.; Moyes, S.; Kēpa, M.; Connolly, M.; Jatrana, S.; Kerse, N. Dietary protein intake may reduce hospitalisation due to infection in Māori of advanced age: LiLACS NZ. *Aust. New Zealand J. Public Health* **2015**, *39*, 390–395. [CrossRef]

32. Bacon, C.J.; Kerse, N.; Hayman, K.J.; Moyes, S.A.; The, R.O.; Kepa, M.; Dyall, L. Vitamin D status of Maori and non-Maori octogenarians in New Zealand: A cohort study (LiLACS NZ). *Asia Pac. J. Clin. Nutr.* **2016**, *25*, 885–897.

33. North, S.M.; Wham, C.A.; Teh, R.; Moyes, S.A.; Rolleston, A.; Kerse, N. High nutrition risk related to dietary intake is associated with an increased risk of hospitalisation and mortality for older Māori: LiLACS NZ. *Aust. New Zealand J. Public Health* **2018**, *42*, 375–381. [CrossRef]

34. Teh, R.; Kerse, N.; Mendonca, N.; Menzies, O.; Hill, T.; Jagger, C. Dietary Protein and Transitions Between Frailty States and to Death in Advanced Age: LiLACS NZ. *Innov. Aging* **2020**, *4*, 239–240. [CrossRef]

35. Ram, A.; Kerse, N.; Moyes, S.; Rolleston, A.; Wham, C. Protein Intake, Distribution and Food Sources in Adults of Advanced Age: Life and Living in Advanced Age: A Cohort Study in New Zealand (LiLACS NZ). *Multidiscip. Digit. Publ. Inst. Proc.* **2019**, *37*, 10. [CrossRef]

36. Wham, C.; Maxted, E.; Dyall, L.; Teh, R.; Kerse, N. Korero te kai o te Rangatira: Nutritional wellbeing of Māori at the pinnacle of life. *Nutr. Diet.* **2012**, *69*, 213–216. [CrossRef]

37. A, R. *Protein Intake, Distribution, Sources, Adequacy and Determinants in Māori and Non-Māori Octogenarians: Life and Living in Advanced Age: A Cohort Study in New Zealand (LiLACS NZ)*; Massey University: Albany, New Zealand, 2019.

38. Bennett, B. *Investigation of Protein Intakes of Māori in Advanced Age*; Massey University: Albany, New Zealand, 2013.

39. Adamson, A.; Davies, K.; Wham, C.; Kepa, M.; Foster, E.; Jones, A.; Mathers, J.; Granic, A.; Teh, R.; Moyes, S.; et al. Assessment of Dietary Intake in Three Cohorts of Advanced Age in Two Countries: Methodology Challenges. *J. Nutr. Health Aging* **2023**, *27*, 59–66. [CrossRef]

40. Bradley, J.; Simpson, E.; Poliakov, I.; Matthews, J.N.S.; Olivier, P.; Adamson, A.J.; Foster, E. Comparison of INTAKE24 (an Online 24-h Dietary Recall Tool) with Interviewer-Led 24-h Recall in 11–24 Year-Old. *Nutrients* **2016**, *8*, 358. [CrossRef]

41. Dietary Intake Data Were Collected Using Intake24.org (NZ 2018): An Open Source Dietary Assessment Research Tool, Freely Available to Researchers, Maintained and Developed by the Nutrition Measurement Platform, MRC Epidemiology Unit, University of Cambridge, in collaboration with Open Lab, Newcastle University. Available online: https://www.mrc-epid.cam.ac.uk/research/measurement-platform/dietary-assessment/intake24/ (accessed on 16 January 2023).

42. Tay, E.; Barnett, D.; Leilua, E.; Kerse, N.; Rowland, M.; Rolleston, A.; Waters, D.; Edlin, R.; Connolly, M.; Hale, L.; et al. The Diet Quality and Nutrition Inadequacy of Pre-Frail Older Adults in New Zealand. *Nutrients* **2021**, *13*, 2384. [CrossRef]

43. Washburn, R.; McAuley, E.; Katula, J.; Mihalko, S.; Boileau, R. The physical activity scale for the elderly (PASE): Evidence for validity. *J. Clin. Epi.* **1999**, *52*, 643–651. [CrossRef]

44. Essink-Bot, M.-L.; Krabbe, P.; Bonsel, G.J.; Aaronson, N.K. An Empirical Comparison of Four Generic Health Status Measures. *Med. Care* **1997**, *35*, 522–537. [CrossRef]

45. Guralnik, J.M.; Simonsick, E.M.; Ferrucci, L.; Glynn, R.J.; Berkman, L.F.; Blazer, D.G.; Scherr, P.A.; Wallace, R.B. A Short Physical Performance Battery Assessing Lower Extremity Function: Association with Self-Reported Disability and Prediction of Mortality and Nursing Home Admission. *J. Gerontol.* **1994**, *49*, M85–M94. [CrossRef]

46. Dyall, L.; Kēpa, M.; The, R.; Mules, R.; Moyes, S.A.; Wham, C.; Kerse, N. Cultural and social factors and quality of life of Māori in advanced age. Te puawaitanga o ngā tapuwae kia ora tonu–Life and living in advanced age: A cohort study in New Zealand (LiLACS NZ). *N. Z. Med. J.* **2014**, *127*, 62–79.

47. Bennett, B.; Wham, C.; Teh, R.; Moyes, S.; Kepa, M.; Maxted, E.; Kerse, N. Protein intake by Māori of advanced age. *MAI J. A New Zealand J. Indig. Sch.* **2017**, *6*, 99–115. [CrossRef]

48. National Health and Medical Research Council; Australian Government Department of Health and Ageing; New Zealand Ministry of Health. *Nutrient Reference Values for Australia and New Zealand*; National Health and Medical Research Council: Canberra, Australia, 2006.

49. Fulgoni, V.L., III. Current protein intake in America: Analysis of the National Health and Nutrition Examination Survey, 2003–2004. *Am. J. Clin. Nutr.* **2008**, *87*, 1554S–1557S. [CrossRef]
50. van Kuijk, M.; Smith, M.B.; Ferguson, C.A.; Kerse, N.M.; The, R.; Gribben, B.; Thomson, W.M. Dentition and nutritional status of aged New Zealanders living in aged residential care. *Oral Dis.* **2021**, *27*, 370–377. [CrossRef]
51. Morais, J.A.; Chevalier, S.; Gougeon, R. Protein turnover and requirements in the healthy and frail elderly. *J. Nutr. Health Aging* **2006**, *10*, 272–283. [PubMed]
52. Mendonça, N.; Hill, T.R.; Granic, A.; Davies, K.; Collerton, J.; Mathers, J.C.; Siervo, M.; Wrieden, W.L.; Seal, C.J.; Kirkwood, T.B.L.; et al. Macronutrient intake and food sources in the very old: Analysis of the Newcastle 85+ Study. *Br. J. Nutr.* **2016**, *115*, 2170–2180. [CrossRef] [PubMed]
53. Berrazaga, I.; Micard, V.; Gueugneau, M.; Walrand, S. The Role of the Anabolic Properties of Plant-versus Animal-Based Protein Sources in Supporting Muscle Mass Maintenance: A Critical Review. *Nutrients* **2019**, *11*, 1825. [CrossRef] [PubMed]
54. Gorissen, S.H.M.; Witard, O.C. Characterising the muscle anabolic potential of dairy, meat and plant-based protein sources in older adults. *Proc. Nutr. Soc.* **2017**, *77*, 20–31. [CrossRef] [PubMed]
55. Guralnik, J.M.; Ferrucci, L.; Simonsick, E.M.; Salive, M.E.; Wallace, R.B. Lower-Extremity Function in Persons over the Age of 70 Years as a Predictor of Subsequent Disability. *N. Engl. J. Med.* **1995**, *332*, 556–562. [CrossRef]
56. Xu, F.; Cohen, S.A.; Greaney, M.L.; Earp, J.E.; Delmonico, M.J. Longitudinal Sex-Specific Physical Function Trends by Age, Race/Ethnicity, and Weight Status. *J. Am. Geriatr. Soc.* **2020**, *68*, 2270–2278. [CrossRef]
57. Granic, A.; Mendonça, N.; Sayer, A.A.; Hill, T.R.; Davies, K.; Adamson, A.; Siervo, M.; Mathers, J.C.; Jagger, C. Low protein intake, muscle strength and physical performance in the very old: The Newcastle 85+ Study. *Clin. Nutr.* **2017**, *37*, 2260–2270. [CrossRef]
58. Rozzini, R.; Frisoni, G.B.; Bianchetti, A.; Zanetti, O.; Trabucchi, M. Physical Performance Test and Activities of Daily Living Scales in the Assessment of Health Status in Elderly People. *J. Am. Geriatr. Soc.* **1993**, *41*, 1109–1113. [CrossRef]
59. Reuben, D.B.; Siu, A.L.; Kimpau, S. The Predictive Validity of Self-Report and Performance-based Measures of Function and Health. *J. Gerontol.* **1992**, *47*, M106–M110. [CrossRef]
60. Manini, T.M.; Beavers, D.P.; Pahor, M.; Guralnik, J.M.; Spring, B.; Church, T.S.; King, A.C.; Folta, S.C.; Glynn, N.W.; Marsh, A.P.; et al. Effect of Physical Activity on Self-Reported Disability in Older Adults: Results from the LIFE Study. *J. Am. Geriatr. Soc.* **2017**, *65*, 980–988. [CrossRef]
61. Cruz-Jentoft, A.J.; Landi, F.; Schneider, S.M.; Zúñiga, C.; Arai, H.; Boirie, Y.; Chen, L.-K.; Fielding, R.A.; Martin, F.C.; Michel, J.-P.; et al. Prevalence of and interventions for sarcopenia in ageing adults: A systematic review. Report of the International Sarcopenia Initiative (EWGSOP and IWGS). *Age Ageing* **2014**, *43*, 748–759. [CrossRef]
62. Chale, A.; Cloutier, G.; Hau, C.; Phillips, E.; Dallal, G.; Fielding, R. Efficacy of whey protein supplementation on resistance exercise-induced changes in lean mass, muscle strength, and physical function in mobility-limited older adults. *J. Gerontol. Ser. A Biol. Sci. Med. Sci.* **2013**, *68*, 682–690. [CrossRef]
63. McMaughan, D.J.; Oloruntoba, O.; Smith, M.L. Socioeconomic Status and Access to Healthcare: Interrelated Drivers for Healthy Aging. *Front. Public Health* **2020**, *8*, 231. [CrossRef]
64. Kwon, D.H.; Park, H.A.; Cho, Y.G.; Kim, K.W.; Kim, N.H. Different Associations of Socioeconomic Status on Protein Intake in the Korean Elderly Population: A Cross-Sectional Analysis of the Korea National Health and Nutrition Examination Survey. *Nutrients* **2019**, *12*, 10. [CrossRef]
65. Kerse, N.; Teh, R.; Moyes, S.A.; Broad, J.; Rolleston, A.; Gott, M.; Kepa, M.; Wham, C.; Hayman, K.; Jatrana, S.; et al. Cohort Profile: Te Puawaitanga o Nga Tapuwae Kia Ora Tonu, Life and Living in Advanced Age: A Cohort Study in New Zealand (LiLACS NZ). *Int. J. Epidemiol.* **2015**, *44*, 1823–1832. [CrossRef]
66. The, R.; Kerse, N.; Kepa, M.; Doughty, R.N.; Moyes, S.; Wiles, J.; Dyall, L. Self-rated health, health related behaviours and medical conditions of Māori and non-Māori in advanced age: LiLACS NZ. *N. Z. Med. J.* **2014**, *127*, 13–29.
67. *Methodology Report for the 2008/09 New Zealand Adult Nutrition Survey*; University of Otago, Ministry of Health: Wellington, New Zealand, 2011.
68. Thompson, F.E.; Byers, T. Dietary assessment resource manual. *J. Nutr.* **1994**, *124*, s2245–s2317. [CrossRef]

Article

Oral Administration of Protease-Soluble Chicken Type II Collagen Ameliorates Anterior Cruciate Ligament Transection–Induced Osteoarthritis in Rats

Nan-Fu Chen [1,2,†], Yen-You Lin [3,†], Zhi-Kang Yao [4,5], Chung-Chih Tseng [2], Yu-Wei Liu [5], Ya-Ping Hung [6], Yen-Hsuan Jean [7,*] and Zhi-Hong Wen [5,8,*]

[1] Division of Neurosurgery, Department of Surgery, Kaohsiung Armed Forces General Hospital, Kaohsiung 80284, Taiwan; g1078020008@mail.802.org.tw
[2] Institute of Medical Science and Technology, National Sun Yat-sen University, Kaohsiung 80424, Taiwan; caviton@g-mail.nsysu.edu.tw
[3] Department of Sports Medicine, China Medical University, Taichung 40402, Taiwan; chas6119@mail.cmu.edu.tw
[4] Department of Orthopedics, Kaohsiung Veterans General Hospital, Kaohsiung 81362, Taiwan; akang329@vghks.gov.tw
[5] Department of Marine Biotechnology and Resources, National Sun Yat-sen University, Kaohsiung 80424, Taiwan; yoweiwei@g-mail.nsysu.edu.tw
[6] R&D Department, Taiyen Biotech Co., Ltd., Tainan 70263, Taiwan; tsi201@tybio.com.tw
[7] Department of Orthopedic Surgery, Pingtung Christian Hospital, Pingtung 90059, Taiwan
[8] Institute of BioPharmaceutical Sciences, National Sun Yat-sen University, Kaohsiung 80424, Taiwan
[*] Correspondence: 02027@ptch.org.tw (Y.-H.J.); wzh@mail.nsysu.edu.tw (Z.-H.W.); Tel.: +886-8-7368686 (Y.-H.J.); +886-7-5252000 (ext. 5038) (Z.-H.W.)
[†] These authors contribute equally to this work.

Citation: Chen, N.-F.; Lin, Y.-Y.; Yao, Z.-K.; Tseng, C.-C.; Liu, Y.-W.; Hung, Y.-P.; Jean, Y.-H.; Wen, Z.-H. Oral Administration of Protease-Soluble Chicken Type II Collagen Ameliorates Anterior Cruciate Ligament Transection–Induced Osteoarthritis in Rats. *Nutrients* **2023**, *15*, 3589. https://doi.org/10.3390/nu15163589

Academic Editor: Tyler Barker

Received: 27 July 2023
Revised: 8 August 2023
Accepted: 14 August 2023
Published: 16 August 2023

Abstract: This study investigated whether oral supplementation with protease-soluble chicken type II collagen (PSCC-II) mitigates the progression of anterior cruciate ligament transection (ACLT)–induced osteoarthritis (OA) in rats. Eight-week-old male Wistar rats were randomly assigned to the following groups: control, sham, ACLT, group A (ACLT + pepsin-soluble collagen type II collagen (C-II) with type I collagen), group B (ACLT + Amano M–soluble C-II with type I collagen), group C (ACLT + high-dose Amano M–soluble C-II with type I collagen), and group D (ACLT + unproteolyzed C-II). Various methods were employed to analyze the knee joint: nociceptive tests, microcomputed tomography, histopathology, and immunohistochemistry. Rats treated with any form of C-II had significant reductions in pain sensitivity and cartilage degradation. Groups that received PSCC-II treatment effectively mitigated the ACLT-induced effects of OA concerning cancellous bone volume, trabecular number, and trabecular separation compared with the ACLT alone group. Furthermore, PSCC-II and unproteolyzed C-II suppressed ACLT-induced effects, such as the downregulation of C-II and upregulation of matrix metalloproteinase-13, tumor necrosis factor-α, and interleukin-1β. These results indicate that PSCC-II treatment retains the protective effects of traditional undenatured C-II and provide superior benefits for OA management. These benefits encompass pain relief, anti-inflammatory effects, and the protection of cartilage and cancellous bone.

Keywords: anterior cruciate ligament transection; type II collagen; protease; nociception; cartilage; inflammation

1. Introduction

Osteoarthritis (OA) is the most common disorder of the musculoskeletal system and a leading cause of joint dysfunction and disability worldwide [1]. OA is generally caused by aging or mechanical-induced dysfunction of biological factors, resulting in an imbalance in cartilage homeostasis. This imbalance causes the degradation of the extracellular matrix (ECM), hyaluronic acid, and proteoglycans in cartilage tissue. It also leads to fibrillation and erosion of the articular surface, chondrocyte death, matrix calcification, and vascular invasion [2]. Excluding the water content, the ECM of articular cartilage comprises collagen (60%), proteoglycans (25–35%), and other noncollagenous proteins (15–20%). Type II collagen (C-II) accounts for 80% of the total collagen and provides mechanical stability to cartilage [3]. Without timely supplementation with nutraceuticals, excessive collagen degradation by matrix metalloproteinases (MMPs) can damage cartilage tissue [4,5].

The current primary treatment for OA involves reducing inflammation, alleviating pain and discomfort, and improving the structure of collagen or temporarily delaying the progression of the disease [6]. Recently, many studies have investigated whether supplementation with nutraceuticals containing undenatured C-II can alleviate OA progression [7]. In a retrospective 2020 study, Hasan et al. found that undenatured C-II improved synovitis and cartilage degradation in humans, horses, dogs, and rodents with OA [8]. Chicken sternal cartilage is a common source for preparing undenatured C-II. In traditional undenatured C-II preparations, the triple helix structure of collagen is retained, and it resembles that of C-II in human cartilage tissue. However, one study reported that its high molecular weight may result in poor absorption [9]. The processing of undenatured C-II ensures the preservation of protein glycosylation and telopeptides, which may participate in immune responses [8,10]. Studies have revealed that native glycosylation on C-II can cause the excessive activation of T cells [11,12]. Degrading or modifying the glycosylation on C-II significantly reduced the incidence, time of onset, and severity of C-II-induced rheumatoid arthritis (RA) in mouse models [11,12]. Moreover, clinical studies have revealed that T cells exhibited a strong response to C-II in patients with RA, and this immune response was correlated with disease severity [13–15]. These results highlight the risk of an excessive immune response to the structural components (e.g., carbohydrates or telopeptides) of native C-II, which may increase joint inflammation.

In cartilage tissue, type IX collagen is linked to C-II through polysaccharide cross-linking, forming a robust fibril structure [16]. Therefore, proteases such as pepsin can degrade the polysaccharide side chains on collagen, increasing the release of C-II from the fibril structure and enhancing the extraction efficiency [17]. Removing the telopeptide region from C-II through protease cleavage can increase water solubility and potentially reduce immune responses [18]. Although undenatured C-II has oral tolerance properties, which are less than the threshold of immune response activation, preserved post-translational modifications or structures may trigger adverse immune reactions in the human body [19,20]. The addition of proteases during the extraction process can disrupt the interaction between type IX collagen and C-II. It can reduce the immune response risks associated with the natural structure of C-II. The present study primarily investigated the protective effects of undenatured C-II obtained through protease addition during the extraction of chicken sternal cartilage on rats with OA.

2. Materials and Methods

2.1. Preparation of Protease-Soluble C-II

Cleaned chicken sternal cartilage without adhering tissue was donated by Taiyen Biotech in Tainan, Taiwan. The cartilage was cut into small pieces (1–2 mm^3) and treated with 20 mM ethylenediaminetetraacetic acid (EDTA, pH 7.5) at 20 °C for 18 h. The cartilage pieces were cleaned with deionized water, and EDTA-free pieces were stored at −20 °C until use. Sternal cartilage was soaked in 50 mM acetic acid for 30 days and then homogenized at 10,000 rpm on a homogenizer (X40/38 E3, Ystral, Ballrechten-Dottingen, Germany) for 30 min on ice at 4 °C. The acid-soaked cartilage was extracted with 1% pepsin (Sigma-Aldrich, St. Louis, MO, USA) and 1% protease M (Amano M, Amano Enzyme Co., Ltd., Tokyo, Japan) or without protease at 15 °C for 72 h while stirring. The mixture was filtered to remove undissolved particles and then lyophilized. The pepsin-soluble, Amano M–soluble, and unproteolyzed C-II contained 74.5, 86.9, and 62.5 mg/g hydroxyproline for 596, 695.2, and 500 mg/g collagen. The presence of epitopes in undenatured C-II was measured using the Chondrex Type II Collagen Detection Kit in accordance with the manufacturer's protocol (Chondrex, Redmond, WA, USA). The results revealed that the percentage of epitopes in pepsin-soluble C-II, Amano M–soluble C-II, and unproteolyzed C-II was 68.4%, 80.5%, and 28.0%, respectively. Pepsin-soluble C-II and Amano M–soluble C-II were fortified with bovine collagen peptide (type I collagen, MW 300–8000 Da, Nippi, Fujinomiya, Shizuoka Prefecture, Japan) to achieve ratios of type I collagen to C-II of 5.25 to promote the dispersion of C-II fibrils.

2.2. Animal Preparation

For this study, 8-week-old male Wistar rats were procured from BioLASCO Taiwan (Taipei, Taiwan). The animal room was set to a photoperiod of 12-h light–dark cycle, with the humidity and temperature maintained at 50–55% and $24 \pm 1\ °C$, respectively, using an air conditioning system throughout the experimental period. During the experiment, the rats weighed approximately 300 ± 10 g and were between 9 and 10 weeks old. Regardless of the experimental mode, all surgical procedures and feeding of the rats were completed after they were anesthetized with 2.5% isoflurane (catalog no. 08547, Panion & BF Biotech, Taoyuan, Taiwan). The handling and experimental use of animals conformed to the Guiding Principles for the Care and Use of Animals of the American Physiological Society, and the study protocol was approved by the Institutional Animal Care and Use Committee of National Sun Yat-sen University (approval number: 10725).

2.3. Establishment of Animal Models

Animal models of anterior cruciate ligament transection (ACLT)–induced OA were established using the methods proposed by Stoop et al. (2001) and Yang et al. (2014) [21,22]. In this study, first, the baseline activity of the rats was measured; the rats were then anesthetized, and their right knee joint was shaved. Methanol was used for sterilization, and the knee joint capsule of each rat was opened through a medial parapatellar incision. The patella was dislocated laterally, and the knee joint was fully flexed to expose the cruciate ligament. An incision was made anterior to the medial collateral ligament, and the anterior drawer test was implemented to verify the adequacy of the cut before surgical suturing. In the sham group, the knee joint capsule was opened, but no incision was made anterior to the medial collateral ligament [21]. After surgery, the rats were injected with cefazolin (20 mg/kg) to prevent wound infection and were returned to their cages to recover for eight weeks. Weekly testing was implemented to analyze differences between the experimental and control groups regarding pain, inflammation, and knee swelling. C-II treatment was administered after significant pain, inflammation, and swelling differences were observed between the experimental and control groups.

2.4. Experimental Groups

The rats were randomly assigned to six experimental groups.
Control: naïve rats ($n = 8$).
Sham: ACL was exposed but not transected ($n = 8$).
ACLT: ACLT of the right knee ($n = 8$).
Group A (ACLT + pepsin-soluble C-II fortified with type I collagen): Rats undergoing ACLT were orally administered 0.25 mg/kg/day collagen (containing 0.04 mg/kg/day C-II and 0.21 mg/kg/day type I collagen), which was dissolved in 1 mL of ultrapure water, once daily for 17 consecutive weeks beginning eight weeks after ACLT ($n = 8$).
Group B (ACLT + Amano M–soluble C-II fortified with type I collagen): Rats undergoing ACLT were orally administered 0.25 mg/kg/day collagen (containing 0.04 mg/kg/day C-II and 0.21 mg/kg/day type I collagen), which was dissolved in 1 mL of ultrapure water, once daily for 17 consecutive weeks beginning eight weeks after ACLT ($n = 8$).
Group C (ACLT + high-dose Amano M–soluble C-II fortified with type I collagen): Rats undergoing ACLT were orally administered 0.75 mg/kg/day collagen (containing 0.12 mg/kg/day C-II and 0.63 mg/kg/day type I collagen), which was dissolved in 1 mL

of ultrapure water, once daily for 17 consecutive weeks beginning eight weeks after ACLT ($n = 8$).

Group D (ACLT + unproteolyzed C-II): Rats undergoing ACLT were orally administered 0.24 mg/kg/day unproteolyzed collagen (containing 0.12 mg/kg/day C-II), which was homogeneously suspended in 1 mL of ultrapure water, once daily for 17 consecutive weeks beginning eight weeks after ACLT ($n = 8$).

2.5. Analysis of Pain Behavior of Rats

2.5.1. Weight-Bearing Distribution in Hind Legs

The difference in weight-bearing distribution between the hind leg with ACLT-induced joint degeneration and the contralateral leg is a pain indicator in OA [23]. A dual-channel weight averager (Singa Technology, Taoyuan, Taiwan) was used to assess the weight-bearing distribution of the rats' hind legs. First, each rat was placed on the sloped channel with its hind legs resting on two weight-averaging platform pads. The researchers ensured the rat was stationary and its posture stable. The measurement button on the instrument was pressed to record the distribution of the animal's body weight on each paw for 3 s. The results are presented as the difference between the amount of weight placed on the uninjured paw (i.e., left paw) and the amount placed on the injured paw (i.e., right paw) [24]. The baseline activity levels of the rats were measured weekly post-surgery from weeks 1 to 24.

2.5.2. Secondary Mechanical Allodynia

Allodynia is pain caused by normally non-noxious stimuli. This study assessed mechanical allodynia by employing von Frey hair monofilaments (27 inches; Touch-Test Sensory Evaluators, NC12775, NorthCoast Medical, Morgan Hill, CA, USA) with stiffness ranging between 2.0 and 28.8 g in combination with Dixon's up-and-down method. The hair monofilaments were applied to the plantar surface of each hind paw in ascending order of stiffness, and whether the rats exhibited a reflex response was observed. If no response occurred, von Frey hairs of increasing stiffness levels were applied until the rat exhibited a reflex response, and the stiffness level was recorded [24]. Mechanical allodynia testing was implemented to measure reflex responses in the rats at baseline and weekly from weeks 1 to 24.

2.5.3. Knee Swelling

Changes in knee swelling can reflect the severity of inflammation in the knee joint. To assess the severity of inflammation in the knee joint, the circumference of each rat's knee joint was measured before surgery (baseline) and weekly from weeks 1 to 24 post-surgery. After the rats were anesthetized with 2.5% isoflurane, a Vernier scale (calipers, AA847R, Aesculap, Tuttlingen, Germany) was used to measure the widths of the rats' knee joints, and the changes in width were recorded [22].

2.6. Sample Collection and Fixation

After the animal behavior experiments, the rats were euthanized for further analysis. Each rat was anesthetized with 2.5% isoflurane, and the chest was opened using surgical instruments. The rats were injected with 600 mL of 4% phosphate-buffered saline (PBS; 137 mM NaCl, 2.68 mM KCl, 10 mM Na_2HPO_4, and 1.76 mM KH_2PO_4; pH = 7.2; storage temperature = 4 °C) and heparin (0.2 U/mL) by using a perfusion needle inserted through the ventricle and into the left aorta with a pump. An incision was made in the right atrium to enable blood outflow, achieving full-body blood replacement. Subsequently, 4% paraformaldehyde (at a storage temperature of 4 °C) was injected in situ for tissue fixation. Finally, the knee joint tissue was extracted using surgical instruments. The extracted tissue samples were maintained in 10% neutral buffered formalin at 4 °C for 3 to 4 days, during which time the solution was replaced every two days to maintain its fixative effect.

2.7. Micro-Computed Tomography Scanning Analysis

This study adapted the research methods of Bagi et al. [25] with modification. Before completing the bone samples' microcomputed tomography (micro-CT) scans (SkySacn 1076, Bruker, Antwerp, Belgium), the Taiwan Mouse Clinic was commissioned to reconstruct 3D animal skeleton models and implement data analysis of the reconstructed images. Micro-CT scanning, which was implemented with a resolution of 35 µm (isotropic voxel size: 35 µm), was used to analyze the 3D microstructure parameters of statistical significance, including trabecular separation, trabecular number, bone mineral density (BMD), tissue volume, bone volume, and the ratio of bone volume to tissue volume.

2.8. Sample Embedding

The fixated tissues were immersed in decalcification solution (100-g EDTA disodium salt dehydrate/1000-mL PBS) at room temperature for 2 to 3 weeks, during which the solution was replaced every two days to maintain its effectiveness for removing calcium deposits. The fixated and decalcified tissue samples were placed into tissue cassettes and processed using an automatic tissue dehydration machine (Tissue-Tek, Sakura Finetek Japan, Tokyo, Japan). The tissue samples were treated (in the following order) with 35% alcohol, 75% alcohol, 85% alcohol, 85% alcohol, 95% alcohol, 95% alcohol, 100% alcohol, 100% alcohol, 90% xyline/alcohol, 100% xyline, soft paraffin, and hard paraffin. The tissue dehydration and paraffin embedding process lasted 18 h. Finally, an embedding center (CSA Embedding Center EC780-1; EC780-2, Pomona, CA, USA) was used to embed the tissues into tissue blocks. A paraffin slicing machine (Microm, HM340E, Kentwood, MI, USA) was then used to slice the tissue blocks into 1-µm slices, which were stained.

2.9. Histological Staining Analysis

To evaluate the cartilage tissue samples, the grading system for safranin-O/fast green staining proposed by the Osteoarthritis Research Society International (OARSI) was employed. Semi-quantitative analysis was implemented using the OARSI grading system (i.e., six histological grades) while considering the rats' OA recovery stages (i.e., four histological stages). The total scores of the samples were calculated as the product of the historical grade and the histological stage and ranged from 1 (normal articular cartilage) to 24 (no repair) [26]. Synovial tissue was analyzed using the tissue assessment method proposed by Krenn et al. (2006), which involves hematoxylin and eosin (H&E) staining [27]. The grading criteria for the synovial lining layer were the presence of subsynovial fibroblasts (0–3 points), synovial hyperplasia (0–3 points), and red blood cell infiltration (0–3 points). The grading criteria for subsynovial tissue were the formation of granulation tissue (0–3 points), angioplasia (0–3 points), and red blood cell infiltration (0–3 points). The severity of inflammation in synovial tissue was divided into three stages; mild, moderate, and severe inflammation were indicated by scores of 0–6, 7–12, and 13–18, respectively. A higher total score represents greater damage to the knee joint. The glass slides of the stained samples were examined using an optical microscope (DM 6000, Leica, Wetzlar, Germany) equipped with a digital microscope camera (SPOT Idea, Diagnostic Instruments, Sterling Heights, MI, USA).

2.10. Immunohistochemical Staining Analysis

In the immunohistochemical staining experiment, the paraffin slides were first immersed in xylene solution for 20 min and then in ethanol with a concentration ranging from 50% to 100% for 30 s. Finally, the paraffin slides were subjected to proteinase K (20 mM; Sigma-Aldrich, St. Louis, MO, USA) digestion for 20 min. Each slide was rinsed twice with tris-buffered saline with Tween (TTBS), encircled with a Dako pen (Dako, S2002), and immersed in hydrogen peroxide for 8 min. The slides were again rinsed twice with TTBS, immersed in the ABC Kit blocking buffer (Vectastain ABC Kit, Vector Laboratories, Burlingame, CA, USA), and oscillated at room temperature for 30 min. The samples were subsequently incubated with a primary antibody specific to the target protein and oscillated overnight at 4 °C. The sections were then incubated with specific primary antibodies, namely anti-C-II antibodies (1:200; catalog no. 234187; Calbiochem, San Diego, CA, USA), anti-matrix metallopeptidase 13 (MMP13) antibodies, (1:100; catalog no. ab39012; Abcam, Cambridge, UK), anti-interleukin-1β antibodies (IL-1β 1:200; catalog no. ab9722; Abcam), and anti-tumor necrosis factor-α (TNF-α antibodies (1:150; catalog no. ab6671; Abcam). The slides were then rinsed twice with TTBS, immersed in the aforementioned blocking buffer, and oscillated at room temperature for 30 min. Next, the slides were incubated with a secondary antibody (Vector Laboratories, Burlingame, CA, USA) in the ABC Kit, oscillated at room temperature for 80 min and in fresh TTBS solution three times for 10 min each. They were subsequently immersed in the aforementioned blocking buffer, oscillated at room temperature for 30 min, and washed once with fresh TTBS. The samples were then incubated with DAB staining solution (DAB Peroxidase Substrate Kit, Vector Laboratories) for 5 min, rinsed twice, soaked in Mayer's hematoxylin for 90 s, rinsed for 5 min, and soaked sequentially in ethanol solutions with concentrations of 50%, 70%, 95%, and 100% for 20 s each. Finally, the slides were immersed in xylene solution for 1 min before being sealed with a coverslip. The slides were imaged under an optical microscope (DM 6000B, Leica, Wetzlar, Germany) equipped with the aforementioned digital microscope camera (SPOT Idea 5.0 Mp 635 Color Digital Camera, Diagnostic Instruments, Sterling Heights, MI, USA).

2.11. Data Analysis

The experimental data are presented as the mean ± standard error of the mean. For statistical analyses, we calculated the variation between groups by a one-way analysis of variance (ANOVA), examined by a post hoc Tukey test. We defined the statistical significance as $p < 0.05$. Statistical analyses were performed using SigmaPlot Version 12.0 (Systat Software, Inc., San Jose, CA, USA).

3. Results

3.1. Effect of (Oral Protease-Soluble Chicken C-II) PSCC-II on ACLT-Induced Weight-Bearing Deficits

A dual-channel weight averager was employed to determine the weight-bearing distribution of the rats' hind legs. Figure 1A presents the difference in ACLT-induced weight-bearing deficits between the rats' hind legs. The results revealed a significant difference in the weight-bearing distribution of the hind legs between the ACLT and sham groups from weeks 1 to 24 post-surgery. Among the rat models of ACLT-induced OA, the differences in the weight-bearing distribution of the hind legs in groups A, B, C, and D were significantly lower than that in the ACLT group from weeks 12–24, 12–24, 11–24, and 11–24 post surgery, respectively. In summary, significant improvements in the ACLT-induced weight-bearing deficit in the hind legs were detected in treatment groups A, B, C, and D.

3.2. Effect of PSCC-II on ACLT-Induced Mechanical Allodynia

Figure 1B presents the effects of PSCC-II on ACLT-induced mechanical allodynia. The results revealed that the paw withdrawal threshold (g) of the ACLT group was significantly lower than that of the sham group post-surgery. Among the groups with ACLT-induced OA, the paw withdrawal thresholds (g) of groups A, B, C, and D were significantly higher than that of the ACLT group from weeks 13–24 post-surgery. In summary, treatment groups A, B, C, and D had reduced considerably mechanical allodynia caused by ACLT-induced OA. Notably, the treatment effect in group C was higher than in the other groups but did not attain statistical significance.

3.3. Effect of PSCC-II on ACLT-Induced Knee Joint Swelling

The knee joint width of the hind legs was measured using a Vernier scale. The results, presented in Figure 1C, revealed that the difference in the knee joint width of the ACLT group was significantly greater than that of the sham group from weeks 2 to 24 post-surgery. Among the groups with ACLT-induced OA, the differences in the knee joint width of groups A, B, C, and D were significantly smaller than that of the ACLT group from weeks 12–24, 12–24, 12–22, and 13–24 postsurgery, respectively. In summary, treatment groups A, B, C, and D had significantly reduced swelling of the knee joint caused by ACLT-induced OA.

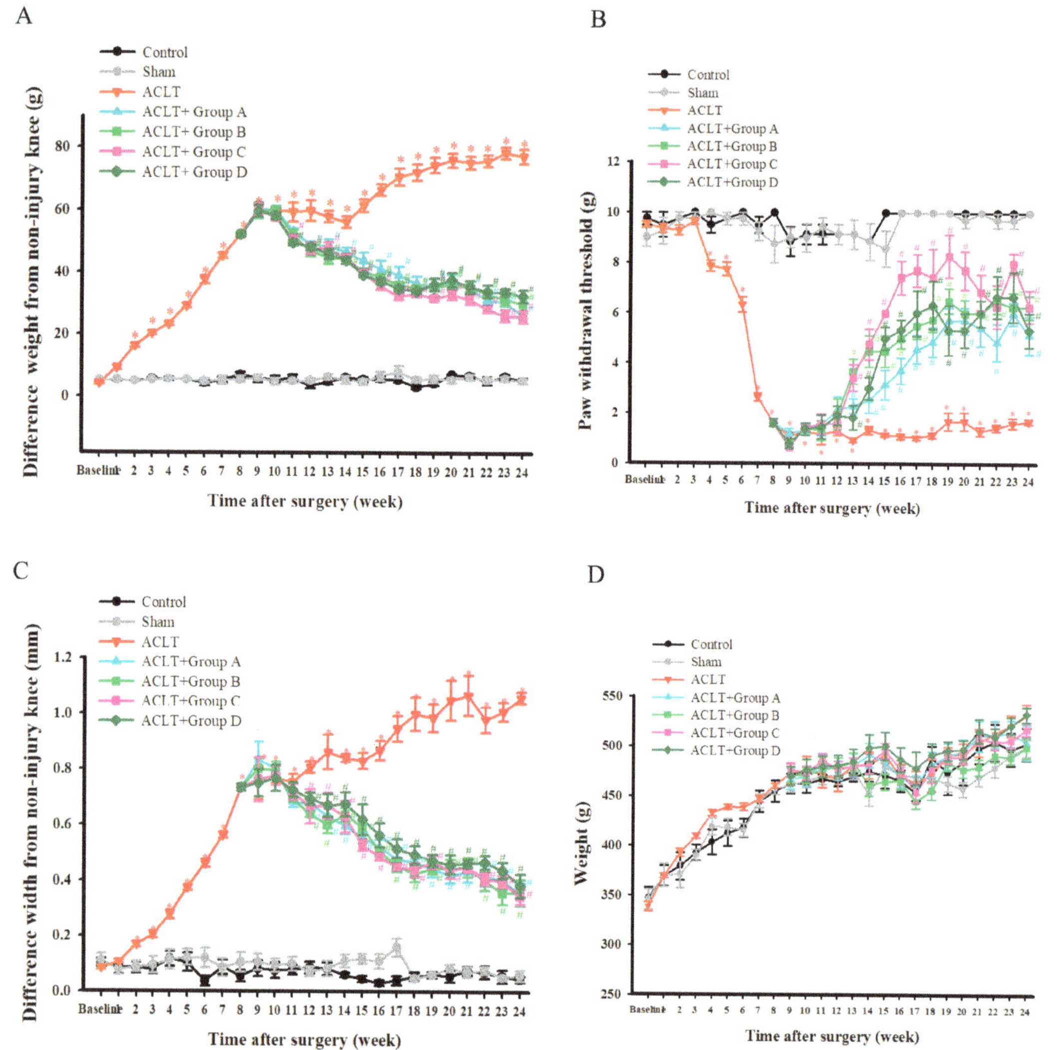

Figure 1. Effects of PSCC-II on ACLT-Induced OA-Related Phenomena. Effects of PSCC-II on ACLT-induced weight-bearing deficits in hindlegs (**A**), mechanical allodynia (**B**), knee joint swelling (**C**), and body weight (**D**) over time. Data are presented as mean ± standard error of the mean. * denotes comparison with sham group ($p < 0.05$); # denotes comparison with ACLT group ($p < 0.05$).

3.4. Effect of PSCC-II on Body Weight

Figure 1D presents the effects of PSCC-II on body weight. The weights of all groups increased over time. The results revealed no significant differences in body weight between the ACLT and sham groups from weeks 0 to 24 post-surgery. No significant differences were found in body weight between the treatment groups (A, B, C, and D) and the ACLT group from weeks 9 to 24 post-surgery. In summary, no significant differences in body weight were observed between groups A, B, C, and D.

3.5. Micro-CT Analysis of Knee Bone Structure in Rats Undergoing ACLT after PSCC-II Treatment

Figure 2A presents 2D micro-CT images of knee joints displaying the tibial cancellous bone. The images reveal greater tibial cancellous bone loss in the ACLT group than in the sham group. The restoration effects in groups A and D were not greater than those in the ACLT group; however, the restoration effects in groups B and C were markedly greater than those in the ACLT group. Micro-CT scanning was employed to quantitatively analyze bone volume (Figure 2B), trabecular separation (Figure 2C), trabecular number (Figure 2D), and BMD (Figure 2E) in the tibial cancellous bone. These parameters were evaluated using reconstructed 3D images of the tibial metaphysis. The results indicated that the ACLT group had significantly lower cancellous bone volume, lower trabecular number, and increased trabecular separation than the sham group. Treatment groups A, B, and C had significantly greater restoration effects than the ACLT group with respect to cancellous bone volume, trabecular number, and trabecular separation; however, the results did not indicate a significant restoration effect in group D. Figure 3 presents the reconstructed 3D images of the subchondral bone (SB) for the control, sham ACLT, and ACLT plus treatment groups. Table 1 presents the quantitative analysis of tissue volume, bone volume, the ratio of bone volume to tissue volume, and BMD of the reconstructed 3D images of the SB. Analysis of the reconstructed 3D images revealed no significant differences between groups. In summary, PSCC-II treatment administered to groups A, B, and C effectively mitigated cancellous bone loss caused by ACLT-induced OA.

Table 1. Analysis of 3D Micro-CT scans of Reconstructed Images of Tibial Subchondral Bone.

Parameter	Unit	Control	Sham	ACLT	ACLT + Group A	ACLT + Group B	ACLT + Group C	ACLT + Group D
Tissue vol.	mm^3	0.492 ± 0.001	0.490 ± 0.001	0.491 ± 0.001	0.490 ± 0.000	0.490 ± 0.000	0.490 ± 0.000	0.490 ± 0.000
Bone vol.	mm^3	0.31 ± 0.01	0.32 ± 0.02	0.31 ± 0.02	0.37 ± 0.02	0.31 ± 0.02	0.37 ± 0.02	0.32 ± 0.04
BV/TV	%	63.92 ± 2.05	66.30 ± 3.238	62.24 ± 3.06	75.02 ± 4.53	63.88 ± 3.89	75.35 ± 4.85	66.06 ± 8.09
BMD	g/cm^3	0.726 ± 0.014	0.711 ± 0.020	0.691 ± 0.013	0.743 ± 0.013	0.684 ± 0.016	0.767 ± 0.023	0.687 ± 0.04

BV/TV, bone vol./tissue vol.; BMD, bone mineral density.

3.6. Effect of PSCC-II on Synovial Tissue Inflammation in Knee Joints Subjected to ACLT

Figure 4A–H present the pathological H&E staining results. Synovial hyperplasia and red blood cell infiltration were significantly greater in the ACLT group than in the sham group. The rats in the ACLT group had severe knee joint inflammation. The rats in group A had mild synovial inflammation, and synovial hyperplasia and neutrophil infiltration were significantly less severe in group A than in the ACLT group. Similar results were observed in groups B, C, and D. PSCC-II treatment effectively alleviated synovial hyperplasia and red blood cell infiltration in knee joints subjected to ACLT in groups A, B, C, and D.

3.7. Effect of PSCC-II on Articular Cartilage Degradation in Knee Joints Subjected to ACLT

Figure 4I–O presents the safranin-O/fast green staining of knee articular cartilage. Cartilage histopathology was further assessed using the OARSI histological scoring system (Figure 4P). The results revealed that the ACLT group had greater damage to the surface of cartilage tissue, greater loss of chondrocytes, and lower staining intensity of cartilage than the sham group. These pathological changes in articular cartilage were reduced in the ACLT plus treatment groups (i.e., A, B, C, and D). The OARSI score was significantly higher in the ACLT group (9.75 ± 0.94) than in the sham group (0.00 ± 0.00). Groups A, B, C, and D had significantly lower OARSI scores than the ACLT group (2.57 ± 0.53, 0.63 ± 0.26, 0.43 ± 0.20, 3.83 ± 0.65, respectively). Therefore, PSCC-II treatment effectively alleviated articular cartilage degradation in the rats with ACLT-induced OA in groups A, B, C, and D. PSCC-II treatment almost completely counteracted ACLT-induced articular cartilage degradation in groups C.

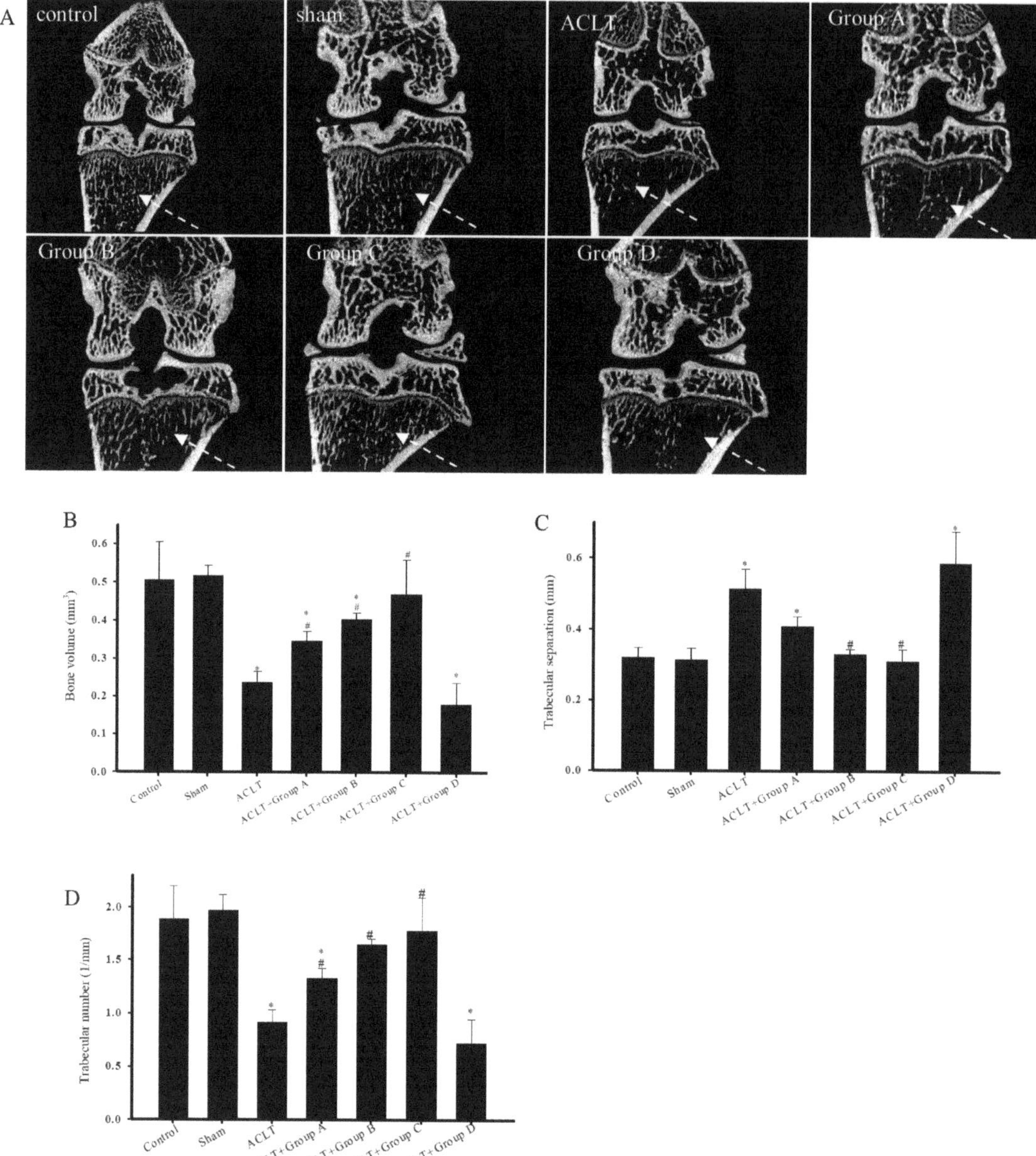

Figure 2. Micro-CT scans of Reconstructed Images of Cancellous Bone and Quantitative Analysis. Micro-computed tomography (micro-CT) was used to analyze the structure of the knee tibia in ACLT after treatment. (**A**) 2D micro-CT scans of reconstructed images of cancellous bone. Dotted arrows indicate image sections for comparing cancellous bone volume. Quantitative analysis of (**B**) bone volume (mm^3), (**C**) trabecular separation (mm), and (**D**) trabecular number (1/mm). Data are presented as mean ± standard error of the mean. * denotes comparison with sham group ($p < 0.05$); # denotes comparison with ACLT group ($p < 0.05$).

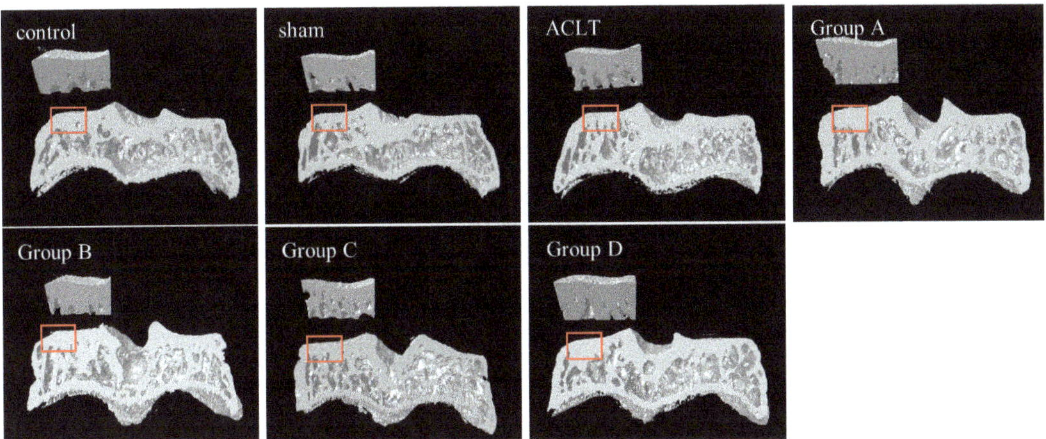

Figure 3. 3D Micro-CT scans of Reconstructed Images of Tibial Metaphysis. Upper left sections marked with red rectangles were used for analysis. No obvious differences were observed between the sham group, ACLT group, and treatment groups A, B, C, and D.

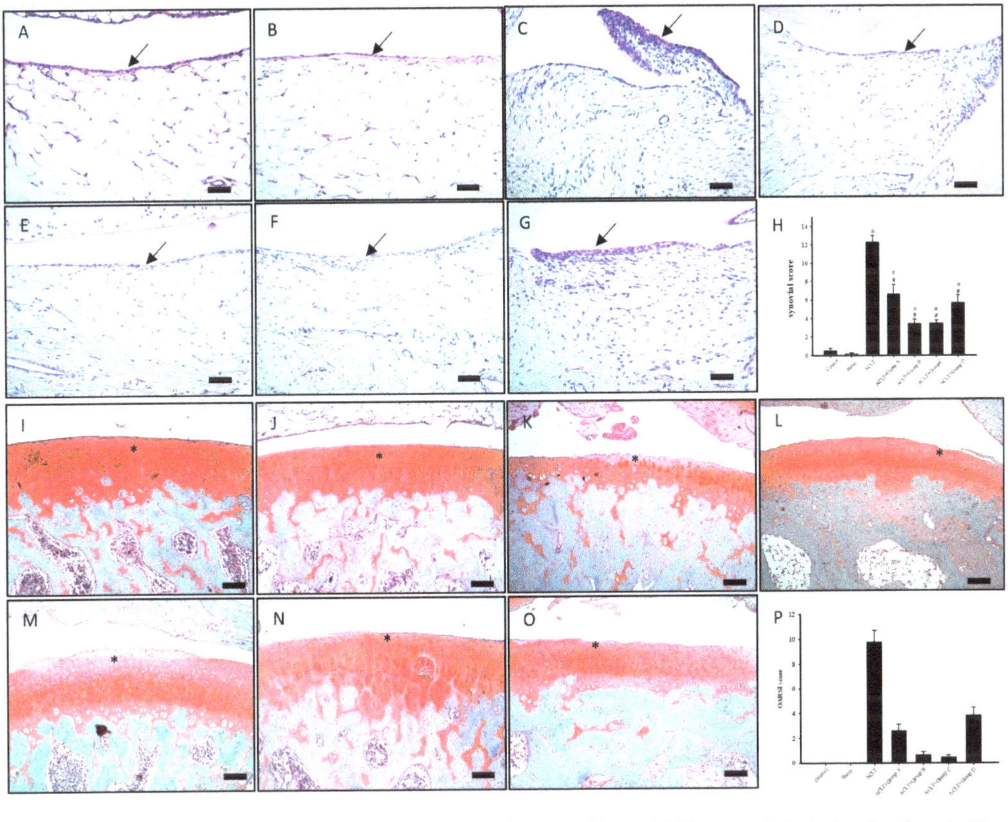

Figure 4. Histopathological Evaluation of Synovial Tissue and Articular Cartilage in Knee Joints Subjected to ACLT after Treatment. (**A,I**) indicate control group; (**B,J**) indicate sham group; (**C,K**) indicate ACLT group; (**D,L**) indicate group A; (**E,M**) indicate group B; (**F,N**) indicate group C; and (**G,O**) indicate group D. Synovial tissue samples were stained with hematoxylin and eosin stain, and cell infiltration, synovial fibrosis, angiogenesis, and synovial hypertrophy (indicated by arrows)

were observed in each group. (**H**) Quantitative synovial scores for various synovial tissue samples. Cartilage degradation (indicated by an asterisk) was stained using safranin-O/fast green staining. (**P**) Quantitative histopathological changes in knee joints were evaluated using the Osteoarthritis Research Society International scoring system. Scale bar = 100 μm. Data are presented as mean ± standard error of the mean. * denotes comparison with sham group ($p < 0.05$); # denotes comparison with ACLT group ($p < 0.05$).

3.8. Effect of PSCC-II on Expression of C-II and MMP-13 in Articular Cartilage in Knee Joints Subjected to ACLT

Figure 5A–H presents the effect of PSCC-II on C-II immunoreactivity in knee articular cartilage. C-II immunoreactivity was lower in the ACLT group than in the sham group. C-II expression was increased in the ACLT plus PSCC-II groups (i.e., groups A, B, and C) and group D compared with that in the ACLT group. Quantitative analysis of C-II expression in articular cartilage revealed significantly decreased immunoreactivity in the ACLT group (0.49 ± 0.08 folds) compared with the sham group (0.97 ± 0.10). The results also revealed significantly increased immunoreactivity in groups A, B, C, and D (1.74 ± 0.11, 1.98 ± 0.11, 2.05 ± 0.14, 1.44 ± 0.14 folds, respectively) compared with the ACLT group. Moreover, all treatment groups exhibited significantly increased immunoreactivity compared with the sham group. In summary, the decreased ACLT-induced C-II expression in articular cartilage was mitigated significantly in groups A, B, C, and D. Figure 5I–P presents the effects of PSCC-II on MMP-13 immunoreactivity in knee articular cartilage. MMP-13 immunoreactivity was higher in the ACLT group than in the sham group. All treatment groups exhibited decreased *MMP-13* expression compared with the ACLT group. Quantitative analysis of *MMP-13* expression in articular cartilage revealed significantly increased immunoreactivity in the ACLT group (48.76% ± 2.59%) compared with the sham group (8.55% ± 1.44%). Immunoreactivity was significantly lower in groups A, B, C, and D (25.67% ± 3.91%, 28.34% ± 1.28%, 20.68% ± 1.42%, 22.93% ± 2.90%, respectively) compared with in the ACLT group. In summary, PSCC-II treatment in groups A, B, and C and the treatment in group D effectively mitigated the decreased *MMP-13* expression caused by ACLT.

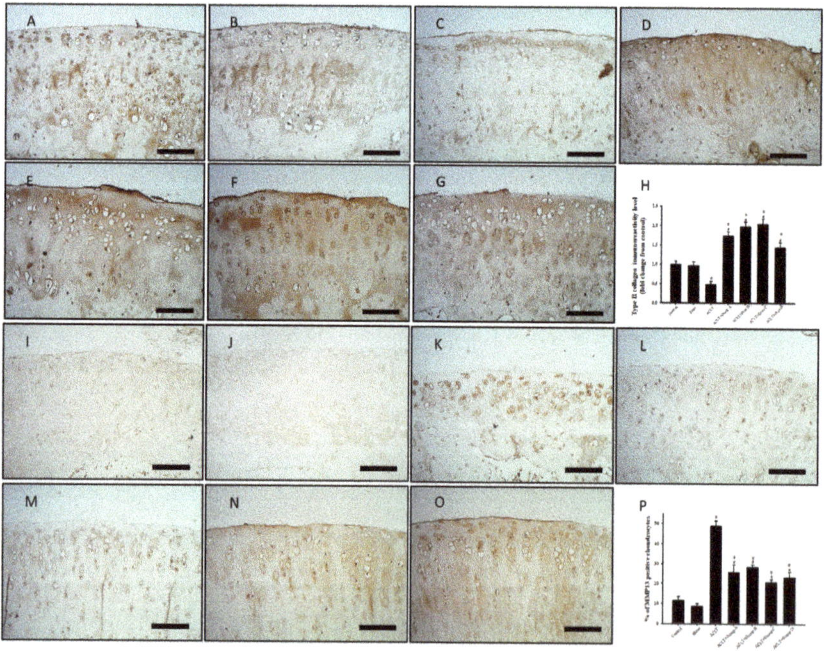

Figure 5. Expression of *C-II* and *MMP-13* in Cartilage Tissue with ACLT after Treatment. (**A,I**) indicate control group; (**B,J**) indicate sham group; (**C,K**) indicate ACLT group; (**D,L**) indicate

group A; (**E,M**) indicate group B; (**F,N**) indicate group C; and (**G,O**) indicate group D. (**A–G**) Immuno-histochemical staining of type II collagen (C-II; brown area) in knee joint sections. (**H**) Quantitative analysis of C-II expression in cartilage. (**I–P**) Immunohistochemical staining of MMP-13 in joint sections. Brown areas indicate immunoreactive cells. (**P**) Quantitative analysis of MMP-13–positive cells in cartilage. Scale bar =100 μm. Data are presented as mean ± standard error of the mean. * denotes comparison with sham group ($p < 0.05$); # denotes comparison with ACLT group ($p < 0.05$).

3.9. Effect of PSCC-II on Expression of TNF-α and IL-1β in Chondrocytes in Articular Cartilage after ACLT

Figure 6A–P present the effects of PSCC-II on the immunoreactivity of TNF-α and IL-1β in chondrocytes within articular cartilage. The immunoreactivity of TNF-α and IL-1β was higher in the ACLT group than in the control and sham groups. The results revealed the decreased expression of TNF-α and IL-1β in groups A, B, C, and D compared with that in the ACLT group. Quantitative analysis of TNF-α and IL-1β expression revealed significantly increased immunoreactivity in the ACLT group compared with that in the sham group. Immunoreactivity was significantly decreased in the treatment groups compared with the ACLT group. In summary, PSCC-II treatment in groups A, B, and C and the treatment in group D effectively inhibited the reduction in TNF-α and IL-1β expression caused by ACLT. Notably, the results indicated no significant difference in the immunoreactivity of TNF-α and IL-1β between the sham group and group C.

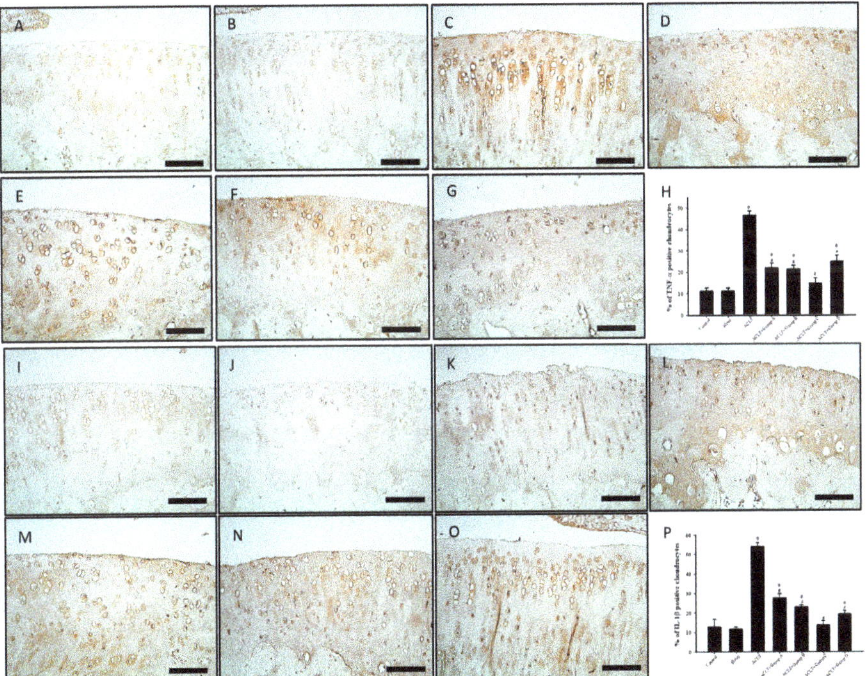

Figure 6. Expression of *TNF-α* and *IL-1β* in Cartilage Tissue with ACLT after Treatment. (**A,I**) indicate control group; (**B,J**) indicate sham group; (**C,K**) indicate ACLT group; (**D,L**) indicate group A; (**E,M**) indicate group B; (**F,N**) indicate group C; and (**G,O**) indicate group D. Brown areas indicate immunoreactive cells. (**A–G**) Immunohistochemical staining of TNF-α in cartilage. (**H**) Quantitative analysis of TNF-α–positive cells. (**I–P**) Immunohistochemical staining of IL-1β in cartilage. (**P**) Quantitative analysis of TNF-α–positive cells. Scale bar = 100 μm. Data are presented as mean ± standard error of the mean. * denotes comparison with sham group ($p < 0.05$); # denotes comparison with ACLT group ($p < 0.05$).

4. Discussion

This study investigated the potential protective effects of orally administered PSCC-II against nociception, histopathological cartilage degradation, and the expression of proinflammatory cytokines in rats with ACLT-induced OA. Additionally, we employed micro-CT to assess the SB microstructure in the tibial knee region. The results revealed that the oral administration of PSCC-II and undenatured C-II effectively reduced nociception sensitivity in rats with ACLT-induced OA. These supplements mitigated weight-bearing deficits, mechanical allodynia, knee joint swelling, cartilage degradation, synovitis, and the expression of ECM degradative enzymes and proinflammatory cytokines in knee articular cartilage. Notably, PSCC-II prevented microstructural changes induced by ACLT in the proximal tibial metaphysis, such as reduced bone volume and trabecular number and increased trabecular separation. However, undenatured C-II did not prevent these changes.

Rats with ACLT-induced OA are a widely used model for studying OA. ACLT in rats leads to knee joint inflammation, neutrophil infiltration, SB remodeling, and osteophyte formation. All these factors can exacerbate the occurrence of OA [28]. Cartilage degradation is simulated in the ACLT-induced OA model, as ACL injury causes long-term joint instability. The progression of this OA model is slow; however, the study observed molecular changes in cartilage tissue, synovitis, and SB sclerosis similar to those in human OA, which may be conducive to drug research [29]. Another study suggested that the experimental model of ACLT alone may be more suitable for evaluating the potential of drugs to alleviate OA [30]. The results of the present study revealed that treatment with oral PSCC-II (groups A, B, and C) and undenatured C-II (group D) significantly inhibited ALCT-induced weight-bearing deficits, mechanical allodynia, and knee joint swelling in rats with OA (Figure 1A–C). PSCC-II and undenatured C-II did not adversely affect body weight (Figure 1D).

Studies have reported that protein glycosylated side chains may play a crucial role in the overexcitation of T cells and contribute to the immune response [11,31]. Studies employing C-II-induced arthritis (CIA) animal model have contended that C-II is a cartilage-specific autoantigen and that its glycosylated side chains can promote T cell overactivation [32,33]. CIA is a common animal model used to study RA. Studies have verified that naïve glycosylated C-II is more likely to induce arthritis and cause changes in articular cartilage and SB than non-glycosylated C-II. Removing or modifying carbohydrates from C-II could reduce the incidence, onset time, and severity of CIA in animals [11,12]. Clinical studies have revealed that 3–27% of patients with RA have serum antibodies against C-II; their results also revealed a positive correlation between T cell responses to C-II and disease severity [15,34]. Undenatured C-II is an untreated bundle of native C-II protein fibers obtained by purifying animal cartilage. The procollagen form of C-II contains three identical alpha chains; each has extensions called non-collagenous telopeptides and retains its glycoprotein side chain structure [35]. One study revealed that collagen-IX and C-II exhibit polysaccharide cross-linking within cartilage tissue, forming a robust fibril structure [16]. The use of proteases can increase extraction efficiency because they target and degrade the side chains of polysaccharides on collagen molecules, facilitating the release of C-II from the fibril structure [17]. However, telopeptides of procollagen are insoluble and may contribute to the immune response [10]. Modification of glycosylated side chains of C-II can also alleviate the excessive activation of the immune system [8]. Therefore, the proteolytic action of enzymes in C-II may involve the modification of glycosylated side chains and the removal of telopeptides, thereby enhancing solubility and reducing adverse immune reactions. In our unpublished experimental results, the molecular weight of undenatured C-II was approximately 300 kDa, whereas that of PSCC-II was approximately 280–300 kDa. We speculate that the telopeptides and glycosylated structures on the C-II fiber bundles of PSCC-II obtained from enzymatic proteolysis may change, potentially leading to reduced autoimmunity. This speculation merits additional investigation.

Studies have widely leveraged micro-CT's high-resolution and 3D reconstruction capabilities for evaluating animal and human OA histopathology [36]. Changes in the

SB structure in joint regions, a histopathological characteristic of OA, can contribute to the load-bearing pressure experienced by the body after cartilage degeneration [37]. In OA, the SB undergoes structural changes, such as decreased bone volume and trabecular number and increased trabecular separation [38]. Structural changes in the SB interact synergistically with articular cartilage degradation; thus, alterations in the SB structure can lead to secondary cartilage injury and degradation. Cartilage damage or loss can alter the load-bearing capacity of the underlying SB [38]. The trabecular bone, a component of SB, has a layered, spongy, and porous structure and crucially contributes to the load-bearing capacity of the body; its mechanical supporting properties are influenced by the mineral content and type I collagen [39]. In 1998, Fazzalari and Parkinson found a significant decrease in trabecular number and increased trabecular separation of the SB in patients with OA [40]. In 2008, Kadri et al. observed a significant decrease in bone volume and an increase in trabecular separation in mice with OA [41]. Bagi et al. employed a rat model in which a partial medial meniscectomy tear was surgically induced to trigger OA development; that study also revealed reduced trabecular bone volume in the SB [25]. Our experimental results align with the findings of these relevant studies. The ACLT-induced OA group exhibited a significant decrease in bone volume and trabecular number (Figure 2B,D) and a significant increase in trabecular separation (Figure 2C). Oral administration of PSCC-II (i.e., groups A, B, and C) mitigated these structural changes in the SB, and the results of group D were superior to those of groups A and B. However, undenatured C-II (group D) treatment did not significantly affect the changes in bone volume, trabecular number, and trabecular separation caused by ACLT.

Type I collagen comprises 90% of the total protein in the SB, and it is influenced by factors such as mechanical loading, pathological insults, and bone types [42–44]. Studies have revealed that patients with osteogenesis imperfecta (OI) primarily exhibit mutations in the collagen type 1 alpha 1 chain (*COL1A1*) gene or the collagen type 1 alpha 2 chain (*COL1A2*) gene, which are associated with the lower trabecular number, trabecular thickness, bone mass, and connectivity in the SB. These findings indicate a significant positive correlation between structural changes in the SB and the expression of type I collagen [45]. Wu et al. (2020) asserted that type I collagen is unevenly distributed and is significantly reduced in the SB of patients with OA. This decrease in type I collagen is accompanied by a decrease in trabecular bone volume and an increase in trabecular separation [46]. In the present study, groups A, B, and C received PSCC-II treatment. During the preparation process of PSCC-II, in addition to proteolytic C-II, type I collagen was added to dissolve PSCC-II. The type I collagen dosage in group C was 0.63 mg/kg/day, higher than in groups A and B (0.21 mg/kg/day). The results revealed that the decrease in bone volume caused by ACLT was more effectively mitigated in group C than in groups B and C (Figure 2B). Therefore, in addition to the proteolytic C-II employed to reduce the risk of adverse immune reactions, the additional type I collagen may counteract the ACLT-induced structural changes in the SB. We predict that PSCC-II can mitigate changes in the SB in patients with OA and can be applied to prevent OI, which should be investigated in future studies.

As early as 1982, Goldenberg et al. revealed an association between OA and synovitis of the knee joint [47]. Cartilage degradation is promoted by synovitis-induced proinflammatory cytokines, such as TNF-α, IL-1β and IL-6, and extracellular matrix-degrading enzymes (i.e., MMPs), resulting in OA progression [48]. In a 2020 retrospective study, Hasan et al. contended that undenatured C-II alleviates synovitis and cartilage degradation in humans, horses, dogs, and rodents with OA [8]. In the present study, the rats with ACLT-induced OA exhibited ACLT-induced synovitis characterized by synovial tissue thickening and increased infiltration of blood cells. The rats also showed cartilage degradation, with losses in proteoglycans and C-II. All these phenomena were attenuated by orally administered PSCC-II (groups A, B, and C) and undenatured C-II (group D; Figure 4). These treatments also significantly inhibited the expression of proinflammatory IL-1β and TNF-α proteins and the ECM-degrading protein MMP13 in cartilage tissue after ACLT (Figures 5P and 6H,P). The pro-inflammatory cytokines IL-1β and TNF-α inhibit the synthesis of major extracellular structural proteins, such as C-II and aggrecan. They simultaneously stimulate chondrocytes to produce MMPs (e.g., MMP-1, MMP-3, and MMP-13), disintegrin and metalloproteinase with thrombospondin motifs, which lead to ECM degradation. This results in cartilage degradation and exacerbates OA progression [49–51].

In summary, this study demonstrated that the oral administration of PSCC-II (groups A, B, and C) and undenatured C-II (group D) enhanced the quantity of C-II in the ECM by suppressing the expression of proinflammatory cytokines and MMP-13. In a 2009 study, cartilage and synovial tissue from patients with OA were cocultured; that study revealed that the administration of type I collagen upregulated the expression of proteoglycans, C-II, and the anti-inflammatory cytokine IL-10, and downregulated the expression of the proinflammatory cytokines IL-1β and TNF-α [52]. Dar et al. (2017) verified that the daily oral administration of hydrolyzed type I collagen mitigated the degradation of cartilage tissue and the expression of MMP13 and TNF-α in mice with meniscal plus ligamentous injury-induced OA, and hydrolyzed type I collagen also inhibited synovial thickening [53]. In the present study, the PSCC-II groups (groups A, B, and C) and the undenatured C-II group exhibited improved ACLT-induced OA symptoms, but the effect was greater in the PSCC-II groups. We suggest that PSCC-II exerts chondroprotective effects similar to those of undenatured C-II. It may also counteract cartilage damage and ECM degradation because it contains additional type I collagen.

Currently, three configurations of C-II are used for nutraceutical applications to prevent OA: natural insoluble undenatured C-II, protease-soluble undenatured C-II, and hydrolyzed C-II. Natural insoluble undenatured C-II can be derived from chicken [25], porcine [54], and salmon nasal cartilages [55], and it is used as a nutraceutical ingredient under names such as UC-II (Lonza, Collavant of Bioiberica, S.A.U., Esplugues de Llobregat, Barcelona) and SCP-II (Guzen Development, Walnut Creek, CA, USA). Protease (pepsin)-soluble undenatured C-II, such as EXT-II (Ryusendo Co., Ltd., Tokyo, Japan) [55], is extracted from chicken sternal cartilage. The C-II hydrolysate is primarily obtained from insoluble undenatured collagen fibrils from chicken sternal cartilage, such as in BioCell Collagen (Biocell Technology, Irvine, CA, USA) [56]. In PSCC-II, only pepsin-soluble undenatured C-II extracted from chicken sternal cartilage is available on the nutraceutical market. The use of marine pepsin-soluble undenatured C-II sourced from the skulls and cartilage of Nile tilapia and sturgeon is in the research stage [57,58]. Epidemiological studies have revealed that OA is a complex disease characterized by chronic joint pain, inflammation, synovitis, bone remodeling, and cartilage loss [59,60]. Orally administered undenatured C-II can alleviate some of the aforementioned symptoms, but fewer studies have investigated the changes in the SB caused by OA. In addition, glycosylated side chains are retained in undenatured C-II during the preparation process, which may induce an adverse immune response. PSCC-II is obtained through the enzymatic degradation of chicken sternal cartilage. In this process, pepsine or protease M and type I collagen aid in the homogenization of C-II in the final product. PSCC-II plus type I collagen may reduce adverse immune responses and mitigate structural changes in the SB in OA. C-II must be processed under acidic conditions before extraction from animal cartilage. In the present study, pepsin and protease M maintained their protease activity in acidic environments. However, pepsin, an enzyme derived from mammalian sources, is associated with the risk of transmitting certain diseases, such as transmissible spongiform encephalopathies or bovine spongiform encephalopathies. Therefore, pepsin use as a food additive has not been approved by the European Union or food safety authorities in other countries. By contrast, protease M, which is derived from the fungus *Aspergillus oryzae*, is regarded as being safer than pepsin. It has been approved as a food additive in certain countries, such as Japan and Taiwan. In the future, we propose that protease M-processed PSCC-II can be administered to patients with OA as a nutritional supplement to prevent fractures and reduce the incidence of osteoporosis.

5. Conclusions

Our results conclude that PSCC-II retains the protective effects of traditional undenatured C-II and provides superior benefits for OA management. These benefits encompass pain relief, anti-inflammatory effects, and the protection of cartilage and cancellous bone.

Author Contributions: Conceptualization, N.-F.C., Z.-H.W., Y.-Y.L. and Y.-H.J.; PSCC-II and undenatured C-II preparation, Y.-P.H.; Methodology, Y.-Y.L., Y.-P.H. and Y.-W.L.; Formal Analysis, Y.-W.L. and Y.-Y.L.; Writing-Original Draft Preparation, N.-F.C., C.-C.T., Z.-K.Y. and Z.-H.W.; Writing-Review and Editing, Z.-H.W. and Y.-H.J.; Supervision, Z.-H.W. All authors have read and agreed to the published version of the manuscript.

Funding: The study received a research grant from Taiyen Biotech Co., Ltd., Taiwan (N107; N112013).

Institutional Review Board Statement: Institutional Animal Care and Use Committee of National Sun Yat-sen University (approval number: 10725).

Informed Consent Statement: Not applicable.

Data Availability Statement: Correspondence and requests for materials should be addressed to Z.-H.W. and Y.-P.H.

Acknowledgments: This manuscript was edited by Wallace Academic Editing.

Conflicts of Interest: Y.-P.H. is an employee of Taiyen Biotech Co., Ltd., Taiwan. Expect PSCC-II and undenatured C-II preparation. The study was run and managed independently by National Sun Yat-sen University. The funders were not involved in the study design, collection, analysis, and interpretation of data, the writing of this article, or the decision to submit it for publication. Research does not endorse any brand or product, nor does it have any financial interests with any supplement manufacturer or distributor.

References

1. Pereira, D.; Ramos, E.; Branco, J. Osteoarthritis. *Acta Med. Port.* **2015**, *28*, 99–106. [CrossRef] [PubMed]
2. He, Y.; Li, Z.; Alexander, P.G.; Ocasio-Nieves, B.D.; Yocum, L.; Lin, H.; Tuan, R.S. Pathogenesis of Osteoarthritis: Risk Factors, Regulatory Pathways in Chondrocytes, and Experimental Models. *Biology* **2020**, *9*, 194. [CrossRef] [PubMed]
3. Gentili, C.; Cancedda, R. Cartilage and bone extracellular matrix. *Curr. Pharm. Des.* **2009**, *15*, 1334–1348. [CrossRef] [PubMed]
4. Bella, J.; Hulmes, D.J. Fibrillar Collagens. *Subcell. Biochem.* **2017**, *82*, 457–490. [CrossRef] [PubMed]
5. Troeberg, L.; Nagase, H. Proteases involved in cartilage matrix degradation in osteoarthritis. *Biochim. Biophys. Acta* **2012**, *1824*, 133–145. [CrossRef]
6. Watt, F.E.; Gulati, M. New Drug Treatments for Osteoarthritis: What is on the Horizon? *Eur. Med. J. Rheumatol.* **2017**, *2*, 50–58. [CrossRef] [PubMed]
7. Honvo, G.; Lengele, L.; Charles, A.; Reginster, J.Y.; Bruyere, O. Role of Collagen Derivatives in Osteoarthritis and Cartilage Repair: A Systematic Scoping Review with Evidence Mapping. *Rheumatol. Ther.* **2020**, *7*, 703–740. [CrossRef] [PubMed]
8. Gencoglu, H.; Orhan, C.; Sahin, E.; Sahin, K. Undenatured Type II Collagen (UC-II) in Joint Health and Disease: A Review on the Current Knowledge of Companion Animals. *Animals* **2020**, *10*, 697. [CrossRef]
9. De Almagro, M.C. The use of collagen hydrolysates and native collagen in osteoarthritis. *Am. J. Biomed. Sci. Res.* **2020**, *6*, 530–532. [CrossRef]
10. Takaoka, K.; Koezuka, M.; Nakahara, H. Telopeptide-depleted bovine skin collagen as a carrier for bone morphogenetic protein. *J. Orthop. Res.* **1991**, *9*, 902–907. [CrossRef]
11. Anderton, S.M. Post-translational modifications of self antigens: Implications for autoimmunity. *Curr. Opin. Immunol.* **2004**, *16*, 753–758. [CrossRef]
12. Michaelsson, E.; Malmstrom, V.; Reis, S.; Engstrom, A.; Burkhardt, H.; Holmdahl, R. T cell recognition of carbohydrates on type II collagen. *J. Exp. Med.* **1994**, *180*, 745–749. [CrossRef] [PubMed]
13. Desai, B.V.; Dixit, S.; Pope, R.M. Limited proliferative response to type II collagen in rheumatoid arthritis. *J. Rheumatol.* **1989**, *16*, 1310–1314. [PubMed]
14. Stuart, J.M.; Huffstutter, E.H.; Townes, A.S.; Kang, A.H. Incidence and specificity of antibodies to types I, II, III, IV, and V collagen in rheumatoid arthritis and other rheumatic diseases as measured by 125I-radioimmunoassay. *Arthritis Rheum.* **1983**, *26*, 832–840. [CrossRef] [PubMed]
15. Kim, W.U.; Kim, K.J. T cell proliferative response to type II collagen in the inflammatory process and joint damage in patients with rheumatoid arthritis. *J. Rheumatol.* **2005**, *32*, 225–230.
16. Eyre, D.R.; Pietka, T.; Weis, M.A.; Wu, J.J. Covalent cross-linking of the NC1 domain of collagen type IX to collagen type II in cartilage. *J. Biol. Chem.* **2004**, *279*, 2568–2574. [CrossRef]

17. Van der Rest, M.; Mayne, R. Type IX collagen proteoglycan from cartilage is covalently cross-linked to type II collagen. *J. Biol. Chem.* **1988**, *263*, 1615–1618. [CrossRef]

18. Beck, K.; Brodsky, B. Supercoiled protein motifs: The collagen triple-helix and the alpha-helical coiled coil. *J. Struct. Biol.* **1998**, *122*, 17–29. [CrossRef]

19. Park, K.S.; Park, M.J.; Cho, M.L.; Kwok, S.K.; Ju, J.H.; Ko, H.J.; Park, S.H.; Kim, H.Y. Type II collagen oral tolerance; mechanism and role in collagen-induced arthritis and rheumatoid arthritis. *Mod. Rheumatol.* **2009**, *19*, 581–589. [CrossRef]

20. Martínez-Puig, D.; Costa-Larrión, E.; Rubio-Rodríguez, N.; Gálvez-Martín, P. Collagen Supplementation for Joint Health: The Link between Composition and Scientific Knowledge. *Nutrients* **2023**, *15*, 1332. [CrossRef]

21. Stoop, R.; Buma, P.; van der Kraan, P.M.; Hollander, A.P.; Billinghurst, R.C.; Meijers, T.H.; Poole, A.R.; van den Berg, W.B. Type II collagen degradation in articular cartilage fibrillation after anterior cruciate ligament transection in rats. *Osteoarthr. Cartil.* **2001**, *9*, 308–315. [CrossRef]

22. Yang, P.Y.; Tang, C.C.; Chang, Y.C.; Huang, S.Y.; Hsieh, S.P.; Fan, S.S.; Lee, H.P.; Lin, S.C.; Chen, W.F.; Wen, Z.H.; et al. Effects of tibolone on osteoarthritis in ovariectomized rats: Association with nociceptive pain behaviour. *Eur. J. Pain* **2014**, *18*, 680–690. [CrossRef] [PubMed]

23. Bove, S.E.; Laemont, K.D.; Brooker, R.M.; Osborn, M.N.; Sanchez, B.M.; Guzman, R.E.; Hook, K.E.; Juneau, P.L.; Connor, J.R.; Kilgore, K.S. Surgically induced osteoarthritis in the rat results in the development of both osteoarthritis-like joint pain and secondary hyperalgesia. *Osteoarthr. Cartil.* **2006**, *14*, 1041–1048. [CrossRef] [PubMed]

24. Fernihough, J.; Gentry, C.; Malcangio, M.; Fox, A.; Rediske, J.; Pellas, T.; Kidd, B.; Bevan, S.; Winter, J. Pain related behaviour in two models of osteoarthritis in the rat knee. *Pain* **2004**, *112*, 83–93. [CrossRef] [PubMed]

25. Bagi, C.M.; Berryman, E.R.; Teo, S.; Lane, N.E. Oral administration of undenatured native chicken type II collagen (UC-II) diminished deterioration of articular cartilage in a rat model of osteoarthritis (OA). *Osteoarthr. Cartil.* **2017**, *25*, 2080–2090. [CrossRef]

26. Pritzker, K.P.; Gay, S.; Jimenez, S.A.; Ostergaard, K.; Pelletier, J.P.; Revell, P.A.; Salter, D.; van den Berg, W.B. Osteoarthritis cartilage histopathology: Grading and staging. *Osteoarthr. Cartil.* **2006**, *14*, 13–29. [CrossRef]

27. Krenn, V.; Morawietz, L.; Burmester, G.R.; Kinne, R.W.; Mueller-Ladner, U.; Muller, B.; Haupl, T. Synovitis score: Discrimination between chronic low-grade and high-grade synovitis. *Histopathology* **2006**, *49*, 358–364. [CrossRef]

28. Louboutin, H.; Debarge, R.; Richou, J.; Selmi, T.A.; Donell, S.T.; Neyret, P.; Dubrana, F. Osteoarthritis in patients with anterior cruciate ligament rupture: A review of risk factors. *Knee* **2009**, *16*, 239–244. [CrossRef]

29. Kuyinu, E.L.; Narayanan, G.; Nair, L.S.; Laurencin, C.T. Animal models of osteoarthritis: Classification, update, and measurement of outcomes. *J. Orthop. Surg. Res.* **2016**, *11*, 19. [CrossRef]

30. Hayami, T.; Pickarski, M.; Zhuo, Y.; Wesolowski, G.A.; Rodan, G.A.; Duong, L.T. Characterization of articular cartilage and subchondral bone changes in the rat anterior cruciate ligament transection and meniscectomized models of osteoarthritis. *Bone* **2006**, *38*, 234–243. [CrossRef]

31. Bagchi, D.; Misner, B.; Bagchi, M.; Kothari, S.C.; Downs, B.W.; Fafard, R.D.; Preuss, H.G. Effects of orally administered undenatured type II collagen against arthritic inflammatory diseases: A mechanistic exploration. *Int. J. Clin. Pharmacol. Res.* **2002**, *22*, 101–110.

32. Von Delwig, A.; Altmann, D.M.; Isaacs, J.D.; Harding, C.V.; Holmdahl, R.; McKie, N.; Robinson, J.H. The impact of glycosylation on HLA-DR1-restricted T cell recognition of type II collagen in a mouse model. *Arthritis Rheum.* **2006**, *54*, 482–491. [CrossRef] [PubMed]

33. Bäcklund, J.; Treschow, A.; Bockermann, R.; Holm, B.; Holm, L.; Issazadeh-Navikas, S.; Kihlberg, J.; Holmdahl, R. Glycosylation of type II collagen is of major importance for T cell tolerance and pathology in collagen-induced arthritis. *Eur. J. Immunol.* **2002**, *32*, 3776–3784. [CrossRef] [PubMed]

34. Raza, K.; Mullazehi, M.; Salmon, M.; Buckley, C.D.; Ronnelid, J. Anti-collagen type II antibodies in patients with very early synovitis. *Ann. Rheum. Dis.* **2008**, *67*, 1354–1355. [CrossRef] [PubMed]

35. Shoulders, M.D.; Raines, R.T. Collagen structure and stability. *Annu. Rev. Biochem.* **2009**, *78*, 929–958. [CrossRef] [PubMed]

36. Bouxsein, M.L.; Boyd, S.K.; Christiansen, B.A.; Guldberg, R.E.; Jepsen, K.J.; Muller, R. Guidelines for assessment of bone microstructure in rodents using micro-computed tomography. *J. Bone Min. Res.* **2010**, *25*, 1468–1486. [CrossRef] [PubMed]

37. Radin, E.L.; Rose, R.M. Role of subchondral bone in the initiation and progression of cartilage damage. *Clin. Orthop. Relat. Res.* **1986**, *213*, 34–40. [CrossRef]

38. Li, G.; Yin, J.; Gao, J.; Cheng, T.S.; Pavlos, N.J.; Zhang, C.; Zheng, M.H. Subchondral bone in osteoarthritis: Insight into risk factors and microstructural changes. *Arthritis Res. Ther.* **2013**, *15*, 223. [CrossRef]

39. Oftadeh, R.; Perez-Viloria, M.; Villa-Camacho, J.C.; Vaziri, A.; Nazarian, A. Biomechanics and mechanobiology of trabecular bone: A review. *J. Biomech. Eng.* **2015**, *137*, 010802–01080215. [CrossRef]

40. Fazzalari, N.L.; Parkinson, I.H. Femoral trabecular bone of osteoarthritic and normal subjects in an age and sex matched group. *Osteoarthr. Cartil.* **1998**, *6*, 377–382. [CrossRef]

41. Kadri, A.; Ea, H.K.; Bazille, C.; Hannouche, D.; Liote, F.; Cohen-Solal, M.E. Osteoprotegerin inhibits cartilage degradation through an effect on trabecular bone in murine experimental osteoarthritis. *Arthritis Rheum.* **2008**, *58*, 2379–2386. [CrossRef] [PubMed]

42. Yamauchi, M.; Sricholpech, M. Lysine post-translational modifications of collagen. *Essays Biochem.* **2012**, *52*, 113–133. [CrossRef]

43. Shiiba, M.; Arnaud, S.B.; Tanzawa, H.; Uzawa, K.; Yamauchi, M. Alterations of collagen matrix in weight-bearing bones during skeletal unloading. *Connect. Tissue Res.* **2001**, *42*, 303–311. [CrossRef]

44. Fonseca, H.; Moreira-Goncalves, D.; Coriolano, H.J.; Duarte, J.A. Bone quality: The determinants of bone strength and fragility. *Sports Med.* **2014**, *44*, 37–53. [CrossRef]

45. Nijhuis, W.H.; Eastwood, D.M.; Allgrove, J.; Hvid, I.; Weinans, H.H.; Bank, R.A.; Sakkers, R.J. Current concepts in osteogenesis imperfecta: Bone structure, biomechanics and medical management. *J. Child. Orthop.* **2019**, *13*, 1–11. [CrossRef] [PubMed]

46. Wu, Z.; Wang, B.; Tang, J.; Bai, B.; Weng, S.; Xie, Z.; Shen, Z.; Yan, D.; Chen, L.; Zhang, J.; et al. Degradation of subchondral bone collagen in the weight-bearing area of femoral head is associated with osteoarthritis and osteonecrosis. *J. Orthop. Surg. Res.* **2020**, *15*, 526. [CrossRef]

47. Goldenberg, D.L.; Egan, M.S.; Cohen, A.S. Inflammatory synovitis in degenerative joint disease. *J. Rheumatol.* **1982**, *9*, 204–209. [PubMed]

48. Mathiessen, A.; Conaghan, P.G. Synovitis in osteoarthritis: Current understanding with therapeutic implications. *Arthritis Res. Ther.* **2017**, *19*, 18. [CrossRef]

49. Kapoor, M.; Martel-Pelletier, J.; Lajeunesse, D.; Pelletier, J.P.; Fahmi, H. Role of proinflammatory cytokines in the pathophysiology of osteoarthritis. *Nat. Rev. Rheumatol.* **2011**, *7*, 33–42. [CrossRef]

50. Chow, Y.Y.; Chin, K.Y. The Role of Inflammation in the Pathogenesis of Osteoarthritis. *Mediat. Inflamm.* **2020**, *2020*, 8293921. [CrossRef]

51. Wiegertjes, R.; van de Loo, F.A.J.; Blaney Davidson, E.N. A roadmap to target interleukin-6 in osteoarthritis. *Rheumatology* **2020**, *59*, 2681–2694. [CrossRef] [PubMed]

52. Furuzawa-Carballeda, J.; Munoz-Chable, O.A.; Barrios-Payan, J.; Hernandez-Pando, R. Effect of polymerized-type I collagen in knee osteoarthritis. I. In vitro study. *Eur. J. Clin. Investig.* **2009**, *39*, 591–597. [CrossRef] [PubMed]

53. Dar, Q.A.; Schott, E.M.; Catheline, S.E.; Maynard, R.D.; Liu, Z.; Kamal, F.; Farnsworth, C.W.; Ketz, J.P.; Mooney, R.A.; Hilton, M.J.; et al. Daily oral consumption of hydrolyzed type 1 collagen is chondroprotective and anti-inflammatory in murine posttraumatic osteoarthritis. *PLoS ONE* **2017**, *12*, e0174705. [CrossRef] [PubMed]

54. Di Cesare Mannelli, L.; Micheli, L.; Zanardelli, M.; Ghelardini, C. Low dose native type II collagen prevents pain in a rat osteoarthritis model. *BMC Musculoskelet. Disord.* **2013**, *14*, 228. [CrossRef]

55. Tomonaga, A.; Takahashi, T.; Tanaka, Y.T.; Tsuboi, M.; Ito, K.; Nagaoka, I. Evaluation of the effect of salmon nasal proteoglycan on biomarkers for cartilage metabolism in individuals with knee joint discomfort: A randomized double-blind placebo-controlled clinical study. *Exp. Ther. Med.* **2017**, *14*, 115–126. [CrossRef]

56. Hector, L.; Lopez, S.H.; Sandrock, J.E.; Kedia, A.W.; Ziegenfuss, T.N. Effects of BioCell Collagen® on connective tissue protection and functional recovery from exercise in healthy adults: A pilot study. *J. Int. Soc. Sports Nutr.* **2014**, *11*, P48.

57. Hou, C.; Li, N.; Liu, M.; Chen, J.; Elango, J.; Rahman, S.U.; Bao, B.; Wu, W. Therapeutic Effect of Nile Tilapia Type II Collagen on Rigidity in CD8(+) Cells by Alleviating Inflammation and Rheumatoid Arthritis in Rats by Oral Tolerance. *Polymers* **2022**, *14*, 1284. [CrossRef]

58. Luo, Q.B.; Chi, C.F.; Yang, F.; Zhao, Y.Q.; Wang, B. Physicochemical properties of acid- and pepsin-soluble collagens from the cartilage of Siberian sturgeon. *Environ. Sci. Pollut. Res. Int.* **2018**, *25*, 31427–31438. [CrossRef]

59. De Lange-Brokaar, B.J.; Ioan-Facsinay, A.; Yusuf, E.; Visser, A.W.; Kroon, H.M.; Andersen, S.N.; Herb-van Toorn, L.; van Osch, G.J.; Zuurmond, A.M.; Stojanovic-Susulic, V.; et al. Degree of synovitis on MRI by comprehensive whole knee semi-quantitative scoring method correlates with histologic and macroscopic features of synovial tissue inflammation in knee osteoarthritis. *Osteoarthr. Cartil.* **2014**, *22*, 1606–1613. [CrossRef]

60. Haseeb, A.; Haqqi, T.M. Immunopathogenesis of osteoarthritis. *Clin. Immunol.* **2013**, *146*, 185–196. [CrossRef]

nutrients

Article

The Gastroprotective Effect of Walnut Peptides: Mechanisms and Impact on Ethanol-Induced Acute Gastric Mucosal Injury in Mice

Yutong Yuan [1,2,†], Xinyi Wang [1,2,†], Yumeng Wang [2], Yaqi Liu [1], Liang Zhao [1,*], Lei Zhao [1] and Shengbao Cai [3,*]

[1] Beijing Engineering and Technology Research Center of Food Additives, School of Food and Health, Beijing Technology and Business University, Beijing 100048, China; s20223061183@cau.edu.cn (Y.Y.); xaesi23@163.com (X.W.); liuyq2617@163.com (Y.L.); zhaolei@th.btbu.edu.cn (L.Z.)

[2] Beijing Key Laboratory of Functional Food from Plant Resources, College of Food Science and Nutritional Engineering, China Agricultural University, Beijing 100083, China; goodbai0427@163.com

[3] Faculty of Food Science and Engineering, Kunming University of Science and Technology, Kunming 650500, China

* Correspondence: liangzhao@btbu.edu.cn (L.Z.); caikmust2013@kmust.edu.cn (S.C.)

† These authors contributed equally to this work.

Abstract: The objective of this research was to explore the protective impact of walnut peptides (WP) against ethanol-induced acute gastric mucosal injury in mice and to investigate the underlying defense mechanisms. Sixty male BALB-c mice were divided into five groups, and they were orally administered distilled water, walnut peptides (200 and 400 mg/kg bw), and omeprazole (20 mg/kg bw) for 24 days. Acute gastric mucosal injury was then induced with 75% ethanol in all groups of mice except the blank control group. Walnut peptides had significant protective and restorative effects on tissue indices of ethanol-induced gastric mucosal damage, with potential gastric anti-ulcer effects. Walnut peptides significantly inhibited the excessive accumulation of alanine aminotransferase (ALT), aspartate transferase (AST), and malondialdehyde (MDA), while promoting the expression of reduced glutathione (GSH), total antioxidant capacity (T-AOC), glutathione disulfide (GSSG), and mouse epidermal growth factor (EGF). Furthermore, the Western blot analysis results revealed that walnut peptides significantly upregulated the expression of HO-1 and NQO1 proteins in the Nrf2 signaling pathway. The defensive impact of walnut peptides on the gastric mucosa may be achieved by mitigating the excessive generation of lipid peroxides and by boosting cellular antioxidant activity.

Keywords: walnut peptides; gastric mucosal injury; LC-MS/MS; alcohol-induced; Western blot

Citation: Yuan, Y.; Wang, X.; Wang, Y.; Liu, Y.; Zhao, L.; Zhao, L.; Cai, S. The Gastroprotective Effect of Walnut Peptides: Mechanisms and Impact on Ethanol-Induced Acute Gastric Mucosal Injury in Mice. *Nutrients* 2023, 15, 4866. https://doi.org/10.3390/nu15234866

Academic Editor: Juergen Schrezenmeir

Received: 26 October 2023
Revised: 19 November 2023
Accepted: 20 November 2023
Published: 22 November 2023

1. Introduction

As the economy grows, the rising popularity of alcoholic drinks coincides with numerous studies indicating that excessive alcohol intake can have detrimental effects on one's well-being [1]. Under normal circumstances, the gastric mucosal barrier is composed of factors such as the mucus–bicarbonate–phospholipid barrier, the epithelial barrier, and the endothelial barrier, which work together to protect the integrity and functionality of the gastric mucosa [2]. When ingested in high concentrations, ethanol can directly erode the gastrointestinal tract, causing acute gastritis and gastric ulcers, which can further result in lesions such as perforation and cancer [3–5].

Proton pump inhibitors (PPI) are currently one of the main drugs used to treat gastroesophageal disorders. Among them, omeprazole, lansoprazole, and others are widely used clinical PPI [6]. In addition to PPI, common medications for treating gastric functional disorders include prokinetic agents and H2-receptor antagonists (H2RA) [7]. However, ample evidence now suggests that the medications employed for managing gastroesophageal disorders might come with certain adverse reactions. For instance, the long-term use of PPI and H2RA can affect the body's absorption of vitamin B12 and increase the risk of developing bacterial peritonitis [8,9]. In summary, it is necessary to seek natural, non-toxic substances that can effectively enhance the gastric mucosal barrier and improve gastrointestinal function, such as polysaccharides and food-derived bioactive peptides; research has shown that wheat peptides and calcitonin-gene-related peptides have protective mechanisms against ethanol-induced gastric ulcers in rats [10–13].

Walnuts are an important nut product, originally hailing from Southeastern Europe, East Asia, and North America [14]. Walnuts currently rank second in global nut production after almonds, whereas walnut kernels are often regarded as the main oilseed crop during processing, while the by-product of walnut meal is ignored [15]. Walnut meal consists of 40% protein, which contains a notably high content of essential amino acids, making it an excellent source of dietary protein intake [16]. In comparison to walnut protein, its hydrolyzed product, walnut peptides, exhibit superior biological functional properties, such as anti-inflammatory [17], antioxidant [18], antibacterial [19], and neuroprotective effects [20]; furthermore, walnut peptides were confirmed to have a regulatory effect on inflammatory bowel disease (IBD) [21]. However, it is currently unclear whether walnut peptides have a positive impact on damaged gastric mucosal barriers.

In this research, we examined the influence of walnut peptides (WP) derived from walnut meal on ethanol-induced gastric mucosal injury in mice, elucidating its mechanism of action. An ethanol-induced gastric mucosal barrier model in mice was established, and gastric mucosal damage was observed and pathologically analyzed. The gastric mucosal injury index, injury inhibition rate, antioxidant indices such as aspartate aminotransferase (AST), alanine aminotransferase (ALT), malondialdehyde (MDA), reduced glutathione (GSH), total antioxidant capacity (T-AOC), glutathione disulfide (GSSG) levels, and inflammatory indices including mouse epidermal growth factor (EGF), myeloperoxidase (MPO), mouse tumor necrosis factor-alpha (TNF-α), and mouse interleukin-1β (IL-1β) were determined for evaluating the protective effect of WP against ethanol-induced gastric mucosal injury. Finally, WP was revealed to mediate its biological effects through antioxidant mechanisms by Western blot, offering a theoretical foundation for utilizing walnut meal resources.

2. Materials and Methods

2.1. Materials and Reagents

Walnut peptides (purity of 99%) were purchased from Shaanxi Xinpai Biotechnology Co., Ltd., Xian, China. Omeprazole was obtained from Renhe (Group) Development Co., Ltd., Beijing, China. MDA, T-AOC, GSH, AST, ALT and GSSG detection kits were purchased from Beijing Solaibao Technology Co., Ltd., Beijing, China. EGF, TNF-α, MPO, and IL-1β ELISA kits were purchased from Wuhan Hualianke Biological Technology Co., Ltd., Wuhan, China. BCA Protein Concentration Determination kits were purchased from Shanghai Biyuntian Biotechnology Co., Ltd., Shanghai, China.

2.2. Animals and Groups

Specific pathogen-free (SPF) 4-week-old male Kunming mice were acquired from Beijing Weitonglihua Experimental Animal Technology Co., Ltd., (Beijing, China). After 7 days of acclimatization, they were randomly divided into five groups, blank control group (CK), model group (MG), positive omeprazole control group (PG, 20 mg/kg bw), low-dose walnut peptide group (LWG, 200 mg/kg bw), and high-dose walnut peptide group (HWG, 400 mg/kg bw), with 12 in each group [13]. All animal procedures were in accordance with the Animal Ethics Committee of the Beijing Key Laboratory of Functional Food from Plant Resources (Permit number: A330-2023-2) and conformed to the guidelines for the care and use of laboratory animals set by the National Institutes of Health. LWG and HWG were given samples at a dose of 10 mL/kg bw daily, while CK and MG were gavaged with an equal dose of distilled water. Mouse body weight was measured every three days, with a fixed weighing time, and the final weight measurement was taken before the last gavage. After the final gavage, the mice were fasted for 24 h to ensure complete digestion and absorption of gastric contents, with free access to water during this period. Except for the CK group, all other groups were given 75% ethanol (0.1 mL/20 g) by gavage to induce acute gastric mucosal injury, while the CK group received an equivalent amount of distilled water [22]. One hour after the gavage, the mice were euthanized by dislocating the cervical spine, and organs (liver, kidney, spleen, gastric) and serum were rapidly collected after enucleating the eyeballs and drawing blood from the neck dislocation site.

2.3. Peptide Components Analysis of Walnut Peptides Based on LC-MS/MS

An appropriate amount of sample was dissolved in 50 mM NH_4HCO_3, and DTT solution was added to the dissolved sample to a final concentration of 10 mmol/L, which was reduced in a 56 °C water bath for 1 h, and alkylated by 50 mM IAM at room temperature in dark for 40 min. The peptide underwent desalting through a self-priming desalting column, and the solvent was removed in a vacuum centrifuge at 45 °C. After reducing alkylation, we used LC-MS/MS (liquid chromatography–tandem mass spectrometry, Thermo Fisher Scientific, Waltham, MA, USA) to analyze the peptide separation and identification, equipped with an Easy-nLC 1200 system (Thermo Fisher Scientific, Waltham, MA, USA) and an Acclaim PepMap RPLC C18 column (size 150 μm × 15 cm, particle size 1.9 μm, Dr. Maisch GmbH, Ammerbuch, Germany). The injection volume was 5.0 μL. The flow rate was 600 nL/min. Solvent A consisted of ultrapure water with 0.1% (v/v) formic acid, while Solvent B comprised acetonitrile with 0.1% (v/v) formic acid. The solvent gradient used for the isolation of walnut peptides was as follows: from 4% B to 8% B in 2 min, from 8% B to 28% B in 43 min, from 28% B to 40% B in 10 min, from 40% B to 95% B in 1 min, and finally maintained at 95% B in 10 min. The walnut peptides were sequenced by Beijing Bio-Tech Pack Technology Company Ltd., (Beijing, China).

2.4. Organ Index Calculation

After euthanasia of the mice, the viscera (liver, kidney, spleen, and gastric) were removed, the fat was removed, and the organs were washed with saline and drained with filter paper to record the weights. The mice's organ index was computed using Equation (1):

$$\text{The organ index (g/g)} = \text{organ mass (g)}/\text{mouse body weight (g)} \tag{1}$$

2.5. Evaluation of Gastric Mucosal Injury

To assess the severity of gastric mucosal lesions, the excised stomachs were dissected along the greater curvature and thoroughly washed with pre-chilled PBS. Subsequently, the emptied and flattened stomach samples were photographed. Following that, the lesion length and width in the flattened stomach samples were measured using a vernier caliper. Briefly, each lesion was scored from 0 to 5 as follows: (a) undetectable lesions (score of zero); (b) bleeding point/pc (score of 1); (c) lesions < 1 mm length (score of 2); (d) lesions of 1–2 mm length (score of 3); (e) lesions of 2–3 mm length (score of 4); (f) lesions > 3 mm length (score of 5), when the width of the lesion exceeded 2 mm, the score was doubled [23]. After washing the gastric specimen, the degree of gastric injury was assessed according to Formulas (2) and (3) as follows:

$$\text{Gastric Mucosal Injury Index} = \text{Bleeding Point Score} + \text{Ulcerated Stripe Score} \tag{2}$$

$$\text{Gastric Mucosal Injury Inhibition Rate} = (\text{Model Group Injury Index} - \text{Test Substance Group Injury Index})/\text{Model Group Injury Index} \times 100\% \tag{3}$$

The most severely affected area of the gastric mucosa from each mouse was dissected to obtain tissue samples measuring approximately 5 mm × 5 mm. These tissue samples were then processed through fixation (4% formalin solution), routine gradient dehydration, embedding, sectioning, slide preparation, and hematoxylin and eosin (HE) staining, The HE staining portion was undertaken by Wuhan Seville Biotechnology Co., Ltd. (Wuhan, China). Subsequently, observations were made under an optical microscope. Microscopic damage was scored from 0 to 14, as follows: (a) loss of mucosal epithelium (0–3 points); (b) upper mucosal edema (0–4 points); (c) hemorrhagic damage (0–4 points); and (d) inflammatory cell influx (0–3 points) [24]. The total score, as per the two evaluation criteria mentioned above, was calculated to express the final gastric mucosal tissue injury score.

2.6. Determination of AST and ALT Levels in Serum

We adhered to the procedural steps outlined in the commercial assay kits for assessing and analyzing the AST and ALT levels in the serum. The enzyme activities for AST and ALT were quantified in U/mL.

2.7. Measurement of Gastric Tissue Antioxidant Capacity

The levels of MDA, T-AOC, GSH, and GSSG were used as indicators of gastric tissue antioxidant capacity. Commercial assay kits (Beijing Soleibao Technology Co., Ltd., Beijing, China) were used to measure these parameters following the manufacturer's instructions. Gastric tissues were minced with scissors and mixed with pre-cooled phosphate buffer solution (PBS, pH = 7.4) at a 1:9 ratio. The mixture was homogenized using a homogenizer to obtain a 10% gastric tissue homogenate (w/v). The homogenate was then centrifuged at 5000 rpm at 4 °C for 15 min. The supernatant was collected, and the levels of MDA, T-AOC, GSH, MPO, and GSSG were determined. MDA levels were expressed in nmol/g, T-AOC levels in μmol/g, GSH levels in μg/mg prot, and GSSG levels in μg/mg prot.

2.8. Measurement of Gastric Tissue Regulatory Mediator Levels

Enzyme-linked immunosorbent assay (ELISA) kits (Wuhan Hualianke Biological Technology Co., Ltd., Wuhan, China) were used to detect the levels of EGF, MPO, TNF-α, and IL-1β in the supernatant of gastric tissue homogenates, following the manufacturer's instructions. EGF, TNF-α, and IL-1β levels were expressed in ng/L, MPO levels in ng/mL.

2.9. Western Blot Assay

Total protein was extracted from gastric tissues by grinding an appropriate amount of tissue in liquid nitrogen. The gastric tissue homogenate was homogenized in a mixture of RIPA lysis buffer (Shanghai Biyuntian Biological Technology Co., Ltd., Shanghai, China) and protease phosphatase inhibitor cocktail (Beijing Solaibao Technology Co., Ltd., Beijing, China). Protein samples underwent separation using 10% SDS-PAGE, were transferred to a polyvinylidene fluoride (PVDF) membrane, and then blocked with 5% skim milk in Tris-buffered saline with Tween 20 (TBST). Subsequently, the membrane was incubated overnight at 4 °C with anti-HO-1, anti-Keap-1, anti-NQO-1, and anti-Nrf2 antibodies. Following three washes with TBST, the membrane was exposed to HRP-conjugated secondary antibodies at room temperature for a duration of 2 h. Images were acquired using the Image Quant LAS 4000 Mini system (GE Healthcare, Chicago, IL, USA), and grayscale values were measured using Image J 2.0 software.

2.10. Statistical Analysis

Statistical analysis was performed using SPSS software (IBM SPSS Statistics 26). Data are presented as the mean ± S.E.M. A one-way analysis of variance (ANOVA) was conducted, followed by Duncan's test for multiple comparisons, with the significance level set at $p < 0.05$. Graphs were created using GraphPad Prism 8 software, and tables were created using Excel 2021 software. Graphical abstracts drawn by Figdraw.

3. Results

3.1. Analysis of Walnut Peptide Amino Acid Sequences

LC-MS/MS was used to identify and analyze the amino acid sequences of the main peptide segments in the walnut peptide samples to determine the composition and identification of the major peptide segments in the samples. Table 1 lists the top eight peptide segments with the highest scores, including EIDIGVPDEVGRL, DLAPTHPIRL, LDRLIPVLE,

SVIQH, NTGSPITVPVGR, SKRPTF, DREIDIGVPDEVGRL, and RDENEKL. Among them, the highest score was 490.0 for EIDIGVPDEVGRL, indicating that the results of peptide segment identification through LC-MS and secondary spectrum quality integration were the most reliable among all peptide segments. In terms of peptide segment abundance, SKRPTF had the highest abundance at 81,398,000, composed of six amino acids, with an *m/z* of 368.211 and a score of 428.9. It had the highest output detected among these eight peptide segments, indicating its high content in the sample. These amino acid sequences may provide potential clues for further research into the mechanisms underlying the impact of walnut peptides on gastric mucosal barrier damage in mice.

Table 1. The amino acid sequence of walnut peptides.

Peptide Sequence	Observed *m/z*	Score	Scan Time	Protein Name	Intensity
EIDIGVPDEVGRL	706.372	490.0	48.0578	A0A2I4EK72	60,222,000
DLAPTHPIRL	566.827	474.2	29.0392	A0A2I4EB92	22,166,000
LDRLIPVLE	534.326	466.0	45.2099	A0A2I4EGJ3	45,179,000
SVIQH	625.331	431.2	7.1603	A0A2I4FTY2	40,419,000
NTGSPITVPVGR	599.334	428.9	25.3499	A0A2I4GH77	25,614,000
SKRPTF	368.211	428.9	8.6987	A0A2I4EFN4	81,398,000
DREIDIGVPDEVGRL	841.940	422.2	43.2557	A0A2I4EK72	13,398,000
RDENEKL	452.230	420.2	7.1808	A0A2I4E5L6	71,960,000

3.2. The Effects of WP on Mouse Body Weight and Organ Indices

During the gavage period, mice in all groups showed no abnormal behavior, and there were no significant changes in food and water intake, or signs of hair loss or death. The quality of the mice directly reflected their growth status, as shown in Figure 1. As a result of fasting, all groups experienced a decline in body weights on Day 24 and the trend in the body weight changes in the LWG and HWG groups were similar to the CK group, suggesting that walnut peptides did not have adverse effects on the body weights or the growth status of the mice. As shown in Table 2, the liver indices of the LWG and HWG groups were 42.15 mg/g and 42.08 mg/g, respectively, which were not significant different compared to the PG group ($p > 0.05$), but higher and significantly different from the CK group ($p < 0.05$). The gastric indexes of the LWG, HWG, and PG groups were 8.92 mg/g, 8.99 mg/g, and 8.67 mg/g, respectively, which were not significantly different compared to the CK and MG groups ($p > 0.05$). However, the gastric index was the lowest in the CK group and highest in the MG group. Similarly, the kidney indexes and spleen indexes of the LWG and HWG groups were not significantly different ($p > 0.05$) compared to the CK group. This indicates that the effect of the walnut peptides on the mice organ coefficients was similar to that of the positive drug and did not cause damage to the internal organs of the mice and did not form a significant effect on the organ coefficients of the mice.

Table 2. Organ indexes of mice in each group. CK group: blank control group; MG group: model group; PG group: positive omeprazole control group (20 mg/kg bw); LWG group: low-dose walnut peptide group (200 mg/kg bw); HWG group: high-dose walnut peptide group (400 mg/kg bw). Different letters represent significant differences between groups at $p < 0.05$.

Group	Liver Index (mg/g)	Kidney Index (mg/g)	Spleen Index (mg/g)	Gastric Index (mg/g)	Starting Weight (g)	Final Weight (g)
CK	39.43 ± 2.32 b	12.17 ± 1.41 ab	29.29 ± 1.50 a	7.85 ± 1.82 a	39.05 ± 0.86 a	43.80 ± 2.04 a
MG	42.81 ± 1.96 a	11.95 ± 1.19 ab	35.28 ± 2.16 a	9.36 ± 1.73 a	38.32 ± 1.15 a	42.25 ± 1.97 a
PG	43.48 ± 3.65 a	13.41 ± 2.97 a	28.25 ± 2.26 a	8.67 ± 2.43 a	40.07 ± 1.61 a	44.43 ± 2.22 a
LWG	42.15 ± 3.10 a	11.61 ± 1.31 b	25.38 ± 0.79 a	8.92 ± 1.24 a	38.93 ± 2.41 a	43.97 ± 2.51 a
HWG	42.08 ± 2.95 a	11.87 ± 1.10 ab	24.40 ± 0.62 a	8.99 ± 1.91 a	39.47 ± 1.69 a	44.53 ± 1.11 a

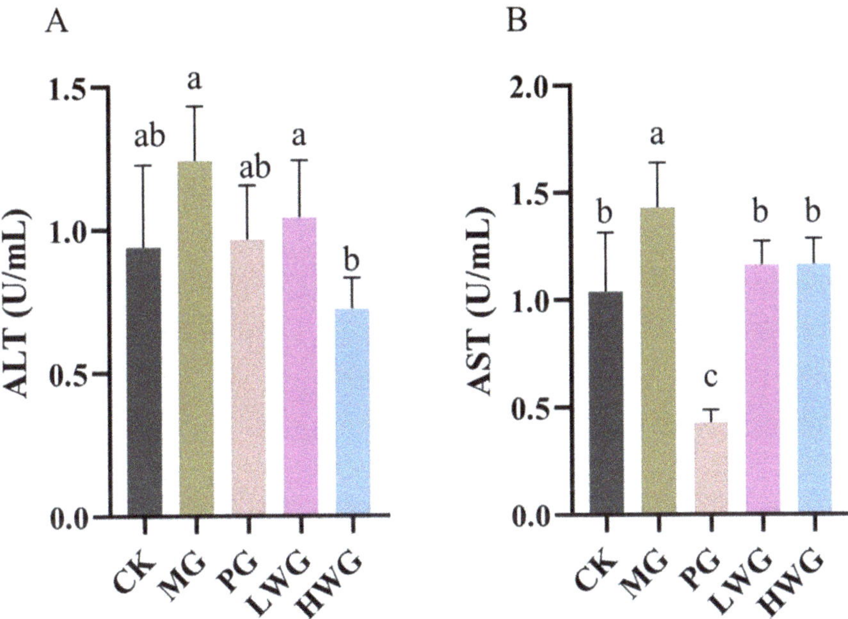

Figure 1. The effects of WP on the serum levels of ALT (**A**) and AST (**B**) in mice. CK group: blank control group; MG group: model group; PG group: positive omeprazole control group (20 mg/kg bw); LWG group: low-dose walnut peptide group (200 mg/kg bw); HWG group: high-dose walnut peptide group (400 mg/kg bw). Different letters represent significant differences between groups at $p < 0.05$.

3.3. The Effects of WP on Serum ALT and AST Levels

The ALT levels were 0.72 U/mL for HWG, 0.94 U/mL for CK, and 0.967 U/mL for PG. According to Figure 1A, it is evident that the ALT levels in HWG were significantly reduced compared to MG ($p < 0.05$) and showed no significant difference compared to CK and PG ($p > 0.05$). The AST levels of LWG and HWG were 1.16 U/mL and 1.17 U/mL, as shown in Figure 1B, compared to the MG group, the AST levels in the WP treatment groups all exhibited a significant decrease from the MG group ($p < 0.05$) and showed no significant difference compared to the CK group ($p > 0.05$).

3.4. The Impact of WP on Gastric Mucosal Tissue Pathological Changes

According to Figure 2, in the CK group, the gastric mucosa of mice appeared intact with normal color and shape, and no obvious signs of hemorrhage, ulceration, or erosion were observed, whereas the gastric mucosa of mice in the MG group showed obvious hemorrhagic damage, with the appearance of multiple large hemorrhagic streaks and hemorrhagic dots, which confirms that the model of ethanol-induced gastric mucosal injury was successfully established. The gastric mucosa of the PG group showed a small number of hemorrhagic spots but no ulceration or erosion, and the degree of injury was significantly reduced compared with that of the MG group, with an inhibition rate of 61.33%; in the LWG and HWG groups, the gastric mucosa of mice also showed a few bleeding spots, but no erosion or ulceration, with an inhibition rate of 58.67% and 66.67%. Compared with the PG group, there was no significant difference in the damage inhibition rate ($p > 0.05$); however, there was a significant difference compared to the MG group ($p < 0.05$). The HE staining results revealed that in the CK group, the gastric mucosal epithelium had a well-organized glandular structure, and the mucosal layer, submucosal layer, and muscular layer appeared intact, well-structured, and free from obvious edema and inflammation infiltration. In the MG group, the glandular structure of the gastric tissue was disrupted, with localized epithelial cell necrosis, shedding, and partial mucosal bleeding, confirming the successful modeling of the gastric mucosal injury induced by ethanol. In the PG, LWG, and HWG groups, there were a noticeable improvement in the status of the gastric mucosal injury. The gastric mucosal injury indices in the PG, LWG, and HWG groups were 2.5, 3.5, and 2, respectively, with no significant difference among these three groups ($p > 0.05$), and significantly different from the MG group ($p < 0.05$).

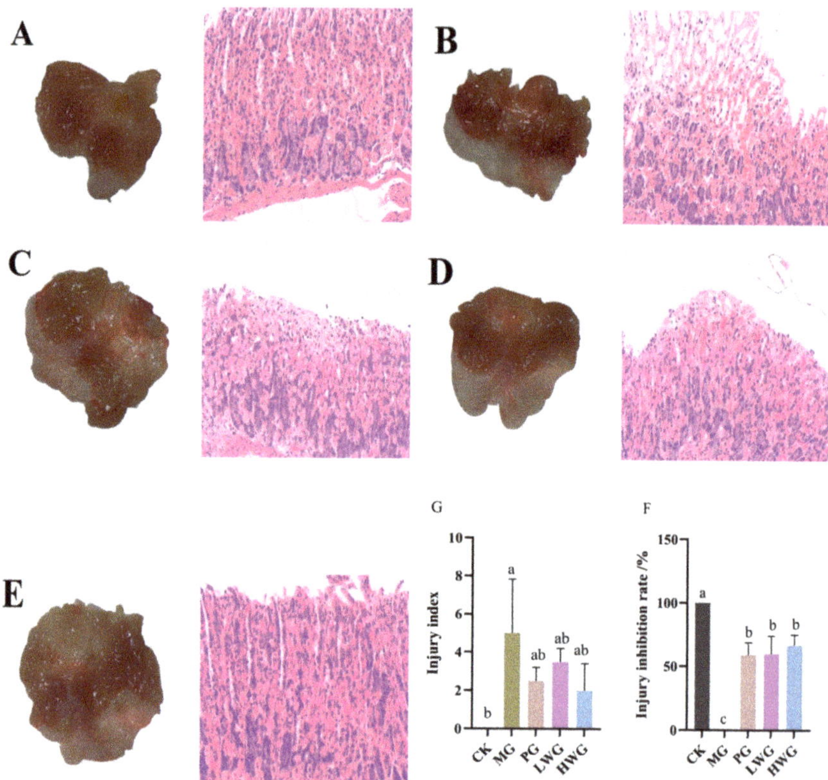

Figure 2. The effects of WP on gastric mucosal injury in mice. From left to right: Observation of gastric mucosal pathological features through gross morphology and HE staining (scale bar: 20 μm). (**A**) CK group. (**B**) MG group. (**C**) PG group. (**D**) LWG group. (**E**) HWG group. (**F**) Injury inhibition rate in each group of mice. (**G**) Injury index in each group of mice. CK group: blank control group; MG group: model group; PG group: positive omeprazole control group (20 mg/kg bw); LWG group: low-dose walnut peptide group (200 mg/kg bw); HWG group: high-dose walnut peptide group (400 mg/kg bw). Different letters represent significant differences between groups at $p < 0.05$.

3.5. The Effects of WP on Gastric Tissue Oxidative Stress

We investigated the effects of WP on relevant oxidative stress markers in ethanol-induced gastric injury in mice using an assay kit. According to Figure 3, we observed that the presence of WP effectively decreased the content of MDA and significantly increased the content of GSH, T-AOC, and GSSG. The GSH contents of the LWG and HWG groups were 83.86 μg/mg prot and 93.26 μg/mg prot, which were not significantly different from the CK group ($p > 0.05$), but formed a significant difference with the MG group ($p < 0.05$). The GSSG content of the HWG group was 0.69 μg/mg prot, which formed a significant difference with the MG group ($p < 0.05$). The T-AOC contents of the LWG, HWG, and CK groups were 11.99 μmol/g, 11.20 μmol/g, and 12.83 μmol/g, and these three results were not significantly different but formed a significant difference with the MG group ($p < 0.05$).

The MDA content of 59.81 nmol/g in the MG group formed a significant difference with the LWG, HWG, and CK groups, which were 25.32 nmol/g, 32.93 nmol/g, and 35.93 nmol/g ($p < 0.05$).

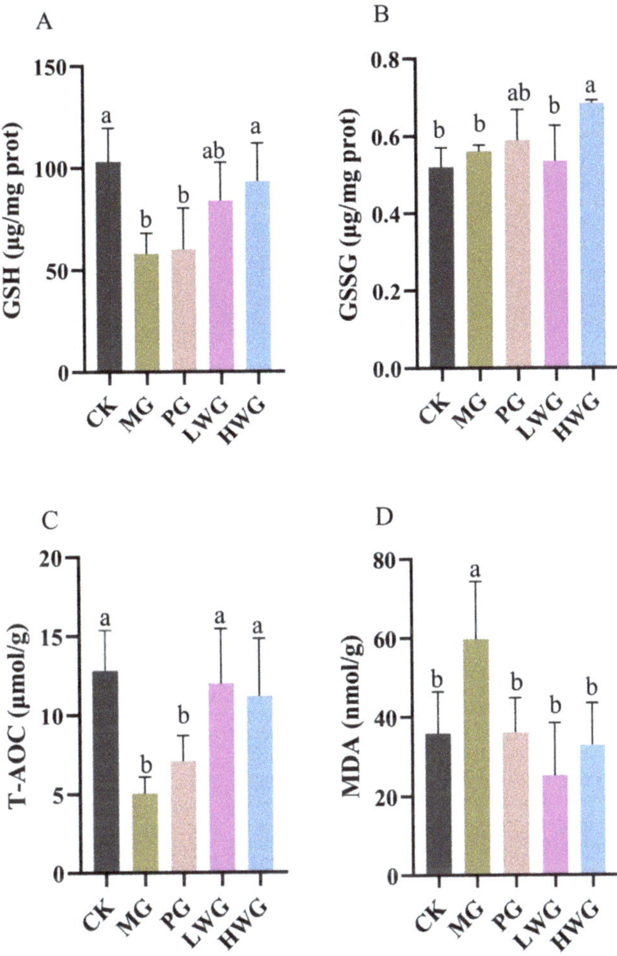

Figure 3. The effects of WP on gastric tissue oxidative stress markers of GSH (**A**), GSSG (**B**), T-AOC (**C**), and MDA (**D**) in mice. CK group: blank control group; MG group: model group; PG group: positive omeprazole control group (20 mg/kg bw); LWG group: low-dose walnut peptide group (200 mg/kg bw); HWG group: high-dose walnut peptide group (400 mg/kg bw). Different letters represent significant differences between groups at $p < 0.05$.

3.6. The Effects of WP on Gastric Tissue Inflammation

We investigated the impact of WP on the gene expression of pertinent inflammatory factors in ethanol-induced gastric injury in mice through ELISA. According to Figure 4, we observed that WP significantly increased the expression level of EGF, which was 99.99 ng/L in the LWG group, 122.89 ng/L in the HWG group, and 103.80 ng/L in the PG group, with no significant difference among the three groups ($p > 0.05$), and was significantly different from the MG group ($p < 0.05$). Although the expression levels of TNF-α and IL-1β in the LWG group were slightly lower than those in the MG group, the differences were not statistically significant ($p > 0.05$). The results of the MPO assay showed no significant differences between the groups ($p > 0.05$).

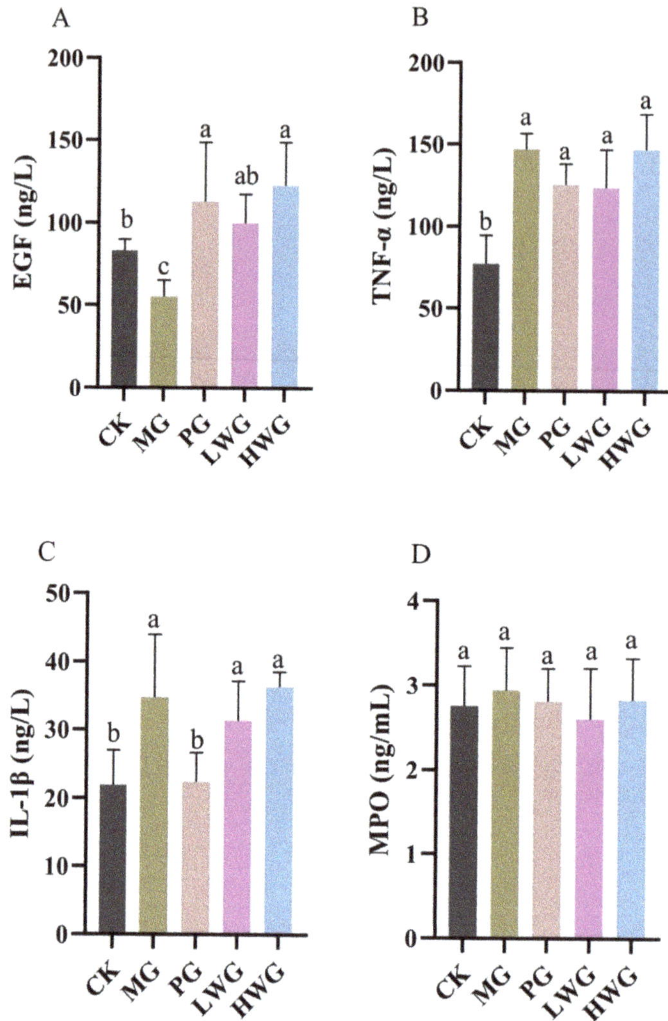

Figure 4. The test results of WP on the expression levels of inflammatory factors in gastric tissue of EGF (**A**), TNF-α (**B**), IL-1β (**C**), and MPO. (**D**) in mice. CK group: blank control group; MG group: model group; PG group: positive omeprazole control group (20 mg/kg bw); LWG group: low-dose walnut peptide group (200 mg/kg bw); HWG group: high-dose walnut peptide group (400 mg/kg bw). Different letters represent significant differences between groups at $p < 0.05$.

3.7. The Effects of WP on the Protein Expression of Inflammatory Factors in Gastric Tissue

The Western blotting (WB) results showed that the Nrf2 expression of the HWG group was 0.82; the difference was not statistically significant compared to the PG group, which was 0.99 ($p > 0.05$), but significantly different from MG group, which was 0.41 ($p < 0.05$), proving that a high dose of WP could effectively up-regulate the expression of Nrf2. The expression levels of Keap1 in all groups were not statistically significant ($p > 0.05$).

To further confirm that Nrf2 is a key target for antioxidant stress and the inhibition of iron metamorphosis, we explored the concentrations of two crucial downstream target genes of Nrf2, namely HO-1 and NQO1. The HO-1 gene expression of the LWG group was 1.27, which was significantly different from that of MG group, which was 0.51 ($p < 0.05$). NQO1 gene expression of the LWG and HWG groups were 1.08 and 1.03, which were significantly different from the MG group content of 0.77 ($p < 0.05$) and not significantly different from the PG group content of 1.23 ($p > 0.05$). In summary: the WB results indicated that the expression of two key downstream target genes, HO-1 and NQO1, was significantly increased in the presence of WP.

4. Discussion

The eight highest scoring peptides, EIDIGVPDEVGRL, DLAPTHPIRL, LDRLIPVLE, SVIQH, NTGSPITVPVGR, SKRPTF, DREIDIGVPDEVGRL, and RDENEKL, were obtained from the LC-MS/MS results. Among them, the highest scoring peptide EIDIGVPDEVGRL has a C-terminal of Leu and an N-terminal of Glu, and this peptide was detected to be derived from the A0A2I4EK72 protein of walnut by searching through the Uniprot Whole Species Sequence Library (https://www.uniprot.org, accessed on 12 September 2023). The most abundant of the eight proteins was a hexapeptide, SKRPTF, with a C-terminal of Phe and an N-terminal of Ser, which was searched to be derived from a walnut-derived RING-type E3 ubiquitin transferase. The determination of the peptide composition and the analysis of the amino acid sequence of the sample can help to provide theoretical and experimental data support for the subsequent active peptides of walnut peptides that contribute significantly to the protection of gastric mucosa. Ethanol exhibits corrosive effects on the gastric mucosa, disrupting the surface mucus layer and mucous cells, thereby weakening the protective role of gastric mucosal defense factors [25]. Research has shown that gavage with 75% ethanol on empty gastric tissue can lead to gastric mucosal vascular occlusion, impaired blood circulation, cell necrosis, and bleeding in mice [26]. These results align with the findings of our study. Furthermore, our research indicates that different doses of walnut peptides can significantly reduce ethanol-induced gastric mucosal damage in mice, as evidenced by a significant decrease in gross observations and pathological tissue damage scores, along with an increase in the injury inhibition rate.

Ethanol-induced liver injury manifests with the release of hepatic enzymes (ALT and AST) into the circulatory system. Elevated serum levels of ALT and AST signify damage to cell membranes and cellular mitochondria in the liver, respectively [27]. The experimental findings revealed that ethanol induced harm to the cell membrane of the liver, but the presence of WP inhibited the excessive release of ALT and AST, thus reducing the damage of alcohol to the liver cells and restoring the AST indexes to almost normal levels, and a high concentration of WP could restore the ALT indexes to almost normal levels, which is consistent with the results of a previous similar study on dendrobium officinale [28].

Gastric tissues undergo a series of oxidative stress responses when stimulated by ethanol. In this experiment, the expression levels of GSH, GSSG, T-AOC, and MDA were used to assess the antioxidant capacity of mice. GSH is the major source of thiol groups in activated gastric mucosal cells and plays a role in maintaining the normal redox state of intracellular protein thiols. It is also an important antioxidant in cellular physiological activities [29]. Glutathione is converted to GSSG by glutathione peroxidase, which is involved in scavenging free radicals. When gastric mucosal cells accelerate the accumulation of hydrogen peroxide and lipid peroxides, the activity of glutathione peroxidase decreases, leading to a decrease in the level of GSSG expression, which can exacerbate the damage [30]. The results showed that ethanol-induced impairment of the gastric mucosal barrier decreased the expression of GSH and GSSG, while the presence of WP promoted the expression of GSH and GSSG, which reduced the excessive damage caused by oxidative stress. T-AOC is one of the main measurements used for assessing and evaluating the level and potential of oxidative stress in certain diseases [31]. The experimental results showed that when ethanol caused gastric mucosal barrier damage, the T-AOC values of mice decreased significantly, and the presence of WP could promote the expression of T-AOC and restore it to normal levels. MDA is an important indicator for assessing lipid peroxidation when reactive oxygen species interact with unsaturated fatty acids [32]. The experimental results showed that the MDA content in mice increased significantly when ethanol caused gastric mucosal barrier damage, and the presence of

WP prevented the excessive accumulation of MDA. It has been demonstrated that wheat peptides can downregulate MDA levels in mice with alcoholic gastric ulcers, and our findings are consistent with them [5,33]. In summary, WP can alleviate the oxidative stress of alcohol-induced gastric mucosal injury by promoting the expression of GSH, GSSG, and T-AOC and reducing the expression of MDA.

When ethanol enters the digestive tract and stimulates gastric mucosal damage, it involves the interaction of various inflammatory cytokines. During the inflammatory process, a large amount of reactive oxygen species is generated at the inflammatory site, exacerbating oxidative stress reactions and causing tissue damage [34]. In this experiment, the expression levels of EGF, IL-1β, TNF-α, and MPO were used to evaluate the anti-inflammatory effects of walnut peptides. EGF is a growth factor that can inhibit gastric acid secretion, promote the proliferation of gastric mucosal epithelial cells, and facilitate tissue healing. When gastric tissue is damaged, the secretion of EGF is inhibited [35]. MPO is a peroxidase produced by neutrophils, which can reflect the number of neutrophils and the degree of inflammation infiltration in the gastric mucosa. It can also indicate the extent of damage caused by inflammation [36]. TNF-α can promote the production of various inflammatory factors and inhibit the blood circulation around the damaged mucosa, thereby delaying healing. Under normal conditions, the expression of TNF-α in tissues is low. When there is damage, the tissue secretes a large amount of TNF-α, worsening the injury [37]. IL-1β is a pro-inflammatory cytokine released by immune cells, and its expression level is strongly associated with intense gastric inflammation, the progression of gastric cancer, and an unfavorable prognosis. This may be because IL-1β has a strong inhibitory effect on parietal cells that induce gastric acid secretion and can drive the overexpression of several pro-inflammatory mediators, thereby altering the gastric mucosa's homeostasis [38,39]. However, this experimental study demonstrated that there was no significant difference in the expression levels of TNF-α, IL-1β, and MPO between the WP experimental group and the MG group. This indicated that the presence of WP did not alleviate the increased expression levels of the inflammatory factors TNF-α, IL-1β, and MPO caused by damage to the gastric mucosal barrier, which proved that WP relied on the increase in the secretion level of EGF to alleviate the effect of ethanol on the gastric mucous membrane damage; this aligns with the findings of a prior analogous study on wheat peptides and fucoidan [40].

An increasing body of experiments demonstrates the significant role of the Nrf2/Keap1 signaling pathway in protecting the gastric mucosa from oxidative stress [41]. HO-1 and NQO1 are key downstream target genes of Nrf2, and studying the gene expression levels of HO-1 and NQO1 can help to validate Nrf2 as a critical target in antioxidant stress [42]. To further investigate the protective effect of WP against ethanol-induced oxidative stress in mouse gastric mucosal histiocytes, the activity of Nrf2, Keap1, HO-1, and NQO1 enzymes related to oxidative stress in the cytoplasm of gastric histiocytes was determined (Figure 5). Experiments have shown that ethanol-induced gastric mucosal damage in mice leads to the suppression of HO-1, NQO1, and Nrf2 expression. However, HO-1 and NQO1 were significantly upregulated in the LWG group compared to the MG group. However, HO-1 and NQO1 were significantly upregulated in the LWG group compared to the MG group. Similarly, Nrf2 in the HWG group had a trend of significant upregulation, with the same trend as in the CK and PG groups. And there was no notable variance in Keap1 expression between the groups. It was demonstrated that the presence of WP could alleviate this inhibition, leading to an enhancement in antioxidant activity and the removal of accumulated lipid peroxides, consequently shielding gastric mucosal cells from harm (Figure 6).

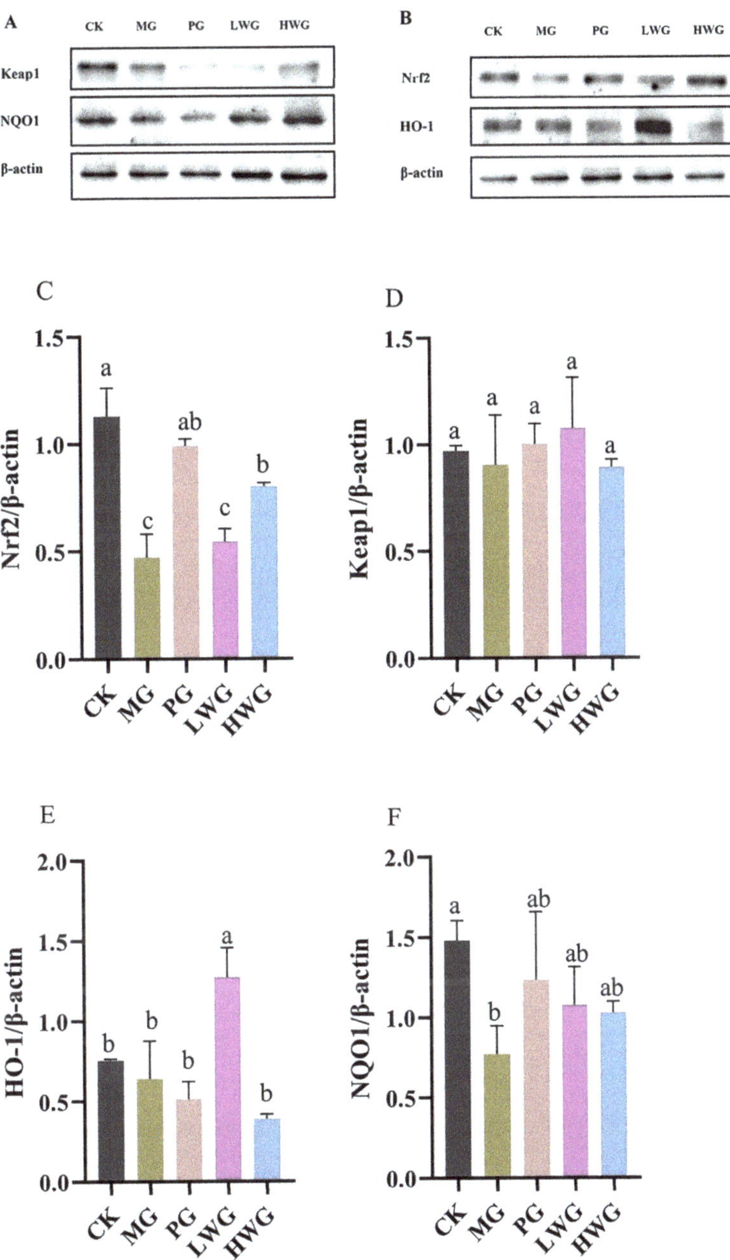

Figure 5. The effects of WP on the protein expression of inflammatory factors in gastric tissue. Protein blot analysis of Nrf2, NQO, Keap1, and HO-1. (**A**,**B**) Protein content analysis of Nrf2, Keap1, HO-1, and NQO1 (**C**–**F**). CK group: blank control group; MG group: model group; PG group: positive omeprazole control group (20 mg/kg bw); LWG group: low-dose walnut peptide group (200 mg/kg bw); HWG group: high-dose walnut peptide group (400 mg/kg bw). Different letters represent significant differences between groups at $p < 0.05$.

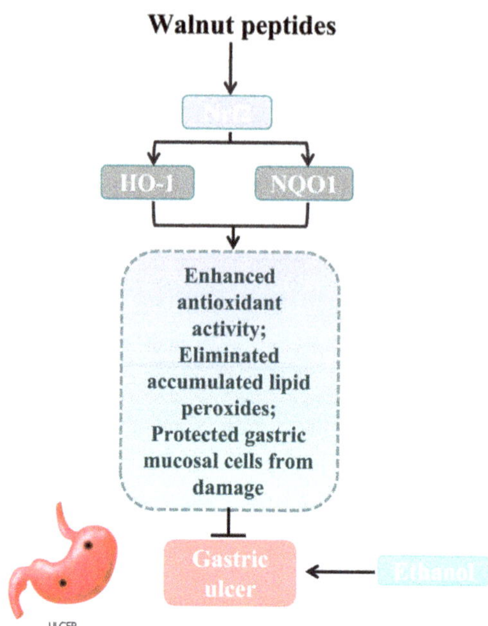

Figure 6. Schematic abstract: WP enhanced the gastric mucosal barrier by activating Nrf2 signaling, as well as increasing the expression of HO-1 and NQO1, thereby ameliorating oxidative stress and ethanol-induced gastric mucosal damage in mice (gastric ulcer image sourced from freepik).

5. Conclusions

In summary, our study demonstrates the potential therapeutic value of walnut peptides in protecting against ethanol-induced gastric mucosal barrier damage in mice. We have investigated the antioxidant and anti-inflammatory mechanisms of walnut peptides during this process. This opens up new avenues for exploring non-toxic and side-effect-free natural products as alternatives to existing drugs that may improve gastric mucosal barrier damage but carry potential risks.

Author Contributions: Y.Y. and X.W. conceived the idea of the study; methodology, Y.W.; Y.L. carried out the statistical analysis; Y.Y. and X.W. drafted the manuscript; L.Z. (Lei Zhao) provided input in the analysis; all authors critically reviewed the manuscript; and L.Z. (Liang Zhao) revised the manuscript for final submission with S.C. All authors have read and agreed to the published version of the manuscript.

Funding: This work was supported by the fund of the Cultivation Project of the Double First-Class Disciplines of Food Science and Engineering, Beijing Technology and Business University (No. BTBUYXTD202207), Discipline Construction—Food Science and Engineering (No. SPKX-202204), the Research Foundation for Youth Scholars of Beijing Technology and Business University (No. QNJJ2022-06), and the Major Science and Technology Project of the Science and Technology Department of Yunnan Province (No. 202202AE090007 and 202202AG050009).

Institutional Review Board Statement: All animal procedures act in accordance with the Animal Ethics Committee of the Beijing Key Laboratory of Functional Food from Plant Resources (Permit number: A330-2023-2) and the guidelines for the care and use of laboratory animals of the National Institutes of Health.

Informed Consent Statement: Not applicable.

Data Availability Statement: Data are contained within the article.

Conflicts of Interest: The authors confirm that they have no conflict of interest with respect to the work described in this manuscript.

References

1. Haghparast, P.; Tchalikian, T.N. Alcoholic Beverages and Health Effects. In *Reference Module in Biomedical Sciences*; Elsevier: Amsterdam, The Netherlands, 2022; p. B978012824315200244X. ISBN 978-0-12-801238-3.
2. Laine, L.; Takeuchi, K.; Tarnawski, A. Gastric Mucosal Defense and Cytoprotection: Bench to Bedside. *Gastroenterology* **2008**, *135*, 41–60. [CrossRef] [PubMed]
3. Zheng, H.; Chen, Y.; Zhang, J.; Wang, L.; Jin, Z.; Huang, H.; Man, S.; Gao, W. Evaluation of Protective Effects of Costunolide and Dehydrocostuslactone on Ethanol-Induced Gastric Ulcer in Mice Based on Multi-Pathway Regulation. *Chem.-Biol. Interact.* **2016**, *250*, 68–77. [CrossRef] [PubMed]
4. Wang, Y.; Su, W.; Zhang, C.; Xue, C.; Chang, Y.; Wu, X.; Tang, Q.; Wang, J. Protective Effect of Sea Cucumber (*Acaudina molpadioides*) Fucoidan against Ethanol-Induced Gastric Damage. *Food Chem.* **2012**, *133*, 1414–1419. [CrossRef]
5. Abdoulrahman, K. Anti-Ulcer Effect of *Ranunculus millefoliatus* on Absolute Alcohol-Induced Stomach Ulceration. *Saudi J. Biol. Sci.* **2023**, *30*, 103711. [CrossRef] [PubMed]
6. Bavishi, C.; DuPont, H.L. Systematic Review: The Use of Proton Pump Inhibitors and Increased Susceptibility to Enteric Infection. *Aliment. Pharmacol. Ther.* **2011**, *34*, 1269–1281. [CrossRef]
7. Ye, H.-Y.; Shang, Z.-Z.; Zhang, F.-Y.; Zha, X.-Q.; Li, Q.-M.; Luo, J.-P. *Dendrobium huoshanense* Stem Polysaccharide Ameliorates Alcohol-Induced Gastric Ulcer in Rats through Nrf2-Mediated Strengthening of Gastric Mucosal Barrier. *Int. J. Biol. Macromol.* **2023**, *236*, 124001. [CrossRef] [PubMed]
8. Lam, J.R.; Schneider, J.L.; Zhao, W.; Corley, D.A. Proton Pump Inhibitor and Histamine 2 Receptor Antagonist Use and Vitamin B_{12} Deficiency. *JAMA* **2013**, *310*, 2435. [CrossRef]
9. Deshpande, A.; Pasupuleti, V.; Thota, P.; Pant, C.; Mapara, S.; Hassan, S.; Rolston, D.D.K.; Sferra, T.J.; Hernandez, A.V. Acid-Suppressive Therapy Is Associated with Spontaneous Bacterial Peritonitis in Cirrhotic Patients: A Meta-Analysis: Acid-Suppressive Therapy and SBP. *J. Gastroenterol. Hepatol.* **2013**, *28*, 235–242. [CrossRef]
10. Al-Sayed, E.; El-Naga, R.N. Protective Role of Ellagitannins from *Eucalyptus citriodora* against Ethanol-Induced Gastric Ulcer in Rats: Impact on Oxidative Stress, Inflammation and Calcitonin-Gene Related Peptide. *Phytomedicine* **2015**, *22*, 5–15. [CrossRef]
11. Lin, Y.; Lv, Y.; Mao, Z.; Chen, X.; Chen, Y.; Zhu, B.; Yu, Y.; Ding, Z.; Zhou, F. Polysaccharides from *Tetrastigma hemsleyanum* Diels et Gilg Ameliorated Inflammatory Bowel Disease by Rebuilding the Intestinal Mucosal Barrier and Inhibiting Inflammation through the SCFA-GPR41/43 Signaling Pathway. *Int. J. Biol. Macromol.* **2023**, *250*, 126167. [CrossRef]
12. Bao, X.; Wu, J. Impact of Food-Derived Bioactive Peptides on Gut Function and Health. *Food Res. Int.* **2021**, *147*, 110485. [CrossRef]
13. Yu, L.; Li, R.; Liu, W.; Zhou, Y.; Li, Y.; Qin, Y.; Chen, Y.; Xu, Y. Protective Effects of Wheat Peptides against Ethanol-Induced Gastric Mucosal Lesions in Rats: Vasodilation and Anti-Inflammation. *Nutrients* **2020**, *12*, 2355. [CrossRef]
14. Gu, M.; Chen, H.-P.; Zhao, M.-M.; Wang, X.; Yang, B.; Ren, J.-Y.; Su, G.-W. Identification of Antioxidant Peptides Released from Defatted Walnut (*Juglans sigillata* Dode) Meal Proteins with Pancreatin. *LWT—Food Sci. Technol.* **2015**, *60*, 213–220. [CrossRef]
15. Sze-Tao, K.W.C.; Sathe, S.K. Walnuts (*Juglans regia* L): Proximate Composition, Protein Solubility, Protein Amino Acid Composition and Proteinin Vitro Digestibility. *J. Sci. Food Agric.* **2000**, *80*, 1393–1401. [CrossRef]
16. Feng, Y.; Wang, Z.; Chen, J.; Li, H.; Wang, Y.; Ren, D.-F.; Lu, J. Separation, Identification, and Molecular Docking of Tyrosinase Inhibitory Peptides from the Hydrolysates of Defatted Walnut (*Juglans regia* L.) Meal. *Food Chem.* **2021**, *353*, 129471. [CrossRef]
17. Qi, Y.; Wu, D.; Fang, L.; Leng, Y.; Wang, X.; Liu, C.; Liu, X.; Wang, J.; Min, W. Anti-Inflammatory Effect of Walnut-Derived Peptide via the Activation of Nrf2/Keap1 Pathway against Oxidative Stress. *J. Funct. Foods* **2023**, *110*, 105839. [CrossRef]
18. Vu, D.C.; Vo, P.H.; Coggeshall, M.V.; Lin, C.-H. Identification and Characterization of Phenolic Compounds in Black Walnut Kernels. *J. Agric. Food Chem.* **2018**, *66*, 4503–4511. [CrossRef]
19. Wang, F.; Pu, C.; Liu, M.; Li, R.; Sun, Y.; Tang, W.; Sun, Q.; Tian, Q. Fabrication and Characterization of Walnut Peptides-Loaded Proliposomes with Three Lyoprotectants: Environmental Stabilities and Antioxidant/Antibacterial Activities. *Food Chem.* **2022**, *366*, 130643. [CrossRef]
20. Xu, X.; Ding, Y.; Liu, M.; Zhang, X.; Wang, D.; Pan, Y.; Ren, S.; Liu, X. Neuroprotective Mechanisms of Defatted Walnut Powder against Scopolamine-Induced Alzheimer's Disease in Mice Revealed through Metabolomics and Proteomics Analyses. *J. Ethnopharmacol.* **2024**, *319*, 117107. [CrossRef]
21. Wang, J.; Wu, T.; Fang, L.; Liu, C.; Liu, X.; Li, H.; Shi, J.; Li, M.; Min, W. Anti-Diabetic Effect by Walnut (*Juglans mandshurica* Maxim.)-Derived Peptide LPLLR through Inhibiting α-Glucosidase and α-Amylase, and Alleviating Insulin Resistance of Hepatic HepG2 Cells. *J. Funct. Foods* **2020**, *69*, 103944. [CrossRef]
22. Huang, Y.; Chen, S.; Yao, Y.; Wu, N.; Xu, M.; Du, H.; Zhao, Y.; Tu, Y. Ovotransferrin Alleviated Acute Gastric Mucosal Injury in BALB/c Mice Caused by Ethanol. *Food Funct.* **2023**, *14*, 305–318. [CrossRef]
23. Wu, J.-Z.; Liu, Y.-H.; Liang, J.-L.; Huang, Q.-H.; Dou, Y.-X.; Nie, J.; Zhuo, J.-Y.; Wu, X.; Chen, J.-N.; Su, Z.-R.; et al. Protective Role of β-Patchoulene from *Pogostemon cablin* against Indomethacin-Induced Gastric Ulcer in Rats: Involvement of Anti-Inflammation and Angiogenesis. *Phytomedicine* **2018**, *39*, 111–118. [CrossRef]
24. Arab, H.H.; Salama, S.A.; Eid, A.H.; Kabel, A.M.; Shahin, N.N. Targeting MAPKs, NF-κB, and PI3K/AKT Pathways by Methyl Palmitate Ameliorates Ethanol-induced Gastric Mucosal Injury in Rats. *J. Cell. Physiol.* **2019**, *234*, 22424–22438. [CrossRef]
25. Amirshahrokhi, K.; Khalili, A.-R. The Effect of Thalidomide on Ethanol-Induced Gastric Mucosal Damage in Mice: Involvement of Inflammatory Cytokines and Nitric Oxide. *Chem.-Biol. Interact.* **2015**, *225*, 63–69. [CrossRef]

26. Zhang, J.; Lu, D.-Y.; Yuan, Y.; Chen, J.; Yi, S.; Chen, B.; Zhao, X. Liubao Insect Tea Polyphenols Prevent HCl/Ethanol Induced Gastric Damage through Its Antioxidant Ability in Mice. *RSC Adv.* **2020**, *10*, 4984–4995. [CrossRef]
27. Pal, S.; Bhattacharjee, A.; Mukherjee, S.; Bhattacharya, K.; Mukherjee, S.; Khowala, S. Effect of *Alocasia indica* Tuber Extract on Reducing Hepatotoxicity and Liver Apoptosis in Alcohol Intoxicated Rats. *BioMed Res. Int.* **2014**, *2014*, 349074. [CrossRef]
28. Jing, Y.; Hu, J.; Zhang, Y.; Sun, J.; Guo, J.; Zheng, Y.; Zhang, D.; Wu, L. Structural Characterization and Preventive Effect on Alcoholic Gastric Mucosa and Liver Injury of a Novel Polysaccharide from *Dendrobium officinale*. *Nat. Prod. Res.* **2022**, 1–8. [CrossRef]
29. Yoo, J.-H.; Park, E.-J.; Kim, S.H.; Lee, H.-J. Gastroprotective Effects of Fermented Lotus Root against Ethanol/HCl-Induced Gastric Mucosal Acute Toxicity in Rats. *Nutrients* **2020**, *12*, 808. [CrossRef]
30. Valcheva-Kuzmanova, S.; Marazova, K.; Krasnaliev, I.; Galunska, B.; Borisova, P.; Belcheva, A. Effect of *Aronia melanocarpa* Fruit Juice on Indomethacin-Induced Gastric Mucosal Damage and Oxidative Stress in Rats. *Exp. Toxicol. Pathol.* **2005**, *56*, 385–392. [CrossRef]
31. Nisha, S.; Bettahalli Shivamallu, A.; Prashant, A.; Yadav, M.K.; Gujjari, S.K.; Shashikumar, P. Role of Nonsurgical Periodontal Therapy on Leptin Levels and Total Antioxidant Capacity in Chronic Generalised Periodontitis Patients—A Clinical Trial. *J. Oral Biol. Craniofacial Res.* **2022**, *12*, 68–73. [CrossRef]
32. Kim, H.-G. Chunggan Extract, a Traditional Herbal Formula, Ameliorated Alcohol-Induced Hepatic Injury in Rat Model. *WJG* **2014**, *20*, 15703. [CrossRef] [PubMed]
33. Yang, X.; Yang, L.; Pan, D.; Liu, H.; Xia, H.; Wang, S.; Sun, G. Wheat Peptide Protects against Ethanol-Induced Gastric Mucosal Damage through Downregulation of TLR4 and MAPK. *J. Funct. Foods* **2020**, *75*, 104271. [CrossRef]
34. Xie, L.; Wu, Y.; Fan, Z.; Liu, Y.; Zeng, J. Astragalus Polysaccharide Protects Human Cardiac Microvascular Endothelial Cells from Hypoxia/Reoxygenation Injury: The Role of PI3K/AKT, Bax/Bcl-2 and Caspase-3. *Mol. Med. Rep.* **2016**, *14*, 904–910. [CrossRef] [PubMed]
35. Chen, W.; Wu, D.; Jin, Y.; Li, Q.; Liu, Y.; Qiao, X.; Zhang, J.; Dong, G.; Li, Z.; Li, T.; et al. Pre-Protective Effect of Polysaccharides Purified from *Hericium erinaceus* against Ethanol-Induced Gastric Mucosal Injury in Rats. *Int. J. Biol. Macromol.* **2020**, *159*, 948–956. [CrossRef]
36. Jeon, Y.-D.; Lee, J.-H.; Lee, Y.-M.; Kim, D.-K. Puerarin Inhibits Inflammation and Oxidative Stress in Dextran Sulfate Sodium-Induced Colitis Mice Model. *Biomed. Pharmacother.* **2020**, *124*, 109847. [CrossRef]
37. Dejban, P.; Eslami, F.; Rahimi, N.; Takzare, N.; Jahansouz, M.; Dehpour, A.R. Involvement of Nitric Oxide Pathway in the Anti-Inflammatory Effect of Modafinil on Indomethacin-, Stress-, and Ethanol-Induced Gastric Mucosal Injury in Rat. *Eur. J. Pharmacol.* **2020**, *887*, 173579. [CrossRef]
38. Filaly, H.E.; Outlioua, A.; Medyouf, H.; Guessous, F.; Akarid, K. Targeting IL-1β in Patients with Advanced *Helicobacter pylori* Infection: A Potential Therapy for Gastric Cancer. *Future Microbiol.* **2022**, *17*, 633–641. [CrossRef]
39. Guo, T.; Qian, J.-M.; Zhao, Y.-Q.; Li, X.-B.; Zhang, J.-Z. Effects of IL-1β on the Proliferation and Apoptosis of Gastric Epithelial Cells and Acid Secretion from Isolated Rabbit Parietal Cells. *Mol. Med. Rep.* **2013**, *7*, 299–305. [CrossRef]
40. Kan, J.; Hood, M.; Burns, C.; Scholten, J.; Chuang, J.; Tian, F.; Pan, X.; Du, J.; Gui, M. A Novel Combination of Wheat Peptides and Fucoidan Attenuates Ethanol-Induced Gastric Mucosal Damage through Anti-Oxidant, Anti-Inflammatory, and Pro-Survival Mechanisms. *Nutrients* **2017**, *9*, 978. [CrossRef]
41. Bhattacharyya, A.; Chattopadhyay, R.; Mitra, S.; Crowe, S.E. Oxidative Stress: An Essential Factor in the Pathogenesis of Gastrointestinal Mucosal Diseases. *Physiol. Rev.* **2014**, *94*, 329–354. [CrossRef]
42. Chen, J.; Zhang, J.; Chen, T.; Bao, S.; Li, J.; Wei, H.; Hu, X.; Liang, Y.; Liu, F.; Yan, S. Xiaojianzhong Decoction Attenuates Gastric Mucosal Injury by Activating the P62/Keap1/Nrf2 Signaling Pathway to Inhibit Ferroptosis. *Biomed. Pharmacother.* **2022**, *155*, 113631. [CrossRef] [PubMed]

Article

Selenium-Enriched Soybean Peptides as Novel Organic Selenium Compound Supplements: Inhibition of Occupational Air Pollution Exposure-Induced Apoptosis in Lung Epithelial Cells

Jian Zhang [1,2], Wenhui Li [1], He Li [1,3,*], Wanlu Liu [1], Lu Li [1] and Xinqi Liu [1,3]

[1] National Soybean Processing Industry Technology Innovation Center, Beijing Technology and Business University, Beijing 100048, China; tsnpzhj@163.com (J.Z.); lwhhuijiop@163.com (W.L.); liuwanlu19990115@163.com (W.L.); lil@btbu.edu.cn (L.L.); liuxinqi@btbu.edu.cn (X.L.)

[2] Department of Nutrition and Health, China Agricultural University, Beijing 100193, China

[3] Beijing Engineering and Technology Research Center of Food Additives, Beijing Technology and Business University, Beijing 100048, China

* Correspondence: lihe@btbu.edu.cn

Abstract: The occupational groups exposed to air pollutants, particularly PM2.5, are closely linked to the initiation and advancement of respiratory disorders. The aim of this study is to investigate the potential protective properties of selenium-enriched soybean peptides (Se-SPeps), a novel Se supplement, in mitigating apoptosis triggered by PM2.5 in A549 lung epithelial cells. The results indicate a concentration-dependent reduction in the viability of A549 cells caused by PM2.5, while Se-SPeps at concentrations of 62.5–500 µg/mL showed no significant effect. Additionally, the Se-SPeps reduced the production of ROS, proinflammatory cytokines, and apoptosis in response to PM2.5 exposure. The Se-SPeps suppressed the PM2.5-induced upregulation of Bax/Bcl-2 and caspase-3, while also restoring reductions in p-Akt in A549 cells. The antiapoptotic effects of Se-SPeps have been found to be more effective compared to SPeps, SeMet, and Na_2SeO_3 when evaluated at an equivalent protein or Se concentration. Our study results furnish evidence that supports the role of Se-SPeps in reducing the harmful effects of PM2.5, particularly in relation to its effect on apoptosis, oxidative stress, and inflammation.

Keywords: selenium-enriched soybean peptides; fine particulate matter; apoptosis; oxidative stress; inflammatory; protective effect

Citation: Zhang, J.; Li, W.; Li, H.; Liu, W.; Li, L.; Liu, X. Selenium-Enriched Soybean Peptides as Novel Organic Selenium Compound Supplements: Inhibition of Occupational Air Pollution Exposure-Induced Apoptosis in Lung Epithelial Cells. *Nutrients* **2024**, *16*, 71. https://doi.org/10.3390/nu16010071

Academic Editor: Ewa Jablonska

Received: 15 October 2023
Revised: 11 December 2023
Accepted: 21 December 2023
Published: 25 December 2023

1. Introduction

In recent years, the accelerated rate of industrialization and urbanization has given rise to a significant worldwide environmental issue: air pollution. Although attempts to regulate air pollution have resulted in some advancements in air quality, there are still specific occupations that necessitate humans working in environments characterized by significant air pollution [1]. In urban settings, the primary origins of PM2.5 are linked to air pollutants from transportation [2], with peak pollutant concentrations in transportation settings reaching up to three times higher than background levels [3]. Consequently, bus drivers, traffic police, sanitation workers, and other occupations are the main exposure groups of PM2.5 [4,5]. Various health problems, particularly the initiation and advancement of respiratory diseases, are closely associated with occupational exposure to PM2.5 [6]. PM2.5 first contacts the lung epithelial cells after entering the human body, inducing stimulation and destruction, increasing the release of ROS, and inducing oxidative damage in lung tissue [7]. In addition, it can further stimulate and induce the release of various proinflammatory cytokines, resulting in inflammation [8]. The damage and shedding of lung epithelial cells can aggravate the toxic effect of PM2.5, thereby leading to more severe inflammatory responses and lung injury. Prolonged exposure to low levels of PM2.5 has been shown to trigger oxidative stress and inflammation in healthy mice, resulting in apoptosis of bronchial and alveolar epithelial cells [9,10]. Zhang et al. demonstrated that PM2.5 prompted the apoptosis of A549 cells by upregulating the Bax/Bcl-2 ratio and the expression levels of caspase-3 [11]. Hence, the initiation of oxidative stress and inflammation by PM2.5 may result in the apoptosis of alveolar epithelial cells. However, utilizing barriers or filtration techniques to lessen particle inhalation is the primary strategy used presently to protect people from PM2.5 exposure [12]. Despite the aforementioned measures, inhaling PM2.5 is frequently unavoidable and has the potential to cause damage to lung cells. Hence, it is imperative to develop novel and effective strategies aimed at preventing or mitigating PM2.5-induced apoptosis and lung damage, as this is essential for safeguarding the health of individuals exposed to occupational hazards.

Natural foods to prevent PM2.5-induced lung damage have attracted a great deal of public interest. Selenium (Se) is considered a crucial microelement for promoting health, primarily owing to its antioxidant and anti-inflammatory properties [13]. Research has shown that administering Se can effectively safeguard rat lungs from acute injuries by activating GSH-Px, diminishing the inflammatory response, and reducing lipid peroxidation [14,15]. Se-containing compounds, which include sodium selenite and selenomethionine, along with selenoproteins and Se nanoparticles, possess the capacity to regulate defense systems against many viral infections, including COVID-19 [16]. Methylselenic acid can also control the cell cycle by influencing the PI3K/AKT/mTOR signaling pathway [17]. Hence, Se exhibits the potential to mitigate lung damage resulting from the inhalation of air pollutants. Se is incorporated into the molecular structure of selenoproteins within the human body, contributing to their biological functions [18]. Organic forms of Se are more effective in fulfilling dietary requirements compared to inorganic forms, due to improvements in bioavailability and low toxicity [19]. In recent years, researchers have become increasingly interested in extracting peptides from Se-enriched, plant-derived foods and determining their anti-inflammatory and antioxidant activities. Several investigations have demonstrated that peptides containing Se have displayed remarkable capabilities in terms of antioxidation and immunoregulation [20,21]. Research findings indicate that Se-enriched soybean peptides (Se-SPeps) possess the ability to mitigate oxidative damage induced by H_2O_2 through the enhancement of GSH-Px activity [22]. Fang et al. revealed that Se-enriched rice peptides exhibited high immunomodulatory activity. Soybean peptides (SPeps) have been proven to have anti-inflammatory and antioxidant properties [23]. The study results demonstrate that SPeps exhibited significant antioxidant properties in safeguarding HepG2 cells from oxidative stress induced by H_2O_2 [24]. In addition, it has been observed that SPeps exhibit a mitigating effect on the inflammatory response within the RAW264.7 cell provoked by LPS [25]. Based upon this literature, it is postulated that Se-SPep have the potential to mitigate the detrimental impacts of PM2.5 on A549 cell apoptosis by regulating oxidative stress and inflammatory responses.

Within this investigation, A549 cells were subjected to the intervention of Se-SPeps and exposed to PM2.5 in vitro. The analysis encompassed the examination of apoptosis, intracellular generation of ROS, secretion of proinflammatory factors, and expression of proteins related to apoptosis. This study aimed to explore the combined mechanism of Se-SPep intervention in countering oxidative stress, inflammatory response, and apoptosis triggered by PM2.5 in A549 cells.

2. Materials and Methods

2.1. Materials

Se-enriched soybeans were acquired from Enshi Se-Run Health Tech Development Co., Ltd. (Enshi, China). The PM2.5 standard reference material (SRM 2786) [26] was procured from the NIST (Gaithersburg, MD, USA). Seleno-DL-methionine (SeMet) and sodium selenite (Na_2SeO_3) were purchased from Sigma-Aldrich Co. (St. Louis, MO, USA). DMEM, FBS, penicillin, and streptomycin were acquired from Gibco (Grand Island, NE, USA). The

ELISA kits for IL-1β, IL-6, and TNF-α were purchased from Multisciences (Hangzhou, China). The antibodies against Bax (AF1270), Bcl-2 (AF6285), caspase-3 (AF1213), Akt (AF1777), p-Akt (Ser473) (AA329), β-Actin (AF5003), and HRP-labeled goat anti-rabbit IgG (H + L) were obtained from Beyotime (Shanghai, China). The Hoechst 33342, ECL kit, CCK-8 kit, and ROS kit were supplied by Beyotime (Shanghai, China). The RIPA lysis buffer, Annexin V-FITC/PI apoptosis kit, and BCA protein assay kit were supplied by Solarbio (Beijing, China).

2.2. Preparation of the Se-SPeps

A previous study described a method for extracting Se-SPeps from Se-enriched soybeans [27]. In summary, the Se-enriched soybeans were subjected to grinding, defatting, and drying processes in order to obtain soybean kernel flour. Using soybean kernel flour as the raw material, Se-enriched soybean protein (Se-SPro) precipitate was obtained through alkali dissolution and acid precipitation. The Se-SPro precipitate was then redissolved, followed by dialysis using a 3500 Da membrane at 4 °C. The dialysate was freeze-dried to produce the Se-SPro. The Se-SPro underwent digestion in a 2:1:1 ratio with alkaline protease, neutral protease, and papain. Proteases, constituting 0.2% of the Se-SPro weight, were introduced and subjected to hydrolysis under optimal conditions at 50 °C for 4 h. Following this, the hydrolysate underwent centrifugation at $3500 \times g$ for 15 min. The resulting supernatant was freeze-dried to yield Se-SPeps. The procedure for SPep preparation mirrored the aforementioned process. The Se concentrations in Se-SPeps and SPeps were determined as 86.03 ± 4.28 mg/kg and 0.06 ± 0.03 mg/kg, respectively, utilizing hydride generation atomic fluorescence spectrometry (LCAFS6500, Beijing Haiguang Instrument Co., Ltd., Beijing, China) following the Chinese national standard GB 5009.93-2010 [28]. The protein concentrations of the Se-SPeps and SPeps were $85.10 \pm 0.21\%$ and $87.41 \pm 0.04\%$, assessed using the Kjeldahl method. Essential amino acids constituted $36.88 \pm 1.23\%$ and $37.18 \pm 1.17\%$ of the total amino acids (Table S1) in Se-SPeps and SPeps, determined using an amino acid analyzer (Biochrom 30+, BioChrom Ltd., Cambridge, UK). Utilizing an ÄKTA pure system (AKTA pure 25, Cytiva, Marlborough, MA, USA) and following the procedure outlined in the Chinese national standard GB/T 22492-2008 [29], the Se-SPeps and SPeps demonstrated molecular weights (Figure S1) below 3000 Da, determined to be 86.42% and 88.46%, respectively.

2.3. Preparation of the PM2.5 Suspension

Briefly, the PM2.5 was obtained from NIST (SRM 2786) with an average particle diameter of 2.8 μm and was collected in 2005 in Prague, Czech Republic. The primary constituents consisted of trace elements, polycyclic aromatic hydrocarbons, and polybrominated diphenyl ether (Table S2). PM2.5 was suspended in DMEM medium without FBS, achieving a final concentration of 1 mg/mL. Subsequently, the suspension underwent sonication for 20 min before administration.

2.4. Cell Lines and Cell Culture

The A549 cells were purchased from the PCRC (Beijing, China). The A549 cells within 10 generations were thawed in 37 °C water bath and transferred to high glucose DMEM medium supplemented with 10% FBS and 1% penicillin/streptomycin. Following this, the cells underwent cultivation and revival within a cell incubator (37 °C, 5% CO_2) for the ensuing cytotoxicity assessment.

2.5. Cytotoxicity of PM2.5 and Se-SPeps

The cell viability of A549 cells was assessed through the CCK-8 assay. Cells were planted in 96-well plates with a density of 5×10^3 cells per well and incubated for 24 h. Next, the cells underwent a 24 h treatment with varying doses of PM2.5 (0, 25, 50, 100, 150, and 200 μg/mL) or Se-SPeps (0, 62.5, 125, 250, 500, 1000, 2000, and 4000 μg/mL), followed by 60 min at a 37 °C incubation with 10 μL CCK-8 solution. The microplate reader (Infinite 200 Pro Nanoquant, Tecan, Männedorf, Switzerland) measured absorbance at 450 nm. Expressing cell viability as a percentage, it compared the absorbance of treated cells to that of untreated cells. Concurrently, an inverted microscope (IX73, OLYMPUS, Tokyo, Japan) was employed to observe and capture images of cell morphology.

2.6. Toxicity Suppression by Se-SPeps

The protective effect of Se-SPeps was examined by plating A549 cells in 96-well plates with a density of 5×10^3 cells per well and allowing them to culture for 24 h. The media was then replaced with DMEM media containing samples of 125, 250, and 500 µg/mL of Se-SPeps, SPeps (the same protein concentration as 250 µg/mL of Se-SPeps), SeMet (the same Se concentration as 250 µg/mL of Se-SPeps), and Na_2SeO_3 (the same Se concentration as 250 µg/mL of Se-SPeps) for 24 h. The Se concentration in the SeMet and Na_2SeO_3 groups was 0.022 µg/mL, which was equivalent to the Se concentration in the 250 µg/mL of Se-SPep group. After removing the samples, PM2.5 was added to the medium at a final concentration of 150 µg/mL and incubated for 24 h. The control group received an equal volume of DMEM, while the model group received an average amount of DMEM only containing PM2.5, and the cell viability test was performed using CKK-8.

2.7. Analysis of Cell Apoptosis

To assess the protective effect of Se-SPeps against PM2.5-induced cell apoptosis, annexin V-FITC/PI double staining was carried out. A549 cells were seeded at a density of 5×10^4 cells per well in 24-well plates, then incubated without or with Se-SPeps, SPeps, SeMet, and Na_2SeO_3 for 24 h followed by PM2.5 treatment as previously described. After Se-SPep and PM2.5 treatments, the cells were suspended in 1 mL of binding buffer, accompanied by 5 µL of annexin V-FITC and 5 µL of PI, in darkness at 37 °C for 20 min. Flow cytometry was used to detect fluorescence immediately after. The treated cells were also resuspended with binding buffer supplemented with Hoechst 33342 for 30 min. A fluorescence microscope was used to monitor the apoptotic cells (IX73, OLYMPUS, Japan).

2.8. Detection of Intracellular ROS

The fluorogenic dye DCFH-DA was used to measure ROS production. The A549 cells were seeded at a density of 5×10^4 cells per well in 24-well plates and then preincubated without or with Se-SPeps, SPeps, SeMet, and Na_2SeO_3 for 24 h followed by PM2.5 treatment as described previously. Following that, cells underwent a PBS solution wash and were subjected to a 30 min incubation with 10 µmol/L DCFA-DA at 37 °C. After incubation, excess DCFH-DA was eliminated by washing the cells with PBS, and 300 µL of PBS was introduced into each well. The flow cytometer (FACSAria III, BD, Franklin Lakes, NJ, USA) was employed to quantify fluorescence intensity, with the excitation wavelength set at 488 nm and the emission wavelength at 525 nm. Additionally, the cells were scrutinized using a fluorescence microscope (IX73, OLYMPUS, Japan).

2.9. Determination of the Cytokines

Supernatants devoid of cells were gathered following treatment with Se-SPeps and PM2.5. The concentrations of IL-1β, IL-6, and TNF-α in the supernatants were determined using ELISA kits.

2.10. Western Blot Analysis

A549 cells were seeded into 24-well plates and subsequently incubated for 24 h with Se-SPeps, SPeps, SeMet, and Na_2SeO_3 before being treated with PM2.5 treatment as previously described. After the treatments, PBS was used to wash A549 cells, followed by lysis on ice in RIPA buffer for 10 min, which included 1% PMSF and a cocktail of phosphatase inhibitors. Whole-cell lysates underwent centrifugation at $10,000 \times g$ for 5 min at 4 °C, leading to the collection of supernatants. The total protein concentration was gauged using the BCA protein quantitative kit. On a 10% polyacrylamide gel, equal amounts of protein (10 µg) were separated and transferred onto PVDF membranes. The membranes were blocked with 5% BSA at room temperature for 1 h. The PVDF membranes were incubated with primary antibodies against β-Actin, Akt, p-Akt, Bax, Bcl-2, and caspase-3 overnight at 4 °C. Following incubation, the membranes were washed three times with TBST and incubated with an HRP-labeled secondary antibody for 1 h at room temperature. After being rewashed with TBST, the bands were developed by an ECL kit. The gray value of the protein bands was analyzed using ImageJ vision 1.8.0 software. The relative expression of the target protein was calculated using the following formula: (the target protein/gray value of β-Actin).

2.11. Statistical Analysis

The mean ± SD represented the data. Statistical analysis was performed utilizing ANOVA followed by Tukey's test with SPSS 23 software. Graphs were generated using Graph Pad Prism 9.2.0 software.

3. Results and Discussion

3.1. Effect of PM2.5 on the Viability of A549 Cells

We investigated the detrimental impacts of PM2.5 on A549 cells. The A549 cells were exposed to different concentrations of PM2.5 (0, 25, 50, 100, 150, and 200 μg/mL) for 24 h. The CCK-8 assay demonstrated that the incubation of A549 cells with 25–200 μg/mL PM2.5 for 24 h reduced their viability in a concentration-dependent manner (Figure 1A), with a significant difference between each dose group and the 0 μg/mL group ($p < 0.05$). At concentrations exceeding 50 μg/mL, the cell viability dropped below 80%, indicating that PM2.5 had a strong toxic effect. At the highest concentration of 200 μg/mL, the cell viability was only 63.05%. The microscopic observation results are shown in Figure 1B. The cells in the 0 μg/mL group have a complete spindle shape and uniform distribution. At concentrations surpassing 150 μg/mL, A549 cells adhered to the PM2.5 particles, and the cell membrane boundary became indiscernible, resulting in evident damage and death. Based on these results, we selected a concentration of PM2.5 of 150 μg/mL, which led to a cell viability of 68.68%. We chose this concentration with the rationale of utilizing a midpoint effective dose, inducing damage to the cell population without triggering a complete collapse.

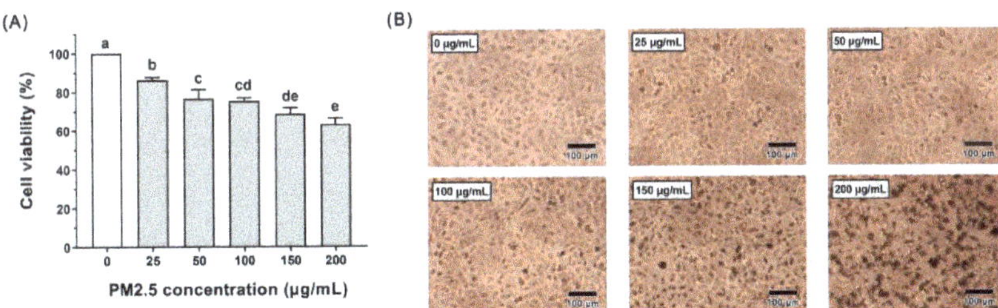

Figure 1. Effect of different concentrations (0, 25, 50, 100, 150, and 200 μg/mL) of PM2.5 pretreatment on A549 cells for 24 h. (**A**) Viability of the A549 cells. (**B**) Cell morphological changes of the A549 cells. Scale bar = 100 μm. Data are shown as the mean ± SD, $n = 8$/group. Statistical analysis was performed using ANOVA followed by Tukey's post hoc test. Different letters over bars indicate statistically significant differences ($p < 0.05$).

3.2. Effect of Se-SPeps on the Viability of A549 Cells

To assess the Se-SPep's potential as a therapeutic intervention against PM2.5-induced toxicity in A549 cells, it was essential to determine a dose range that would not adversely affect the cells. Although Se and peptides exhibit documented benefits to human health, certain reports have indicated potential adverse effects when Se is present in high concentrations [30]. Hence, A549 cells were exposed to varying concentrations of Se-SPeps (0, 62.5, 125, 250, 500, 1000, 2000, and 4000 μg/mL). Subsequently, the cell viability was quantified to determine the optimal concentration for our assays. Illustrated in Figure 2A, the CCK-8 assay revealed that exposure to Se-SPep concentrations ranging from 62.5 to 500 µg/mL for 24 h did not induce a noteworthy impact on the viability of A549 cells. The viability of A549 cells was 99.04%, 96.83%, 95.76%, and 94.37% when the Se-SPep concentration was below 500 µg/mL compared to the 0 µg/mL group, respectively. When the concentration of Se-SPeps was 1000, 2000, and 4000 µg/mL, the cell viability decreased to 72.93%, 67.57%, and 65.18%, respectively. The results demonstrated that Se-SPeps are cytotoxic at high concentrations (more than 1000 µg/mL). Furthermore, to determine the optimal Se-SPep concentration, the cell morphology was examined. Illustrated in Figure 2B, cells in the low-concentration Se-SPep treatment group (less than 500 µg/mL) were in a normal state, and no significant decrease in the cell number was observed compared to the 0 µg/mL group. Treatment with high doses of Se-SPeps (over 1000 µg/mL) resulted in cell damage and decreased cell numbers. Based on the cell viability assessment and cell state observation, 125, 250, and 500 µg/mL were selected as the treatment concentrations of Se-SPeps for subsequent experiments.

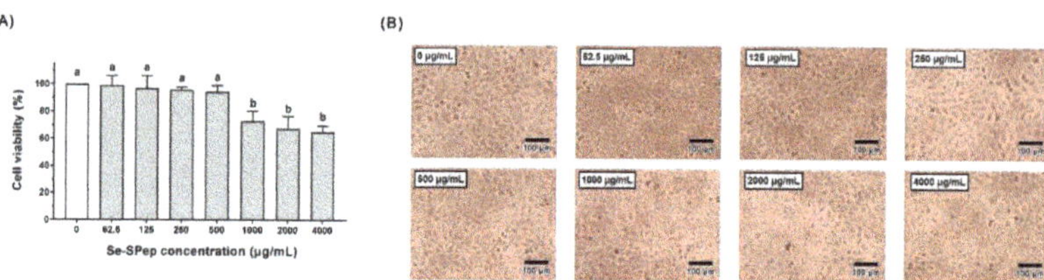

Figure 2. Effect of different concentrations (0, 62.5, 125, 250, 500, 1000, 2000, and 4000 µg/mL) of Se-SPep pretreatment on A549 cells for 24 h. (**A**) Viability of the A549 cells. (**B**) Cell morphological changes of the A549 cells. Scale bar = 100 µm. Data are shown as the mean ± SD, n = 8/group. Statistical analysis was performed using ANOVA followed by Tukey's post hoc test. Different letters over bars indicate statistically significant differences ($p < 0.05$).

3.3. Effect of Se-SPeps on the Viability of A549 Cells Exposed to PM2.5

After the Se-SPep intervention, A549 cells encountered exposure to PM2.5 to explore the protective impact of Se-SPeps on their viability following injury induced by PM2.5. As shown in Figure 3, the 150 µg/mL of PM2.5 significantly decreased the viability of the DMEM-pretreated cells compared to control cells ($p < 0.05$). However, the cell viabilities of Se-SPep- and SPep-pretreated cells, after PM2.5 exposure, were markedly higher than those of DMEM-pretreated cells ($p < 0.05$). The SeMet group exhibited a higher cell viability than the PM2.5 group, yet the disparity lacked statistical significance ($p > 0.05$). Contrarily, the Na$_2$SeO$_3$ group witnessed a significant decrease in cell viability ($p < 0.05$), reaching 15.22%, indicating that Na$_2$SeO$_3$ pretreatment had a significant toxic effect on A549 cells. The findings underscore the Se-SPep's protective influence on viability against the PM2.5-induced injury in A549 cells, with the organic Se form outperforming the inorganic Se form.

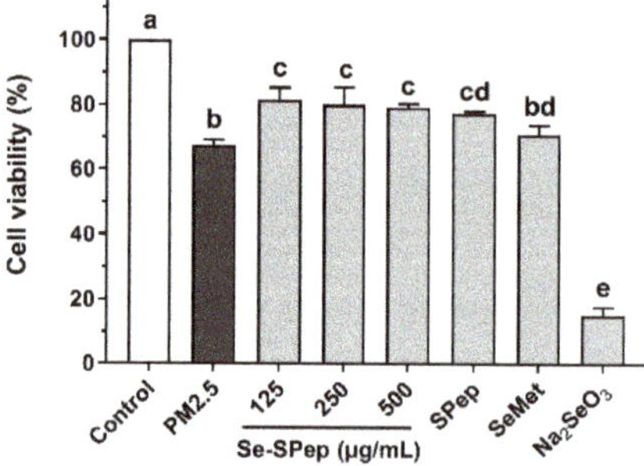

Figure 3. Effect of Se-SPep pretreatment on the cell viability in A549 cells exposed to PM2.5. The Se concentration in the SeMet and Na$_2$SeO$_3$ groups is 0.022 µg/mL, which is equivalent to the Se concentration

in the 250 µg/mL Se-SPep group. Data are shown as the mean ± SD, $n = 8$/group. Statistical analysis was performed using ANOVA followed by Tukey's post hoc test. Different letters over bars indicate statistically significant differences ($p < 0.05$).

3.4. Effect of Se-SPeps on Cell Apoptosis of A549 Cells Exposed to PM2.5

A distinctive feature of PM2.5 toxicity is its ability to directly affect cells, leading to apoptosis [31]. To assess the protective effect of Se-SPeps on PM2.5-induced apoptosis in A549 cells, the Hoechst 33342 and annexin V-FITC assays were employed. Illustrated in Figure 4A, the PM2.5 group exhibited a pronounced increase in potent blue fluorescence compared to the control group. Simultaneously, a noteworthy reduction in blue fluorescence was observed after Se-SPep, SPep, and SeMet interventions compared to the PM2.5 group. A few viable cells were stained in the Na_2SeO_3 group, and a more intense blue fluorescence was observed. As shown in Figure 4B,C, the flow cytometry analysis revealed a notable rise in the apoptotic percentage in the PM2.5 group ($p < 0.05$), reaching 16.81% compared to the control group. This implies a substantial induction of cell apoptosis by PM2.5. In contrast, Se-SPep, SPep, and SeMet treatments significantly decreased the apoptotic cell percentages to 10.64%, 9.39%, 9.21%, 12.62%, and 12.19%, respectively. The Se-SPep group had significantly lower cell apoptosis than the SPep and SeMet groups ($p < 0.05$). Na_2SeO_3 intervention induced a noteworthy elevation in the apoptotic rate of A549 cells, reaching 57.22%. The results showed that Se-SPeps could prevent the apoptosis of A549 cells induced by PM2.5. The antiapoptotic ability of Se-SPeps was greater than that of SPeps and SeMet.

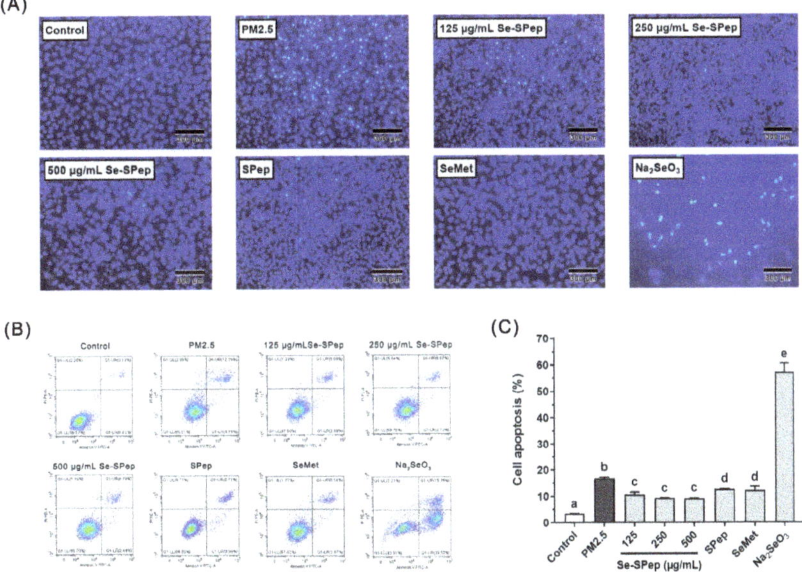

Figure 4. Effect of Se-SPep pretreatment on cell apoptosis in A549 cells exposed to PM2.5. (**A**) Representative photomicrographs of A549 cells stained with Hoechst 33342 fluorescent dye. Scale bar = 300 µm. (**B**) Flow cytometry analysis of cell apoptosis induced by PM2.5. (**C**) The apoptosis ratio induced by PM2.5 after intervention with Se-SPeps. The Se concentration in the SeMet and Na_2SeO_3 groups is 0.022 µg/mL, which is equivalent to the Se concentration in the 250 µg/mL Se-SPep group. Data are shown as the mean ± SD, $n = 3$/group. Statistical analysis was performed using ANOVA followed by Tukey's post hoc test. Different letters over bars indicate statistically significant differences ($p < 0.05$).

3.5. Effect of Se-SPeps on ROS Generation of A549 Cells Exposed to PM2.5

Research findings have offered substantiation that prolonged exposure to PM2.5 can induce respiratory system damage [32,33]. Oxidative stress, a crucial mechanism, is implicated in the induction of damage to the respiratory system, representing a significant factor in the negative impacts of PM2.5 air pollution on respiratory health [34]. Cells produce a large amount of ROS when exposed to PM2.5, causing oxidative damage to tissues and cells, triggering cellular inflammatory reaction and apoptosis [35,36]. Lao et al. reported that oxidative stress occurred in A549 cells after exposure to airborne PM2.5 elements [37]. Further investigations have demonstrated a significant increase in ROS levels in bronchial epithelial cells following exposure to airborne PM2.5 components [38]. In addition, a decreased cell viability caused by PM2.5 exposure was correlated with ROS overproduction.

As shown in Figure 5A, DCFH-DA detection reveals that ROS green fluorescence in the control group is virtually invisible, whereas the green fluorescence in the PM2.5 group surpasses that of the control group significantly. Meanwhile, Figure 5B shows that ROS fluorescence intensity in the PM2.5 group is 1.8 times higher than that in the control group, indicating that PM2.5 induces a strong intracellular oxidative stress response in cells. The intracellular green fluorescence and DCF fluorescence intensities were significantly lower after Se-SPep, SPep, and SeMet intervention compared to the PM2.5 group ($p < 0.05$). The ROS level in the 500 μg/mL of Se-SPep group was markedly lower than that in the SPep group ($p < 0.05$). However, due to apoptosis, both the cell count and fluorescence intensity in the Na_2SeO_3 group were significantly decreased ($p < 0.05$). These results suggest that Se-SPeps, SPeps, and SeMet inhibit the generation of intracellular ROS, protecting A549 cells from oxidative stress generated by PM2.5 exposure and preventing cell apoptosis. Simultaneously, Se-SPeps have a more significant inhibitory effect on PM2.5-induced cellular oxidative stress. This effect was amplified when Se was combined with SPeps, indicating that Se-SPeps can suppress ROS generation, which is beneficial for cell viability.

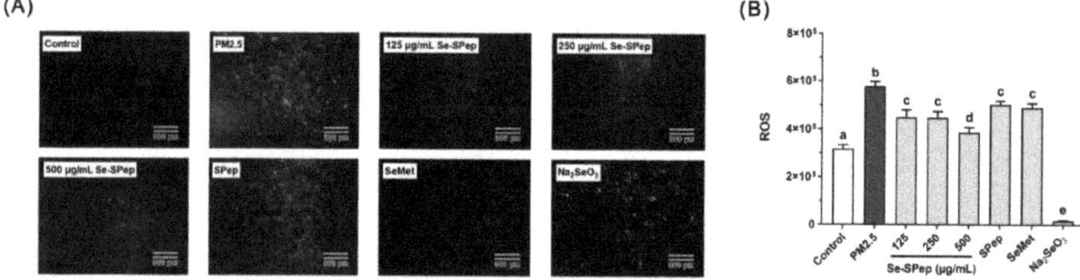

Figure 5. Effect of Se-SPep pretreatment on ROS generation in A549 cells exposed to PM2.5. (**A**) Representative photomicrographs of A549 cells stained with DCFH-DA fluorescent dye. Scale bar = 300 μm. (**B**) Relative fluorescence intensity of intracellular ROS of A549 cells. The Se concentration in the SeMet and Na_2SeO_3 groups is 0.022 μg/mL, which is equivalent to the Se concentration in the 250 μg/mL Se-SPep group. Data are shown as the mean ± SD, $n = 3$/group. Statistical analysis was performed using ANOVA followed by Tukey's post hoc test. Different letters over bars indicate statistically significant differences ($p < 0.05$).

3.6. Effect of Se-SPeps on Proinflammatory Cytokines Release of A549 Cells Exposed to PM2.5

Respiratory diseases exhibit a close association with inflammation. PM2.5 can stimulate lung immune cells to initiate an acute inflammatory response, stimulate normal cells to secrete numerous proinflammatory cytokines, modulate the inflammatory response, and induce inflammatory damage [39,40]. The inflammatory response induced by PM2.5 has been assessed by measuring the levels of expression of various proinflammatory cytokines. Figure 6 illustrates the ability of Se-SPeps to alleviate the release of IL-1β, IL-6, and TNF-α from PM2.5-induced A549 cells. The levels of proinflammatory cytokines IL-1β, IL-6, and TNF-α in the PM2.5 group were significantly increased in comparison to the control group ($p < 0.05$). The findings demonstrate that PM2.5 has the ability to induce multiple proinflammatory cytokines and inflammatory responses in A549 cells, proving that exposure to PM2.5 is immunotoxic to A549 cells. Meanwhile, after Se-SPep and SeMet intervention, the intracellular levels of IL-1β, IL-6, and TNF-α exhibited a significant decrease in comparison to the PM2.5 group ($p < 0.05$). The intervention with SPeps resulted in notably lower levels of IL-6 and TNF-α compared to the PM2.5 group ($p < 0.05$). In particular, the levels of IL-6 in the Se-SPep group exhibited a significant reduction compared to the SPep group ($p < 0.05$), while the levels of IL-1β and TNF-α in the Se-SPep group were lower than those in the SPep group ($p > 0.05$). Concentrations of proinflammatory cytokines in the Na_2SeO_3 group were notably lower due to severe apoptosis.

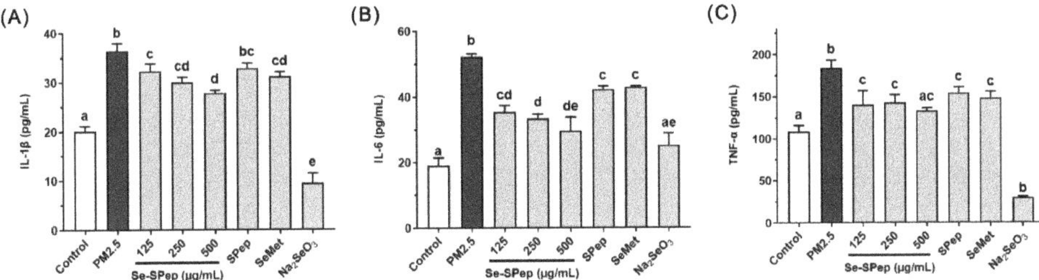

Figure 6. Effect of Se-SPep pretreatment on proinflammatory cytokines secretion in A549 cells exposed to PM2.5. (**A**) IL-1β secretion. (**B**) IL-6 secretion. (**C**) TNF-α secretion. The Se concentration in the SeMet and Na_2SeO_3 groups is 0.022 μg/mL, which is equivalent to the Se concentration in the 250 μg/mL Se-SPep group. Data are shown as the mean ± SD, n = 3/group. Statistical analysis was performed using ANOVA followed by Tukey's post hoc test. Different letters over bars indicate statistically significant differences ($p < 0.05$).

PM2.5 has the potential to induce and exacerbate inflammatory responses in cells, exerting toxic effects on cellular function [41,42]. Several studies have indicated that PM2.5 has the capacity to elevate the secretion of proinflammatory cytokines [43,44], indicating that inflammation is a crucial mechanism of cellular damage in the respiratory system. Xue et al. demonstrated that PM2.5 could significantly promote the release of proinflammatory factors in cells and lead to cell damage [45]. In alignment with the outcomes of this investigation, PM2.5 has the capacity to induce an inflammatory response in A549 cells, elevating the secretion of proinflammatory cytokines such as IL-1β, IL-6, and TNF-α. The results show that after the intervention of Se-SPeps, SPeps, and SeMet, the proinflammatory cytokines decrease significantly compared with the PM2.5 group, proving that the intervention of Se-SPeps, SPeps, and SeMet could inhibit the PM2.5-induced inflammatory damage of cells. In addition, Se-SPeps exhibited a more robust capability to inhibit the production of proinflammatory cytokines compared to SPeps.

3.7. Effect of Se-SPeps on the Mitochondrial Apoptotic Pathway of A549 Cells Exposed to PM2.5

According to our studies, when A549 cells are stimulated by PM2.5, a significant quantity of ROS is created, an inflammatory response is initiated, and proinflammatory cytokines are released, ultimately leading to apoptosis. The modulation of various proteins in the mitochondrial apoptosis pathway governs the intricate process of cellular apoptosis. Currently, the Bcl-2 family is the most studied of the apoptosis pathway-related proteins and is mainly classified as apoptosis-promoting and antiapoptotic [46]. The Bcl-2 family plays a pivotal role in the endogenous apoptosis pathway [47]. It was found that the ratio of Bax to Bcl-2 was the main factor determining the inhibitory effect on apoptosis [48]. An elevation in the Bax to Bcl-2 ratio modifies the potential of the mitochondrial membrane, stimulates the cytosolic release of cytochrome C, and triggers caspase-3 activation, ultimately leading to apoptosis [49]. Research has indicated that the oxidative stress induced by PM2.5 can disturb the antioxidant system and facilitate apoptosis through mitochondria-dependent mechanisms [50]. Therefore, regulating the expression of antiapoptotic and proapoptotic proteins within the Bcl-2 family is crucial for apoptosis control. The protein expression of antiapoptotic protein Bcl-2 and proapoptotic protein Bax in the intervention of Se-SPeps to prevent cell apoptosis was verified. Displayed in Figure 7A–C, the exposure to PM2.5 notably reduced the expression of the antiapoptotic protein Bcl-2 ($p < 0.05$), increased the expression of the proapoptotic protein Bax ($p < 0.05$), and elevated the Bax/Bcl-2 ratio significantly ($p < 0.05$). Zhang et al. proved that PM2.5-induced apoptosis could lead to an elevation in the Bax/Bcl-2 ratio [11]. We also found that PM2.5 could induce an increase in Bax while Bcl-2 levels experienced a significant decrease, highlighting Bax/Bcl-2 as a pathway for PM2.5-induced apoptosis in A549 cells. In the cells treated with Se-SPeps and SPeps, the Bax protein level was significantly decreased ($p < 0.05$) compared to the PM2.5 group, while the Bcl-2 protein level was significantly increased ($p < 0.05$), resulting in a decreased Bax/Bcl-2 ratio ($p < 0.05$). Meanwhile, the Se-SPep group exhibited significantly higher levels of the antiapoptotic protein Bcl-2 compared to the SPep and SeMet groups ($p < 0.05$), along with a lower Bax/Bcl-2 ratio than the SPep and SeMet groups ($p < 0.05$). These findings suggest that Se-SPep and SPep intervention can prevent PM2.5-induced apoptosis in A549 cells by regulating Bax/Bcl-2 protein expression. Secondly, the Se-SPeps have a more solid ability than the SPeps and SeMet to control the expression of Bax/Bcl-2 protein.

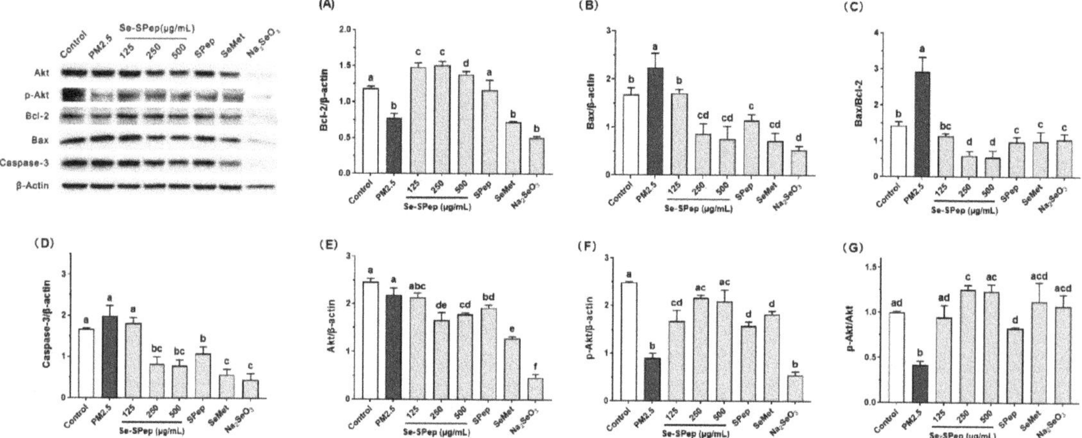

Figure 7. Effect of Se-SPep pretreatment on the mitochondrial apoptotic pathway in A549 cells exposed to PM2.5. The expression of Bcl-2, Bax, caspase-3, Akt, and p-AKT in PM2.5-exposed A549 cells was analyzed using Western blotting. β-Actin was used as the internal control. (**A**) Protein expression of Bcl-2. (**B**) Protein expression of Bax. (**C**) Protein expression of Bax relative to Bcl-2. (**D**) Protein expression of caspase-3. (**E**) Protein expression of Akt. (**F**) Protein expression of p-Akt. (**G**) Protein expression of p-AKT relative to AKT. The Se concentration in the SeMet and Na_2SeO_3 groups is 0.022 µg/mL, which is equivalent to the Se concentration in the 250 µg/mL Se-SPep group. Data are shown as the mean ± SD, $n = 3$/group. Statistical analysis was performed using ANOVA followed by Tukey's post hoc test. Different letters over bars indicate statistically significant differences ($p < 0.05$).

Caspase-3, identified as the "death executor protease", acts as a pivotal player in apoptosis execution, amplifying signals from the caspase promoter and instigating apoptosis [51]. The activation of caspase-3 is considered to be a central link to apoptosis [52]. Illustrated in Figure 7D, the activation level of caspase-3 surged in cells exposed to PM2.5, signifying that PM2.5 exposure heightened apoptosis. However, the expression of caspase-3 significantly decreased in cells treated with Se-SPeps, SPeps, and SeMet compared to the PM2.5 group ($p < 0.05$). These findings imply that Se-SPeps, SPeps, and SeMet can protect A549 cells from apoptosis caused by an increased caspase-3 expression induced by PM2.5 exposure.

Research has indicated that the increased phosphorylation of Akt is related to the inhibition of apoptosis [53]. Akt can phosphorylate target proteins through multiple downstream pathways to play an antiapoptotic role [54]. The activation of this signaling pathway necessitates the initiation of S473 phosphorylation in a hydrophobic sequence [55]. Bcl-2 can depolymerize with phosphorylated BAD when Akt is activated, and free Bcl-2 can play an antiapoptotic role [56]. Figure 7E–G indicates that exposure to PM2.5 significantly decreased the phosphorylation level of Akt ($p < 0.05$). Intervention with Se-SPeps, SPeps, and SeMet markedly elevated the phosphorylation level of Akt ($p < 0.05$) in comparison to the PM2.5 exposure group. These findings suggest that Se-SPeps, SPeps, and SeMet may promote the antiapoptotic effect of A549 cells by increasing Akt phosphorylation. Meanwhile, the Se-SPeps had a more significant effect on Akt phosphorylation than the SPeps ($p < 0.05$), indicating a stronger antiapoptotic effect. The study showed that PM2.5 exposure inhibited Akt phosphorylation in A549 cells, which is consistent with the present findings [57]. These findings suggest that the decreased viability and apoptosis of A549 cells induced by PM2.5 are related to the activation of the mitochondrial apoptosis pathway triggered by oxidative stress and inflammatory response.

4. Conclusions

In this study, we investigated the protective effects of Se-SPeps on PM2.5-induced apoptosis via regulating oxidative stress and inflammatory response. The results demonstrated that Se-SPep intervention could prevent PM2.5-induced apoptosis and maintain regular cell morphology. Simultaneously, Se-SPep intervention can protect cells from the oxidative stress response caused by PM2.5 exposure by inhibiting the generation of intracellular ROS. Regarding the inflammatory response, Se-SPep intervention inhibited the overproduction of the proinflammatory cytokines IL-1β, IL-6, and TNF-α, thereby inhibiting the inflammatory damage caused by PM2.5 and reducing cell apoptosis. The intervention of Se-SPeps can play an antiapoptotic role by promoting the phosphorylation of Akt and the depolymerization of the Bcl-2 protein. Se-SPeps can also inhibit the proapoptotic effect of Bax by inhibiting the expression of the Bax protein. In addition, Se-SPep intervention inhibits the expression of caspase-3, thereby inhibiting its proapoptotic effects. This study also showed that Se-SPeps were more effective than SPeps in regulating the expression of Akt/Bcl-2/Bax pathway proteins to exert antiapoptotic effects. Finally, this study proved that nutritional intervention in the form of Se-SPeps and SeMet was safer compared to the primarily inorganic Se form of Na_2SeO_3. Our study demonstrates the effectiveness of Se-SPeps, a novel organic Se oral supplement, in providing protective effects against occupational air pollution-induced apoptosis of lung epithelial cells.

Supplementary Materials: The following supporting information can be downloaded at: https://www.mdpi.com/article/10.3390/nu16010071/s1, Figure S1: Sephadex G-25 chromatograms of the standard molecular weight samples of Se-SPep (A) and SPep (B); Table S1: The amino acid compositions in Se-SPep and SPep; Table S2: Mean concentration (μg/g) of main compositions detected in PM2.5 sample.

Author Contributions: J.Z.: conceptualization, investigation, formal analysis, and writing—original draft. W.L. (Wenhui Li): investigation and formal analysis. H.L.: conceptualization, resources, formal analysis, supervision, writing—review and editing, and funding acquisition. W.L. (Wanlu Liu): investigation and formal analysis. L.L.: investigation and writing—review and editing. X.L.: conceptualization, resources, supervision, writing—review and editing, and funding acquisition. All authors have read and agreed to the published version of the manuscript.

Funding: This research was funded by the National Key Research and Development Program of China, grant number 2021YFD2100402.

Institutional Review Board Statement: Not applicable.

Informed Consent Statement: Not applicable.

Data Availability Statement: The data presented in this study are available on request from the corresponding author.

Conflicts of Interest: The authors declare no conflict of interest.

References

1. Zhai, S.; Jacob, D.J.; Wang, X.; Shen, L.; Li, K.; Zhang, Y.; Gui, K.; Zhao, T.; Liao, H. Fine particulate matter (PM2.5) trends in China, 2013–2018: Separating contributions from anthropogenic emissions and meteorology. *Atmos. Chem. Phys.* **2019**, *19*, 11031–11041. [CrossRef]
2. Chao, H.; Hsu, J.; Ku, H.; Wang, S.; Huang, H.; Liou, S.; Tsou, T. Inflammatory response and PM2.5 exposure of urban traffic conductors. *Aerosol Air Qual. Res.* **2018**, *18*, 2633–2642. [CrossRef]
3. Morales Betancourt, R.; Galvis, B.; Balachandran, S.; Ramos-Bonilla, J.P.; Sarmiento, O.L.; Gallo-Murcia, S.M.; Contreras, Y. Exposure to fine particulate, black carbon, and particle number concentration in transportation microenvironments. *Atmos. Environ.* **2017**, *157*, 135–145. [CrossRef]
4. Zhang, Y.; Sun, L.; Zhu, C.; Zhang, Y.; Jia, Q.; Li, Z.; Fan, R.; Lyu, X. Personal PM2.5 exposure and the health risk assessment of metal elements in different occupational populations of Jinan. *Environ. Chem.* **2022**, *41*, 2962–2973. [CrossRef]
5. Shakya, K.M.; Rupakheti, M.; Aryal, K.; Peltier, R.E. Respiratory effects of high levels of particulate exposure in a cohort of traffic police in kathmandu, nepal. *J. Occup. Environ. Med.* **2016**, *58*, 218–225. [CrossRef] [PubMed]
6. Mannucci, P.M.; Franchini, M. Health effects of ambient air pollution in developing countries. *Int. J. Environ. Res. Public Health* **2017**, *14*, 1048. [CrossRef] [PubMed]
7. Rui, W.; Guan, L.; Zhang, F.; Zhang, W.; Ding, W. PM2.5-induced oxidative stress increases adhesion molecules expression in human endothelial cells through the ERK/AKT/NF-kappa B-dependent pathway. *J. Appl. Toxicol.* **2016**, *36*, 48–59. [CrossRef]
8. Wang, H.; Shen, X.; Tian, G.; Shi, X.; Huang, W.; Wu, Y.; Sun, L.; Peng, C.; Liu, S.; Huang, Y.; et al. AMPK alpha 2 deficiency exacerbates long-term PM2.5 exposure-induced lung injury and cardiac dysfunction. *Free Radic. Biol. Med.* **2018**, *121*, 202–214. [CrossRef]
9. He, M.; Ichinose, T.; Yoshida, S.; Ito, T.; He, C.; Yoshida, Y.; Arashidani, K.; Takano, H.; Sun, G.; Shibamoto, T. PM2.5-induced lung inflammation in mice: Differences of inflammatory response in macrophages and type II alveolar cells. *J. Appl. Toxicol.* **2017**, *37*, 1203–1218. [CrossRef]
10. Yan, Z.; Wang, J.; Li, J.; Jiang, N.; Zhang, R.; Yang, W.; Yao, W.; Wu, W. Oxidative stress and endocytosis are involved in upregulation of interleukin-8 expression in airway cells exposed to PM2.5. *Environ. Toxicol.* **2016**, *31*, 1869–1878. [CrossRef]
11. Zhang, Y.; Darland, D.; He, Y.; Yang, L.; Dong, X.; Chang, Y. Reduction of PM2.5 toxicity on human alveolar epithelial cells a549 by tea polyphenols. *J. Food Biochem.* **2018**, *42*, e12496. [CrossRef] [PubMed]
12. Cai, D.; He, Y. Daily lifestyles in the fog and haze weather. *J. Thorac. Dis.* **2016**, *8*, 75–77. [CrossRef]
13. Guillin, O.M.; Vindry, C.; Ohlmann, T.; Chavatte, L. Selenium, selenoproteins and viral infection. *Nutrients* **2019**, *11*, 2101. [CrossRef] [PubMed]
14. Amini, P.; Kolivand, S.; Saffar, H.; Rezapoor, S.; Motevaseli, E.; Najafi, M.; Nouruzi, F.; Shabeeb, D.; Musa, A.E. Protective effect of selenium-L-methionine on radiation-induced acute pneumonitis and lung fibrosis in rat. *Curr. Clin. Pharmacol.* **2019**, *14*, 157–164. [CrossRef] [PubMed]
15. Liu, J.; Yang, Y.; Zeng, X.; Bo, L.; Jiang, S.; Du, X.; Xie, Y.; Jiang, R.; Zhao, J.; Song, W. Investigation of selenium pretreatment in the attenuation of lung injury in rats induced by fine particulate matters. *Environ. Sci. Pollut. Res.* **2017**, *24*, 4008–4017. [CrossRef] [PubMed]
16. Mal'tseva, V.N.; Goltyaev, M.V.; Turovsky, E.A.; Varlamova, E.G. Immunomodulatory and anti-inflammatory properties of selenium-containing agents: Their role in the regulation of defense mechanisms against COVID-19. *Int. J. Mol. Sci.* **2022**, *23*, 2360. [CrossRef] [PubMed]
17. Varlamova, E.G.; Turovsky, E.A. The Main Cytotoxic Effects of Methylseleninic Acid on Various Cancer Cells. *Int. J. Mol. Sci.* **2021**, *22*, 6614. [CrossRef] [PubMed]
18. Zhang, J.; Zhou, H.; Li, H.; Ying, Z.; Liu, X. Research progress on separation of selenoproteins/Se-enriched peptides and their physiological activities. *Food Funct.* **2021**, *12*, 1390–1401. [CrossRef]
19. Lyons, M.P.; Papazyan, T.T.; Surai, P.F. Selenium in food chain and animal nutrition: Lessons from nature-review. *Asian-Australas. J. Anim. Sci.* **2007**, *20*, 1135–1155. [CrossRef]
20. Zhang, J.; Zhang, Q.; Li, H.; Chen, X.; Liu, W.; Liu, X. Antioxidant activity of SSeCAHK in HepG2 cells: A selenopeptide identified from selenium-enriched soybean protein hydrolysates. *RSC Adv.* **2021**, *11*, 33872–33882. [CrossRef]
21. Zhang, J.; Gao, S.; Li, H.; Cao, M.; Li, W.; Liu, X. Immunomodulatory effects of selenium-enriched peptides from soybean in cyclophosphamide-induced immunosuppressed mice. *Food Sci. Nutr.* **2021**, *9*, 6322–6334. [CrossRef]
22. Ye, Q.; Wu, X.; Zhang, X.; Wang, S. Organic selenium derived from chelation of soybean peptide-selenium and its functional properties in vitro and in vivo. *Food Funct.* **2019**, *10*, 4761–4770. [CrossRef]
23. Fang, Y.; Pan, X.; Zhao, E.; Shi, Y.; Shen, X.; Wu, J.; Pei, F.; Hu, Q.; Qiu, W. Isolation and identification of immunomodulatory selenium-containing peptides from selenium-enriched rice protein hydrolysates. *Food Chem.* **2019**, *275*, 696–702. [CrossRef]
24. Yi, G.; Din, J.U.; Zhao, F.; Liu, X. Effect of soybean peptides against hydrogen peroxide induced oxidative stress in hepg2 cells via nrf2 signaling. *Food Funct.* **2020**, *11*, 2725–2737. [CrossRef]

25. Pan, F.; Wang, L.; Cai, Z.; Wang, Y.; Wang, Y.; Guo, J.; Xu, X.; Zhang, X. Soybean peptide qrpr activates autophagy and attenuates the inflammatory response in the raw264.7 cell model. *Protein Pept. Lett.* **2019**, *26*, 301–312. [CrossRef]
26. *SRM 2786*; Fine Atmospheric Particulate Matter (Mean Particle Diameter < 4 µm). National Institute of Standards and Technology: Gaithersburg, MD, USA, 2021.
27. Gao, S.; Zhang, J.; Zhang, Q.; Li, W.; Li, H.; Yu, T.; Liu, Q. Preparation and in vivo absorption characteristics of selenium-enriched soybean peptides. *Food Sci.* **2021**, *42*, 165–172. [CrossRef]
28. *GB 5009.93-2010*; National Food Safety Standard Determination of Selenium in Foods. Ministry of Health of the People's Republic of China: Beijing, China, 2010.
29. *GB/T 22492-2008*; Soy Peptides Power. General Administration of Quality Supervision, Inspection and Quarantine of the People's Republic of China: Beijing, China, 2008.
30. Tinggi, U. Essentiality and toxicity of selenium and its status in Australia: A review. *Toxicol. Lett.* **2003**, *137*, 103–110. [CrossRef]
31. Zhou, Q.; Bai, Y.; Gao, J.; Duan, Y.; Lyu, Y.; Guan, L.; Elkin, K.; Xie, Y.; Jiao, Z.; Wang, H. Human serum-derived extracellular vesicles protect A549 from PM2.5-induced cell apoptosis. *Biomed. Environ. Sci.* **2021**, *34*, 40–49. [CrossRef]
32. Pun, V.C.; Kazemiparkouhi, F.; Manjourides, J.; Suh, H.H. Long-term PM2.5 exposure and respiratory, cancer, and cardiovascular mortality in older us adults. *Am. J. Epidemiol.* **2017**, *186*, 961–969. [CrossRef]
33. Qiu, Y.; Wang, G.; Zhou, F.; Hao, J.; Tian, L.; Guan, L.; Geng, X.; Ding, Y.; Wu, H.; Zhang, K. PM2.5 induces liver fibrosis via triggering ROS-mediated mitophagy. *Ecotoxicol. Environ. Saf.* **2019**, *167*, 178–187. [CrossRef]
34. Cui, Y.; Xie, X.; Jia, F.; He, J.; Li, Z.; Fu, M.; Hao, H.; Liu, Y.; Liu, J.Z.; Cowan, P.J.; et al. Ambient fine particulate matter induces apoptosis of endothelial progenitor cells through reactive oxygen species formation. *Cell. Physiol. Biochem.* **2015**, *35*, 353–363. [CrossRef]
35. Deng, X.; Zhang, F.; Rui, W.; Long, F.; Wang, L.; Feng, Z.; Chen, D.; Ding, W. PM2.5-induced oxidative stress triggers autophagy in human lung epithelial A549 cells. *Toxicol. Vitr.* **2013**, *27*, 1762–1770. [CrossRef]
36. Xu, Z.; Zhang, Z.; Ma, X.; Ping, F.; Zheng, X. Effect of PM2.5 on oxidative stress-JAK/STAT signaling pathway of human bronchial epithelial cells. *J. Hyg. Res.* **2015**, *44*, 451–455. [CrossRef]
37. Lao, W.; Bi, T.; Zhou, Y.; Chen, S.; Zhao, X.; Diao, Y. Protective effect of ferulic acid on PM2.5-induced mitochondrial damage in A549 cells. *Food Sci.* **2017**, *38*, 195–200. [CrossRef]
38. Wang, L.; Xu, J.; Liu, H.; Li, J.; Hao, H. PM2.5 inhibits SOD1 expression by up-regulating microRNA-206 and promotes ROS accumulation and disease progression in asthmatic mice. *Int. Immunopharmacol.* **2019**, *76*, 105871. [CrossRef]
39. Fernando, I.P.S.; Jayawardena, T.U.; Kim, H.S.; Lee, W.W.; Vaas, A.P.J.P.; De Silva, H.I.C.; Abayaweera, G.S.; Nanayakkara, C.M.; Abeytunga, D.T.U.; Lee, D.S.; et al. Beijing urban particulate matter-induced injury and inflammation in human lung epithelial cells and the protective effects of fucosterol from *Sargassum binderi* (Sonder ex J. Agardh). *Environ. Res.* **2019**, *172*, 150–158. [CrossRef]
40. Wu, S.; Ni, Y.; Li, H.; Pan, L.; Yang, D.; Baccarelli, A.A.; Deng, F.; Chen, Y.; Shima, M.; Guo, X. Short-term exposure to high ambient air pollution increases airway inflammation and respiratory symptoms in chronic obstructive pulmonary disease patients in Beijing, China. *Environ. Int.* **2016**, *94*, 76–82. [CrossRef]
41. Dagher, Z.; Garcon, G.; Gosset, P.; Ledoux, F.; Surpateanu, G.; Courcot, D.; Aboukais, A.; Puskaric, E.; Shirali, P. Pro-inflammatory effects of Dunkerque city air pollution particulate matter 2.5 in human epithelial lung cells (L132) in culture. *J. Appl. Toxicol.* **2005**, *25*, 166–175. [CrossRef]
42. Dergham, M.; Lepers, C.; Verdin, A.; Billet, S.; Cazier, F.; Courcot, D.; Shirali, P.; Garcon, G. Prooxidant and proinflammatory potency of air pollution particulate matter (PM2.5-0.3) produced in rural, urban, or industrial surroundings in human bronchial epithelial cells (BEAS-2B). *Chem. Res. Toxicol.* **2012**, *25*, 904–919. [CrossRef]
43. Lin, X.; Fan, Y.; Wang, X.; Chi, M.; Li, X.; Zhang, X.; Sun, D. Correlation between tumor necrosis factor-alpha and interleukin-1 beta in exhaled breath condensate and pulmonary function. *Am. J. Med. Sci.* **2017**, *354*, 388–394. [CrossRef]
44. Ogino, K.; Zhang, R.; Takahashi, H.; Takemoto, K.; Kubo, M.; Murakami, I.; Wang, D.; Fujikura, Y. Allergic airway inflammation by nasal inoculation of particulate matter (PM2.5) in NC/Nga mice. *PLoS ONE* **2014**, *9*, 92710. [CrossRef]
45. Xue, Z.; Wang, J.; Yu, W.; Li, D.; Zhang, Y.; Wan, F.; Kou, X. Biochanin A protects against PM2.5-induced acute pulmonary cell injury by interacting with the target protein MEK5. *Food Funct.* **2019**, *10*, 7188–7203. [CrossRef]
46. Cui, Y.F.; Xia, G.W.; Fu, X.B.; Yang, H.; Peng, R.Y.; Zhang, Y.; Gu, Q.Y.; Gao, Y.B.; Cui, X.M.; Hu, W.H. Relationship between expression of Bax and Bcl-2 proteins and apoptosis in radiation compound wound healing of rats. *Chin. J. Traumatol.* **2003**, *6*, 135–138.
47. Yong, F.; Zi, X.; Yi, S.; Fei, P.; Wenjian, Y.; Ning, M.; Muinde Kimatu, B.; Kunlun, L.; Weifen, Q.; Qiuhui, H. Protection mechanism of Se-containing protein hydrolysates from Se-enriched rice on Pb2+-induced apoptosis in PC12 and RAW264.7 cells. *Food Chem.* **2017**, *219*, 391–398. [CrossRef]
48. Wang, H.; Liu, J.; Liu, X.; Liu, Z. Protective effects of blueberry against hydrogen peroxide-induced oxidative stress in HEPG2 cells: Involvement of mitochondrial BCL-2-dependent. *Br. Food J.* **2019**, *121*, 2809–2820. [CrossRef]
49. Babbitt, S.E.; Sutherland, M.C.; Francisco, B.S.; Mendez, D.L.; Kranz, R.G. Mitochondrial cytochrome c biogenesis: No longer an enigma. *Trends Biochem. Sci.* **2015**, *40*, 446–455. [CrossRef]
50. Li, X.; Ding, Z.; Zhang, C.; Zhang, X.; Meng, Q.; Wu, S.; Wang, S.; Yin, L.; Pu, Y.; Chen, R. MicroRNA-1228(*) inhibit apoptosis in A549 cells exposed to fine particulate matter. *Environ. Sci. Pollut. Res.* **2016**, *23*, 10103–10113. [CrossRef]
51. Liu, J.; Liang, S.; Du, Z.; Zhang, J.; Sun, B.; Zhao, T.; Yang, X.; Shi, Y.; Duan, J.; Sun, Z. PM2.5 aggravates the lipid accumulation, mitochondrial damage and apoptosis in macrophage foam cells. *Environ. Pollut.* **2019**, *249*, 482–490. [CrossRef]

52. Xiong, Q.; Ru, Q.; Chen, L.; Tian, X.; Li, C. Mitochondrial dysfunction and inflammatory response in the cytotoxicity of NR8383 macrophages induced by fine particulate matter. *Environ. Toxicol. Pharmacol.* **2017**, *55*, 1–7. [CrossRef]
53. Hers, I.; Vincent, E.E.; Tavare, J.M. Akt signalling in health and disease. *Cell. Signal.* **2011**, *23*, 1515–1527. [CrossRef] [PubMed]
54. Wang, A.S.; Xu, Y.; Zhang, Z.W.; Lu, B.B.; Yin, X.; Yao, A.J.; Han, L.Y.; Zou, Z.Q.; Li, Z.; Zhang, X.H. Sulforaphane protects MLE-12 lung epithelial cells against oxidative damage caused by ambient air particulate matter. *Food Funct.* **2017**, *8*, 4555–4562. [CrossRef] [PubMed]
55. Vicencio, J.M.; Yellon, D.M.; Sivaraman, V.; Das, D.; BoiDoku, C.; Arjun, S.; Zheng, Y.; Riquelme, J.A.; Kearney, J.; Sharma, V.; et al. Plasma exosomes protect the myocardium from ischemia-reperfusion injury. *J. Am. Coll. Cardiol.* **2015**, *65*, 1525–1536. [CrossRef] [PubMed]
56. Siddiqui, W.A.; Ahad, A.; Ahsan, H. The mystery of BCL2 family: Bcl-2 proteins and apoptosis: An update. *Arch. Toxicol.* **2015**, *89*, 289–317. [CrossRef] [PubMed]
57. Li, J.; Zhou, Q.; Yang, T.; Li, Y.; Zhang, Y.; Wang, J.; Jiao, Z. SGK1 inhibits PM2.5-induced apoptosis and oxidative stress in human lung alveolar epithelial A549 cells. *Biochem. Biophys. Res. Commun.* **2018**, *496*, 1291–1295. [CrossRef]

MDPI

St. Alban-Anlage 66

4052 Basel

Switzerland

www.mdpi.com

Nutrients Editorial Office

E-mail: nutrients@mdpi.com

www.mdpi.com/journal/nutrients